Quaternary Dating Methods

Quaternary Dating Methods

MIKE WALKER

Department of Archaeology and Anthropology
University of Wales, Lampeter, UK

John Wiley & Sons, Ltd

Telephone (+44) 1243 779777

Email (for orders and customer service enquiries): cs-books@wiley.co.uk
Visit our Home Page on www.wiley.com

Other Wiley Editorial Offices

John Wiley & Sons Inc., 111 River Street, Hoboken, NJ 07030, USA

Jossey-Bass, 989 Market Street, San Francisco, CA 94103-1741, USA

Wiley-VCH Verlag GmbH, Boschstr. 12, D-69469 Weinheim, Germany

John Wiley & Sons Australia Ltd, 33 Park Road, Milton, Queensland 4064, Australia

John Wiley & Sons (Asia) Pte Ltd, 2 Clementi Loop #02-01, Jin Xing Distripark, Singapore 129809

John Wiley & Sons Canada Ltd, 22 Worcester Road, Etobicoke, Ontario, Canada M9W 1L1

Wiley also publishes its books in a variety of electronic formats. Some content that appears in print may not be available in electronic books.

Library of Congress Cataloging in Publication Data

Walker, M.J.C. (Mike J.C.), 1947–
Quaternary dating methods / Mike Walker.
p. cm.
Includes bibliographical references and index.
ISBN 0-470-86926-7 (hb : acid-free paper) — ISBN 0-470-86927-5 (pbk. : acid-free paper)
1. Geochronometry. 2. Geology, Stratigraphic—Quaternary. 3. Radioactive dating. I. Title.
QE508.W348 2005
551.7′01—dc22

2004029171

British Library Cataloguing in Publication Data

A catalogue record for this book is available from the British Library

ISBN 0-470-86926-7 (HB)
ISBN 0-470-86927-5 (PB)

Typeset in 10/12pt Times by Integra Software Services Pvt. Ltd, Pondicherry, India
Printed and bound in Great Britain by Antony Rowe Ltd, Chippenham, Wiltshire
This book is printed on acid-free paper responsibly manufactured from sustainable forestry in which at least two trees are planted for each one used for paper production.

For John Lowe

Contents

Preface

In a letter to Thomas Manning in 1810, Charles Lamb wrote: 'Nothing puzzles me more than time and space; and yet nothing troubles me less, as I never think about them.' All of us working in the field of Quaternary science would, I suspect, tend to agree with the first part of this statement but take issue over the second. I for one have always been fascinated by time and, in particular, by the way in which we are able to assign ages to events in the distant past. My family and friends have been amused and intrigued in equal measure by me talking, with *apparent* confidence and authority, about the earth being formed 4.5 billion years ago, or the present warm period within which we live lasting 11 500 yrs. 'But how can you be so sure?' is the usual question. One of my aims in writing this book is to show them that there are indeed ways in which we can date the past and, moreover, that we can do so with an ever-increasing sense of assurance. My principal purpose, however, is to describe the various dating techniques that are routinely employed in Quaternary science in a way that is comprehensible to both undergraduate students and interested lay-people alike. I have therefore tried to avoid using mathematical formulae, although in the first chapter I felt it necessary to cover some of the basics of chemistry in order to provide the groundwork for what comes later. I have also orientated the book towards the practical aspects of dating by basing it around specific examples. Hopefully, this approach will appeal to students and others with a non-scientific background but, at the same time, will not appear to those who are fortunate in possessing a stronger scientific pedigree to be 'dumbing down'. Above all, however, my aim is to encourage readers (unlike Charles Lamb) to think a little more about the past and to recognise the importance of being able to frame the momentous events of recent earth and human history within a reasonably secure temporal framework.

Throughout the book I have drawn on a previous volume that I wrote with John Lowe (*Reconstructing Quaternary Environments*, 1997, Addison-Wesley-Longman, London). I make no apologies for this because I know that book has been, and continues to be, widely used at undergraduate and postgraduate levels in both Britain and abroad. I hope that this new book on *Quaternary Dating Methods* will find an equally wide readership. John and I are about to embark on the third edition of *Reconstructing Quaternary Environments* (due 2006), and during the course of preparing that revision, I hope I will be able to reciprocate and that some of the material contained in the following pages will find its way into Lowe and Walker Mark III. The text also includes a large number of references. Some might find that this disrupts the flow of the narrative, but I felt that it was important not only to acknowledge the sources of material upon which I have drawn but, equally importantly, to point the reader in the direction of this work so that those who might be interested in taking matters further will be able to do so.

It is customary in a Preface to express thanks to those who have assisted either directly or indirectly in the production of the book, and I do not intend to depart from that practice. Over the last 15 years or so, I have enjoyed the national and international collaboration, and friendship, of many colleagues, first through the North Atlantic Seaboard Programme of IGCP-253, and more recently through the INTIMATE (Integration of ice-core, marine and terrestrial records) Programme of INQUA (International Quaternary Union). I am particularly appreciative of the time that I have spent at a number of different meetings with, amongst others, Hilary Birks, Sjoerd Bohncke, Svante Björck, Russell Coope, Les Cwynar, Irka Hajdas, Jan Heinemeir, Wim Hoek, Konrad Hughen, Sigfus Johnsen, Karen-Luise Knudsen, Nalan Koç, Thomas Litt, Jørgen Peder Steffensen, Chris Turney, Bas van Geel and Barbara Wohlfarth. My work with the Natural Environmental Research Council, formerly as a member and subsequently as chairman of the NERC Radiocarbon Facilities Committee, and latterly as chairman of the NERC AMS (Accelerator Mass Spectrometry) Strategy Group, has brought me into contact with colleagues at the East Kilbride and Oxford Radiocarbon Dating Laboratories, notably Chris Bronk-Ramsay, Charlotte Bryant, Doug Harkness, Robert Hedges and Tony Fallick, whose company I have enjoyed and from whom I have learned a great deal. I should also like to thank Lin Kay and Chris Franklin at NERC for supporting me in my role as Committee Chairman. Finally, I am grateful to my colleagues in the Department of Archaeology and Anthropology, University of Wales, Lampeter, especially David Austin and John Crowther, for providing such a congenial working environment over the past four years, and to the university itself for allowing me a period of study leave during which much of the first draft of the book was completed.

In writing this book, I have constantly been aware of the fact that I am approaching the material as a member of the user community. I am not an expert in the technical aspects of dating, and hence I have prevailed upon colleagues who know far more about these matters than I ever will to read what I have written and to show me where I have gone wrong. I am deeply indebted to Tim Atkinson, Simon Blockley, Charlotte Bryant, Tony Fallick, Rob Kemp, Olav Lian, Danny McCarroll, James Scourse, Mike Summerfield, Chris Turney and John Westgate for their careful scrutiny and constructive critical appraisal of various sections of the text; I simply could not have completed this book without their assistance. It goes without saying, however, that any remaining errors are my own. Several friends and colleagues have provided me with photographs, for which I am most grateful, and Phil Gibbard and Richard Preece helped considerably in the compilation of Figure 1.4. I should also like to thank Sally Wilkinson, Keily Larkins, Lynette James and the staff in the production department of John Wiley. Last, and by no means least, I would like to express my gratitude to my wife, Gro-Mette, who has not only been a constant source of encouragement, but who has also read the draft text from cover to cover, and has provided many valuable inputs along the way.

One name is missing from the above list. As colleagues within the Quaternary community will know, for more than 30 years I have worked in collaboration with John Lowe. We first met as postgraduate students in the University of Edinburgh and since then we have produced more than 50 joint publications. I have no doubt whatsoever that

John could have written this book and, I suspect, he might well have made a better fist of it. Nevertheless, I hope he will find some of the material in the following pages of interest and that he will enjoy reading it. Not only have John and I been close academic colleagues, but we have also remained firm friends, and in acknowledgement of this I would like to dedicate the book to him.

Mike Walker
October, 2004

1

Dating Methods and the Quaternary

Whatever withdraws us from the power of our senses; whatever makes the past, the distant or the future, predominate over the present, advances us in the dignity of thinking beings.

Samuel Johnson

1.1 Introduction

The Quaternary is the most recent period of the geological record. Spanning the last 2.5 million years or so of geological time[1] and including the **Pleistocene** and **Holocene** epochs,[2] it is often considered to be synonymous with the 'Ice Age'. Indeed, for much of the Quaternary, the earth's land surface has been covered by greatly expanded ice sheets and glaciers, and temperatures during these **glacial** periods were significantly lower than those of the present. But the Quaternary has also seen episodes, albeit much shorter in duration, of markedly warmer conditions, and in these **interglacials** the temperatures in the mid- and high-latitude regions may have exceeded those of the present day. Indeed, rather than being a period of unremitting cold, the hallmark of the Quaternary is the repeated oscillation of the earth's global climate system between glacial and interglacial states.

Establishing the timing of these climatic changes, and of their effects on the earth's environment, is a key element in Quaternary research. Whether it is to date a particular climatic episode, to estimate the rate of operation of past geological or geomorphological processes, or to determine the age of an artefact or cultural assemblage, we need to be able to establish a chronology of events. The aim of this book is to describe, evaluate and exemplify the different dating techniques that are applicable within the field of

Quaternary Dating Methods M. Walker

Quaternary science. It is not, however, a dating manual. Rather, it is a book that is written from the perspective of the user community as opposed to that of the laboratory expert. It is, above all, a book that lays emphasis on the practical side of Quaternary dating, for the principal focus is on examples or case studies. To paraphrase the words of the actor John Cleese, it is intended to show just what Quaternary dating can do for us!

In this chapter, we examine the development of ideas relating to geological time and, in particular, to Quaternary dating. We then move on to consider the ways in which the quality of a date can be evaluated, and to discuss some basic principles of radioactive decay as these apply to Quaternary dating. Finally, we return to the Quaternary with a brief overview of the Quaternary stratigraphic record, and of Quaternary nomenclature and terminology. These sections provide important background information, and both a chronological and stratigraphic context for the remainder of the book.

1.2 The Development of Quaternary Dating

Early approaches to dating the past were closely associated with attempts to establish the age of the earth. Some of the oldest writings on this topic are to be found in the classical literature where the *leitmotif* of much of the Greek writings is the concept of an infinite time, equivalent in many ways to modern day requirements for steady-state theories of the universe (Tinkler, 1985). This position contrasts markedly with that in post-Renaissance Europe where biblical thinking placed the creation of the world around 6000 years ago, and when the end of the universe was predicted within a few hundred years. This restricted chronology for earth history derives from the biblical researches of James Ussher, Archbishop of Armagh, who in 1654 published his considered conclusion, based on Old Testament genealogical sources, that the earth was created on Sunday 23 October 4004 BC, with 'man and other living creatures' appearing on the following Friday. Another momentous event in the Old Testament, the 'great flood', occurred 1656 years after the creation, between 2349 and 2348 BC.

In his magisterial review of the history of earth science, Davies (1969) has observed that although modern researchers have tended to scoff at Ussher's chronology he was, in fact, no fanatical fundamentalist but rather a brilliant and highly respected scholar of his day. It is perhaps for this reason that his chronology had such a pervasive influence on scientific thought, although it is perhaps less clear to modern geologists why it still forms a cornerstone of contemporary creationist 'science'! During the eighteenth and nineteenth centuries, however, with the development of uniformitarianist thinking in geology,[3] the pendulum began to swing once more towards longer timescales for the formation of the earth and for the longevity of operation of geological processes, a view encapsulated by James Hutton's famous observation in his *Theory of the Earth* (1788) that '...we find no vestige of a beginning, no prospect of an end'.

The difficulty was, of course, that pre-twentieth-century scientists had no bases for determining the passage of geological time. One of the earliest attempts to tackle the problem was William McClay's work in 1790 on the retreat of the Niagara Falls escarpment, which led him to propose an age of 55 440 years for the earth (Tinkler, 1985). Others tried a different tack. The nineteenth-century scientist John Joly, for example, calculated the quantity of sodium salt in the world's oceans, as well as the amount added every year from rock erosion, and arrived at a figure of 100 million years for the age of the earth. Increasingly, however, came an awareness that even this extended time frame was simply not long enough to account for the entire history of the earth and, moreover, for organic evolution, a view that was underscored by the publication of Darwin's seminal work *Origin of Species* in 1859. Further challenges to the Ussher timescale and to its successors came from the field of archaeology, with noted antiquarians such as John Evans (and his geological colleague Joseph Prestwich) arguing, on the basis of finds of ancient handaxes, for a protracted period of human occupation extending into a period of antiquity '.... remote beyond any of which we have hitherto found traces' (Renfrew, 1973).

It was into this atmosphere of chronological uncertainty that Louis Agassiz introduced his revolutionary idea of a 'Great Ice Period', which arguably marks the birth of modern Quaternary science. This notion, first propounded in 1837, was initially received with a degree of scepticism by the geological establishment, but the idea not only of a single glaciation but, indeed, of multiple glaciations rapidly gained ground. By the beginning of the twentieth century, most geologists were subscribing to the view that four major glacial episodes had affected the landscapes of both Europe and North America, although the basis for dating these events remained uncertain. An early attempt at establishing a glacial–interglacial chronology was made by the German geologist Albrecht Penck, using the depth of weathering and 'intensity of erosion' in the northern Alpine region of Europe to estimate the duration of interglacial periods. On this basis, an age of 60 000 years was assigned to the Last Interglacial and 240 000 years to the Penultimate Interglacial, the duration of the Quaternary being estimated at 600 000 years (Penck and Bruckner, 1909). An alternative approach using the astronomical timescale based on observed variations in the earth's orbit and axis[4] again arrived at a similar figure, although if older glaciations recorded in the Alpine region were included, the time span of the Quaternary was extended to around 1 million years (Zeuner, 1959). This figure has since been widely quoted and, for the first half of the twentieth century at least, was generally regarded as the best estimate of age for the Quaternary.

At about the time that the Quaternary glacial chronology was being worked out for the European Alps, the first attempts were being made to develop a timescale for the last deglaciation, using laminated or layered sediment sequences which were interpreted as reflecting annual sedimentation cycles. These are known as **varves**, and are still employed as a basis for Quaternary chronology at the present day (section 5.3). Some of the earliest studies were made on the sediments in Swiss lakes and produced estimates of between 16 000 and 20 000 years since the last glacial maximum (Zeuner, 1959), results that are not markedly different from those derived from more recent dating programmes. The seminal work on varved sequences, however, was carried out in Scandinavia where Gerard de Geer (Figure 1.1) developed the world's first high-resolution deglacial chronology in relation to the wasting Fennoscandian ice sheet (section 5.3.3.1). This approach was subsequently applied in North America to date glacial retreat along parts of the southern margin of the last (Laurentide) ice sheet (Antevs, 1931).

The early years of the twentieth century saw the development of another dating technique which is still widely used in Quaternary science, namely **dendrochronology** or **tree-ring dating** (section 5.2). Research on tree rings has a long history, and the relationship between tree rings and climate (a field of study known as **dendroclimatology**) has

Figure 1.1 *Gerard de Geer measuring varves at Beckomberga, Stockholm, in 1931. Varve chronology was the first dating technique to provide a realistic estimate of Quaternary time (photo: Ebba Hult de Geer, courtesy of Lars Brunnberg and Stefan Wastegård)*

intrigued scientists since the Middle Ages. Indeed, some of the earliest writings on this subject can be found in the papers of Leonardo da Vinci (Stallings, 1937). The basics of modern dendrochronology, however, were formulated by the American astronomer Andrew Douglass, who was the first to link simple dendrochronological principles to historical research and to climatology (Schweingruber, 1988). Together with Edmund Schulmann, he founded the world-famous Laboratory for Tree-Ring Research at the University of Arizona in 1937. In Europe, it was not until the end of the 1930s that dendrochronology began to gain a foothold, largely through the work of the German botanist, Bruno Huber. His research laid the foundation for the modern school of German dendrochronology which has remained at the forefront of tree-ring research in Europe to the present day.

The most significant advance in Quaternary chronology, however, came during and immediately after the Second World War, with the discovery that the decay of certain radioactive elements could form a basis for dating. Although measurements had been made more than 30 years earlier on radioactive minerals of supposedly Pleistocene age (Holmes, 1915), it was the pioneering work of Willard Libby and his colleagues that led to the development of radiocarbon dating, and to the establishment of the world's first radiocarbon dating laboratory at the University of Chicago in 1948. During the 1950s and 1960s, other **radiometric methods** were developed that built on technological advances (increasingly sophisticated instrumentation) and an increasing understanding of the nuclear decay process. These included uranium-series and potassium–argon dating (Chapter 3), while a growing appreciation of the effects on minerals and other materials of exposure to radiation led to the development of another family of techniques which includes thermo-luminescence, fission track and electron spin resonance dating (Chapter 4). In the late 1960s and 1970s, advances in molecular biology enabled post-mortem changes in protein structures to be used as a basis for dating (amino acid geochronology), while remarkable developments in coring technology led to the recovery of long-core sequences from ocean sediments and from polar ice sheets, out of which came the first marine and ice-core chronologies. The last two decades of the twentieth century have been characterised by a series of technological innovations that led not only to a further expansion in the range of Quaternary dating techniques, but also to significant improvements in analytical precision. A major advance was the development of accelerator mass spectrometry (AMS), which not only revolutionised radiocarbon dating (Chapter 2), but also made possible the technique of cosmogenic nuclide dating (section 3.4). The last decade has also witnessed the creation of the high-resolution chronologies from the GRIP and GISP2 Greenland ice cores, and from the Vostok and EPICA cores in Antarctica (section 5.5).

These various developments and innovations mean that Quaternary scientists now have at their disposal a portfolio of dating methods that could not have been dreamed of only a generation ago, and which are capable of dating events on timescales ranging from single years to millions of years. The year 2004 sees the 350th anniversary of the publication of the second edition of Ussher's ground-breaking volume on the age of the earth. How he would have reconciled the recent advances in Quaternary dating technology with his 6000-year estimate for the age of the earth is difficult to imagine!

1.3 Precision and Accuracy in Dating

Before going further, it is important to say something about how we can judge the quality of an age determination. Two principal criteria reflect the quality of a date, namely **accuracy** and **precision**, and these apply not only to dates on Quaternary events, but to all age determinations made within the earth, environmental and archaeological sciences. For dating practitioners and for interpreting dates, it is important to understand the meaning and significance of these terms. **Accuracy** refers to the *degree of correspondence* between the true age of a sample and that obtained by the dating process. In other words, it refers to the degree of *bias* in an age measurement. **Precision** relates to the *statistical uncertainty* that is associated with any physical or chemical analysis that is used as a basis for determining age. As we shall see, all dating methods have their own distinctive set of problems, and hence each age measurement will have an element of uncertainty associated with it. These uncertainties tend to be expressed in statistical terms and provide us with an indication of the level of precision of each age determination (Chapter 2).

An example of the distinction between accuracy and precision in the context of a dated sequence is shown in Figure 1.2. In sample A, there is close agreement in terms of mean age between the four dated samples, and the standard errors (indicated by the range bars) are small; however, the dates are 2000–2500 years younger than the 'true age'. These dates are therefore precise, but inaccurate. In sample B, the reverse obtains; the dates cluster around the true age but have wide error bars. Hence they are accurate but imprecise. In sample C, however, the dates are of similar age and have narrow error bars. These age determinations are both accurate and precise, which is the optimal situation in dating.

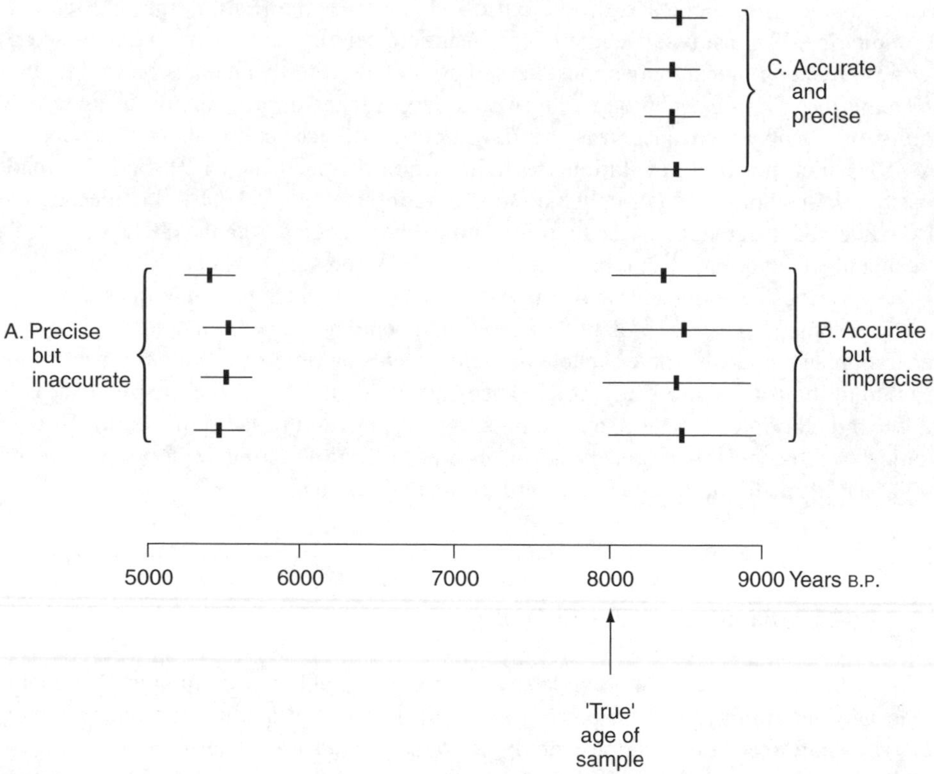

Figure 1.2 *Accuracy and precision in a dated sequence (modified after Lowe and Walker, 1997); see text for details*

1.4 Atomic Structure, Radioactivity and Radiometric Dating

Radiometric dating methods form a significant component of the Quaternary scientist's dating portfolio. Indeed, half of the chapters in this book that deal specifically with dating methods are concerned with radiometric dating. All radiometric techniques are based on the fact that certain naturally occurring elements are unstable and undergo spontaneous changes in their structure and organisation in order to achieve more stable atomic forms. This process, known as **radioactive decay**, is time-dependent, and if the rate of decay for a given element can be determined, then the ages of the host rocks and fossils can be established.

In order to understand the basics of radiometric dating, it is necessary to know something about atomic structure and the radioactive process. Matter is composed of minute particles known as **atoms**, the nuclei of which contain positively charged particles (**protons**), and particles with no electrical charge (**neutrons**), which together make up most of the mass of an atom. Third elements are **electrons**, which are tiny particles of negative charge and negligible mass that spin around the nucleus. Collectively, protons, neutrons and electrons are referred to as **elementary** or **sub-atomic particles**, and for many years were considered to be the fundamental building blocks of matter. With the development of large particle accelerators however, machines that are capable of accelerating samples to such high speeds that matter breaks down into its constituent parts, dozens of new sub-atomic particles have been discovered and current research suggests that atomic matter is made up of elementary particles from two families, **quarks** and **leptons**. Our understanding of electrons (which are members of the lepton family of particles) and their behaviour has also changed. At one time it was believed that electrons orbited the nucleus in shells (or orbitals), similar to the way in which the planets orbit the sun, and that in each of these orbits they had certain energy. However, the situation now appears to be more complex, as modern physics has shown that it is not possible to determine both the location *and* the velocity of a sub-atomic particle.[5] More recent work on atomic structure therefore envisages electrons with a particular energy existing in *volumes of space* around the nucleus, even though their exact location cannot be established. These volumes are known as **atomic orbitals**. The build-up of electrons in atomic orbitals allows scientists to explain many of the physical and chemical properties of elements, and lies at the heart of our modern understanding of chemistry.

When an atom gains or loses electrons, it acquires a net electrical charge, and such atoms are known as **ions**. The electrical charge can be positive or negative; a positive ion is referred to as a **cation** and a negative ion as an **anion**. The nature of the electrical charge is based on the number of protons minus the number of electrons, and is often referred to as the **valence**. Hence, an element with eight protons and eight electrons has a net electrical charge of 0. If it gains two electrons, it has a negative electrical charge (valence = 2^-) and it becomes an anion. If it loses two electrons, it develops a positive charge (valence = 2^+) and becomes a cation. **Ionisation**, which is the process whereby electrons are removed (usually) or added (occasionally) to atoms, is an important element in radiation (see below).

The atoms of each chemical element have a specific **atomic number** and **atomic mass number**. The former refers to the number of protons contained in the nucleus of an atom, while the latter is the number of protons plus neutrons. In other words, the mass number is the total number of particles (**nucleons**) in the nucleus. The atomic number is usually written in subscript on the left-hand side of the symbol for the chemical element (e.g. oxygen – $_8O$; uranium – $_{92}U$), while the atomic mass number is shown in superscript (e.g. ^{16}O; ^{238}U). In some elements, although the number of protons in the nucleus remains the same, the number of neutrons may vary. Elements that possess the same number of protons but different numbers of neutrons are referred to as **isotopes**. The number of electrons is constant for isotopes of each element, and hence they have the same chemical properties, but the isotopes differ in mass, and this will be reflected in change in the atomic mass number. Examples include carbon (^{12}C, ^{13}C, ^{14}C) and oxygen (^{16}O, ^{17}O, ^{18}O). Individual isotopes of an element are referred to as **nuclides**. Most of these are stable; in other words the binding forces created by the electrical charges are sufficient to keep the atomic particles together. In some cases, however, where there are too many or too few neutrons in the nucleus, for example, the nuclides are unstable and this results in a spontaneous emission of particles or energy to achieve a stable state. This is the process of **radiation** (or **radioactive 'decay'**), and such isotopes are known as **radioactive nuclides**.

Unstable nuclei can rid themselves of excess energy in a variety of ways, but the three most common forms are **alpha**, **beta** and **gamma decay**. In **alpha (α) decay**, a nucleus emits an *alpha particle* consisting of two protons and two neutrons, which is a nucleus of helium. Nuclides that emit alpha particles lose both mass and positive charge. The atomic mass number changes to reflect this, and the result is that one chemical element can be created by the decay of others. In **beta (β) decay**, a different kind of particle is ejected – an electron. The emission of a negatively charged electron does not alter mass, hence there is no change in atomic mass number. There is, however, a change in atomic number because the reason for the ejection of the electron is that as the nucleus decays, a neutron transmutes into a proton, and the nucleus must rid itself of some energy and increase its electrical charge. The emission of an electron, with its negative charge and small amount of excess energy, enables this to be achieved. The third common form of radioactivity is **gamma (γ) decay**. Here, the nucleus does not emit a particle, but rather a highly energetic form of electromagnetic radiation. Gamma radiation does not change the number of protons and neutrons in the nucleus, but it does reduce the energy of the nucleus. Gamma rays are not important in most forms of radiometric dating (with the exception of some short-lived isotopes: Chapter 3), but they do contribute to the build-up of luminescent properties in minerals (Chapter 4). In addition, the cosmic rays from deep space that constantly bombard the earth's upper atmosphere, and which initiate the chemical reaction that leads to the formation of radiocarbon (Chapter 2) and other cosmogenic isotopes (Chapter 3), are largely composed of gamma radiation.

An atom that undergoes radioactive decay is termed a **parent nuclide** and the decay product is often referred to as a **daughter nuclide**. Some parent–daughter transformations are accomplished in a single stage, a process known as **simple decay**. Others involve a more complex reaction in which the nuclide with the highest atomic number decays to a stable form through the production of a series of intermediate nuclides, each of which is unstable. This is known as **chain decay** and occurs, for example, in uranium series (section 3.3). The intermediate nuclides that are formed during the course of decay are therefore both the products (or daughters) of previous nuclear transformations and the parents in subsequent radioactive decay. Such nuclides are referred to as **supported**. Where the decay process involves a nuclide that has not, in itself, been created by the decay process, or where that nuclide has been separated from earlier nuclides in the chain through the operation of physical, chemical or biological processes, this is known as **unsupported decay**. The distinction between supported and unsupported decay is considered further in the context of ^{210}Pb (lead-210) dating (section 3.5.1).

Radioactive decay processes are governed by atomic constants. The number of transformations per unit of time is proportional to the number of atoms present in the sample and for each decay pathway there is a **decay constant**. This represents the probability that an atom will decay in a given period of time. Although the radioactive decay of an individual atom is an irregular (stochastic) process, in a large sample of atoms it is possible to establish, within certain statistical limits, the rate at which overall distintegration proceeds. In all radioactive nuclides, the decay is not linear but exponential (e.g. Figure 2.1) and is usually considered in terms of the **half-life**, i.e. the length of time that is required to reduce a given quantity of a parent nuclide to one half. For example, if 1 gm of a parent nuclide is left to decay, after $t_{1/2}$ only 0.5 gm of that parent will remain. It will then take the same period of time to reduce that 0.5 gm to 0.25 gm, and to reduce the 0.25 gm to 0.125 gm, and so on. The half-life concept is fundamental to all forms of radiometric dating.

1.5 The Quaternary: Stratigraphic Framework and Terminology

As we saw above, the Quaternary is conventionally subdivided into **glacial** (cold) and **interglacial** (temperate) stages, with further subdivisions into **stadial** (cool) and **interstadial** (warm) episodes. The distinction between glacials and stadials on the one hand, and interglacials and interstadials on the other, is often blurred, but glacials are generally considered to be cold periods of extended duration (spanning tens of thousands of years) during which temperatures in the mid- and high-latitude regions were low enough to promote extensive glaciation. Stadials are cold episodes of lesser duration (perhaps 10 000 years or less) when cold conditions obtained and when short-lived glacial readvances occurred. Interglacials, on the other hand, were warm periods when temperatures in the mid- and high latitudes were comparable with, or may even have exceeded, those of the present, and whose duration may have been 10 000 years or more. Interstadials, by contrast, were short-lived (typically less than 5000 years) warmer episodes within a glacial stage, during which temperatures did not reach those of the present day. This type of categorisation, which is based on inferred climatic characteristics, is known as **climatostratigraphy** (Lowe and Walker, 1997).

Evidence for former glacial and interglacial conditions (as well as stadial and interstadial environments) has long been recognised in the terrestrial stratigraphic record. Former cold episodes are represented by glacial deposits, by periglacial sediments and structures, and by biological evidence (such as pollen or vertebrate remains) which are indicative of a cold-climate régime. Interglacial and interstadial phases are reflected primarily in the fossil record (pollen, plant macrofossils, fossil insect remains, etc.), or in biogenic sediments that have accumulated in lakes or ponds during a period of warmer climatic conditions. However, because of the effects of erosion, especially glacial erosion, the Quaternary terrestrial stratigraphic record is highly fragmented and, apart from some unusual contexts such as deep lakes in areas that have escaped the direct effects of glaciation, long and continuous sediment records are rarely preserved. During the later twentieth century, therefore, Quaternary scientists turned to the deep oceans of the world, where sedimentation has been taking place continuously over hundreds of thousands of years. Indeed, many ocean sediment records extend in an uninterrupted fashion back through the Quaternary and into the preceding Tertiary period. One of the great technological breakthroughs of the twentieth century was the development of coring equipment mounted on specially designed ships (Figure 1.3) which enabled complete sediment cores to be obtained from the deep ocean floor, sometimes from water depths in excess of 3 km!

What these cores revealed was a remarkable long-term record of oceanographic and, by implication, climatic change. This is reflected in the **oxygen isotope 'signal'** (or trace) in marine microfossils contained within the ocean floor sediments. The variations in the ratio between two isotopes of oxygen, the more common and 'lighter' oxygen-16 (^{16}O) and the rarer 'heavier' oxygen-18 (^{18}O), are indications of the changing isotopic composition of ocean waters between glacial and interglacial stages. As the balance between the two oxygen isotopes in sea water is largely controlled by fluctuations in land ice volume,[6] downcore variations in the oxygen isotope ratio ($\delta^{18}O$) can be read as a record of glacial/interglacial climatic oscillations, working on the principle that ice sheets and glaciers would have been greatly expanded during glacial times but much less extensive during interglacials (Shackleton and Opdyke, 1973). The sequence can therefore be divided into a series of **isotopic stages** (marine oxygen isotope or **MOI stages**) and these are numbered from the top down, interglacial (temperate) stages being assigned odd numbers, while even numbers denote glacial (cold) stages. The record shows that over the course of the

Figure 1.3 *The* Joides Resolution, *a specially commissioned ocean-going drilling ship for coring deep-sea sediments (photo Bill Austin)*

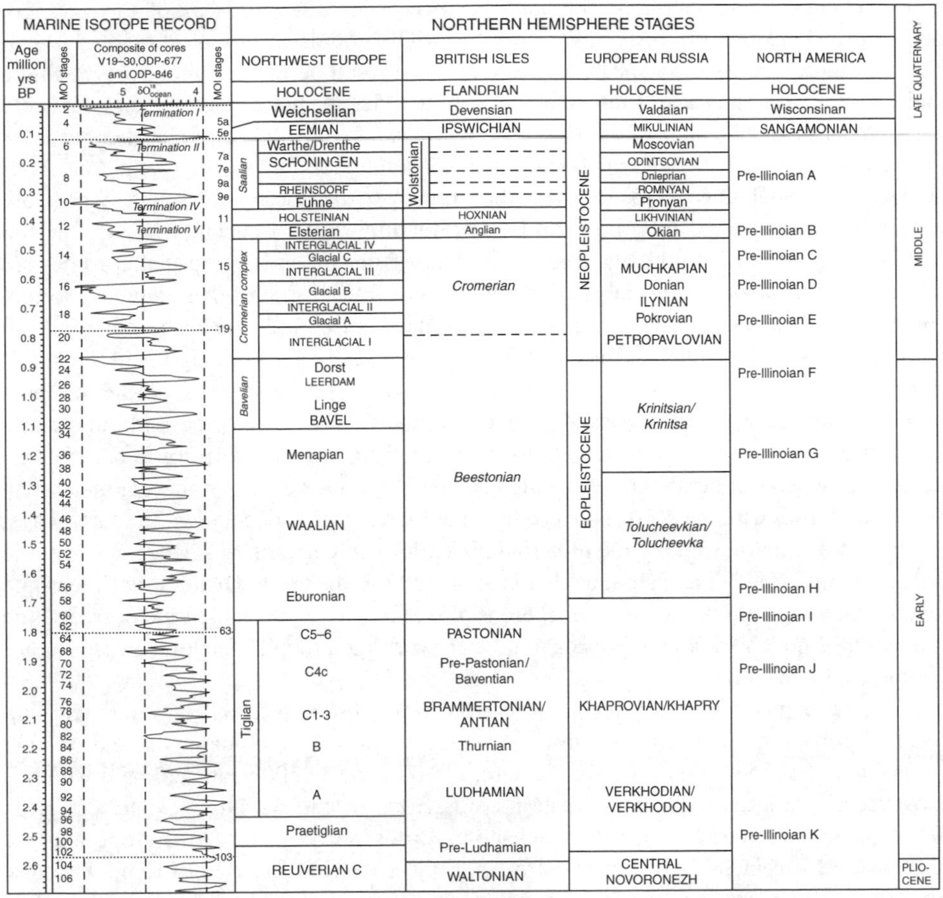

MARINE ISOTOPE RECORD

Age million yrs BP: 0.1, 0.2, 0.3, 0.4, 0.5, 0.6, 0.7, 0.8, 0.9, 1.0, 1.1, 1.2, 1.3, 1.4, 1.5, 1.6, 1.7, 1.8, 1.9, 2.0, 2.1, 2.2, 2.3, 2.4, 2.5, 2.6

MOI stages: 2, 4, 6, 8, 10, 12, 14, 16, 18, 20, 22, 24, 26, 28, 30, 32, 34, 36, 38, 40, 42, 44, 46, 48, 50, 52, 54, 56, 58, 60, 62, 64, 68, 70, 72, 74, 76, 78, 80, 82, 84, 86, 88, 90, 92, 94, 96, 98, 100, 102, 104, 106

Composite of cores V19–30, ODP-677 and ODP-846; 5 $\delta^{18}O_{ocean}$ 4; Termination I; Termination II; Termination IV; Termination V

MOI stages: 5a, 5e, 7a, 7e, 9a, 9e, 11, 15, 19, 63, 103

NORTHERN HEMISPHERE STAGES

NORTHWEST EUROPE	BRITISH ISLES	EUROPEAN RUSSIA	NORTH AMERICA	
HOLOCENE	FLANDRIAN	HOLOCENE	HOLOCENE	LATE QUATERNARY
Weichselian	Devensian	Valdaian	Wisconsinan	
EEMIAN	IPSWICHIAN	MIKULINIAN	SANGAMONIAN	
Saalian: Warthe/Drenthe	*Wolstonian*	NEOPLEISTOCENE: Moscovian	Pre-Illinoian A	MIDDLE
SCHONINGEN		ODINTSOVIAN		
		Dnieprian		
RHEINSDORF		ROMNYAN		
Fuhne		Pronyan		
HOLSTEINIAN	HOXNIAN	LIKHVINIAN		
Elsterian	Anglian	Okian	Pre-Illinoian B	
Cromerian complex: INTERGLACIAL IV	*Cromerian*	MUCHKAPIAN	Pre-Illinoian C	
Glacial C		Donian	Pre-Illinoian D	
INTERGLACIAL III		ILYNIAN		
Glacial B		Pokrovian	Pre-Illinoian E	
INTERGLACIAL II				
Glacial A				
INTERGLACIAL I		PETROPAVLOVIAN		
Bavelian: Dorst	*Beestonian*	EOPLEISTOCENE: *Krinitsian/ Krinitsa*	Pre-Illinoian F	EARLY
LEERDAM				
Linge				
BAVEL				
Menapian			Pre-Illinoian G	
WAALIAN		*Tolucheevkian/ Tolucheevka*		
Eburonian			Pre-Illinoian H	
			Pre-Illinoian I	
Tiglian: C5–6	PASTONIAN	KHAPROVIAN/KHAPRY		
C4c	Pre-Pastonian/ Baventian		Pre-Illinoian J	
C1-3	BRAMMERTONIAN/ ANTIAN			
B	Thurnian			
A	LUDHAMIAN	VERKHODIAN/ VERKHODON		
Praetiglian	Pre-Ludhamian		Pre-Illinoian K	
REUVERIAN C	WALTONIAN	CENTRAL NOVORONEZH		PLIO-CENE

Figure 1.4 *The MOI record based on a composite of deep-ocean cores (V19–30; ODP-677 and ODP-846) (left) and the Quaternary stratigraphy of the northern hemisphere set against this record (right). The marine isotope signal shows the oxygen isotope stages back to 2.6 million years* BP. *In the correlation table, temperate (interglacial) stages are shown in upper case, while cold (glacial) stages are shown in lower case. Complexes which include both temperate and cold stages are in italics (based on Gibbard* et al., *2004)*

past 800 000 years or so, there have been around ten interglacial and ten glacial stages, while over the course of the entire Quaternary, back to 2.5 million years or so, more than 100 isotopic stages have been identified (Shackleton *et al.*, 1990; Figure 1.4, left). This is many more temperate and cold stages than has been recognised in terrestrial sequences, and hence the deep-ocean isotope signal provides a unique proxy record[7] of global climate change. It also constitutes an independent climatostratigraphic scheme against which terrestrial sequences can be compared. This approach is exemplified in a number of case studies discussed in the following pages, while the use of oxygen isotope stratigraphy as a basis for dating Quaternary events is considered in Chapter 7.

The Quaternary terrestrial stratigraphic sequence in different areas of the northern hemisphere, and possible correlatives with the MOI record, is shown on the right-hand side of Figure 1.4. Broadly speaking, the Quaternary can be divided into Early, Middle and Late periods. The **Late Quaternary**, which includes the present interglacial, last cold stage and last interglacial (ca. 0–125 000 years ago), is readily correlated between the various regions, and this warm–cold–warm sequence can be equated with the MOI stratigraphy (MOI stages 1–5). Prior to that, however, the various regional records are less easily correlated. During the **Middle Quaternary**, which encompasses the period from ca. 125 000 to 780 000 years ago, a number of glacial and interglacial episodes are reflected in the various terrestrial stratigraphic records, but several of these have no formal designation. Moreover, some designated warm and cold periods appear to contain both warm and cold stages (often several), while there are clearly gaps or hiatuses in the stratigraphic sequences. As a result, correlation not only between each of the regional sequences but also between these and the MOI 'template' becomes increasingly uncertain. These problems are even more acute during the **Early Quaternary** (prior to ca. 780 000 years ago) where the number of designated stages is even fewer, and both regional correlations and links with the MOI sequence become increasingly speculative. In many ways, Figure 1.4 exemplifies one of the principal difficulties in Quaternary science, namely the lack of a universal dating technique that is applicable to the entire Quaternary time range and to all stratigraphic contexts. The figure is, nevertheless, a useful aide-memoir, and the reader will find it helpful to refer back to it when working through some of the case studies later in the book.

One part of the Quaternary record where there is a broad measure of agreement and where, moreover, there is also closer dating control is the climatic oscillation that occurred at the end of the Last Cold Stage (ca. 15 000 and 11 500 years ago) and which is most clearly reflected in proxy climate records from around the North Atlantic region. This episode is referred to as the **Weichselian Lateglacial** in northern Europe and the **Devensian Lateglacial** in Britain (Figure 1.5, right). It is characterised by rapid warming around 14 800 years ago (the **Bølling-Allerød Interstadial** in Europe; **Lateglacial** or **Windermere Interstadial** in Britain), a significant cooling (**Younger Dryas** or **Loch Lomond Stadial**) around 12 900 years ago, and finally an abrupt climatic amelioration at the onset of the present (**Holocene**) interglacial at ca. 11 500 years ago. In the Greenland (GRIP) ice core (Chapter 5), this climatic oscillation is reflected in a series of clearly defined 'events' in the oxygen isotope record (GS-2; GI-1; GS-1: Figure 1.5, left). Greenland Interstadial 1 is further divided into a series of sub-events, with GI-1a, GI-1c and GI-1e representing warmer intervals, and GI-1b and GI-1d reflecting cooler episodes (Björck *et al.*, 1998; Walker *et al.*, 1999). Whereas the timescale for terrestrial sequences from Britain and northern Europe is based on calibrated radiocarbon years (section 2.6), the Greenland (GRIP) record is in ice-core years (section 5.5). Again, the reader may find it useful to cross-reference some of the later case studies with Figure 1.5.

1.6 The Scope and Content of the Book

In the following chapters, the various dating techniques that are available to Quaternary scientists (Figure 1.6) are introduced, explained and evaluated. The last element is

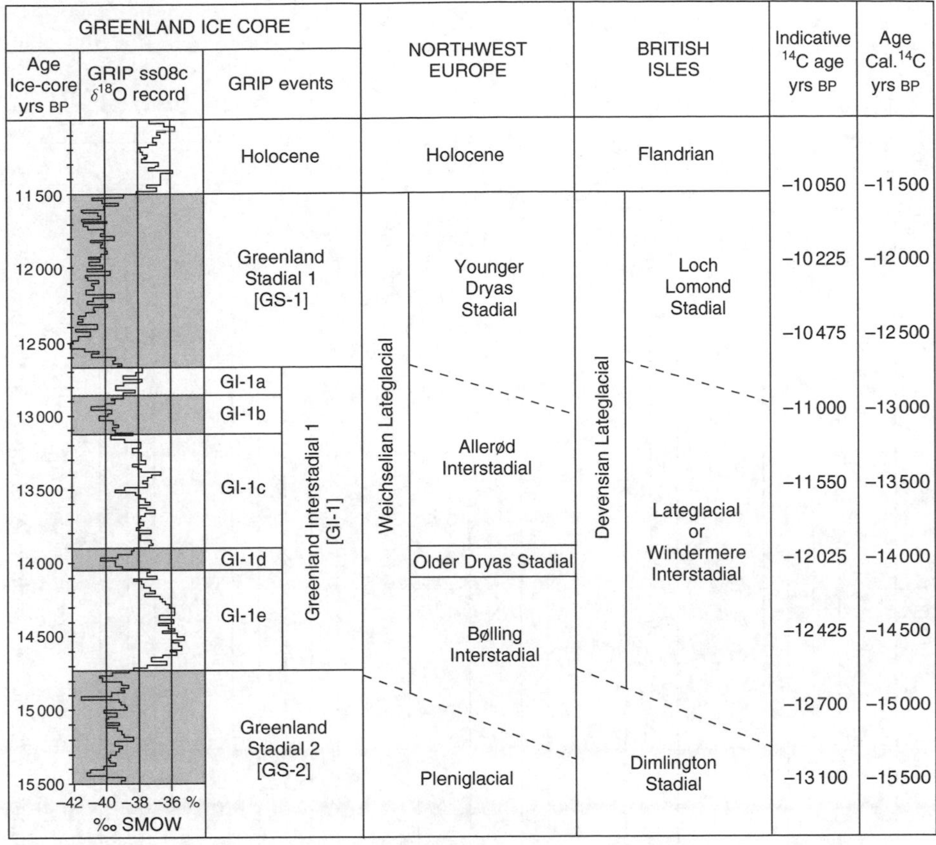

Figure 1.5 *The $\delta^{18}O$ record from the GRIP Greenland ice core showing the Lateglacial event stratigraphy (left), and the stratigraphic subdivision of the Lateglacial in northwest Europe and the British Isles. The isotopic record is based on the GRIP ss08c chronology, and the colder stadial episodes are indicated by dark shading. The radiocarbon ages should be regarded as indicative ages only (partly after Lowe* et al.*, 2001)*

especially important because it is important to understand not only how each method works, but also where and why errors are likely to occur. Some of these may arise from the nature of the sample; others from analytical limitations. Whatever the cause, these will impact on the resultant age determinations. Each section concludes with a number of examples or case studies. These have been carefully selected to show how the different techniques can be employed in Quaternary science and to give an indication of the range of applications of each method.

Chapters 2–5 deal with techniques that enable ages to be determined in years before the present (years BP). In other words, they allow **estimates of age** to be obtained. Chapters 2–4 describe **radiometric dating techniques**, where age is determined from measurements either of radioactive decay of some unstable chemical elements (Chapters 2 and 3), or of the effects of radioactive decay on the crystal structure of certain minerals or fossils (Chapter 4). Chapter 5 reviews a group of methods based on the regular accumulation of

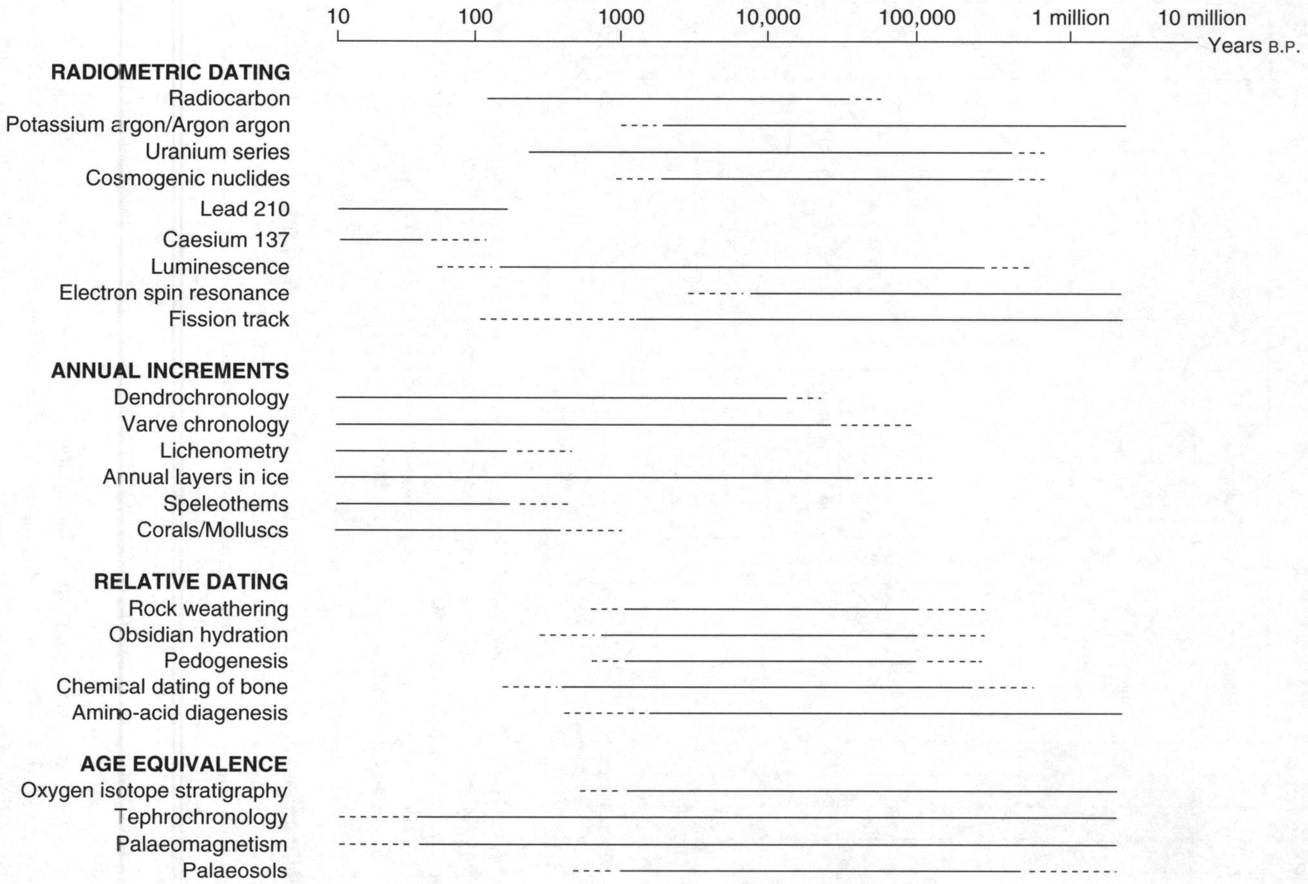

Figure 1.6 *The effective dating ranges of the different techniques discussed in this book*

sediment or biological material through time, and which form **annually banded records**. All of the techniques that enable estimates of age to be made have sometimes been referred to as *absolute dating methods*. This term has not been used here; indeed it has been deliberately avoided because it implies a level of accuracy and precision that can seldom, if ever, be achieved in reality. As we saw above, and as will be amplified in the following discussions, where age estimates are being obtained, errors are unavoidable and hence there will inevitably be an element of uncertainty associated with each age determination. There is, therefore, nothing 'absolute' about a date, and it should not be referred to as such.

In Chapters 6 and 7, two further groups of dating techniques are considered. The first involves the grouping of fossils or sedimentary units which are then ranked in relative order of antiquity; hence, these are known as **relative dating methods**. Some are based on the principles of stratigraphy where relative age can be determined by the position of stratigraphic units in a geological sequence; others use the degree of degradation or chemical alteration (both of which may be time-dependent) on rock surfaces, in soils or in fossils, to establish relative order of age. Chapter 7 considers methods that enable **age equivalence** to be determined, based on the presence of contemporaneous horizons in separate and often quite different stratigraphic sequences. With respect to both relative dating and age-equivalence techniques, where the various stratigraphic units or fossil materials can be dated by one of the age-estimate methods described in Chapters 2–5, it may prove possible to fix the relative or age-equivalent chronologies in time. In other words, they can be calibrated to an independently derived timescale.

Notes

1. The duration of the Quaternary is still a matter for debate, with some authorities arguing for a 'shorter' timescale of around 1.6–1.8 million years, while others subscribe to the view that a longer timescale of 2.5–2.6 million years is more appropriate. There is, perhaps, a majority in favour of the longer chronology and this interpretation has been followed here.
2. The **Pleistocene** epoch ended around 11 500 years ago and was succeeded by the **Holocene**, the warm period in which we live. As the present temperate period is simply the most recent of a number of temperate episodes that form part of a long-term climatic cycle, the last 11 500 years can be seen as part of the Pleistocene (West, 1977). Hence, the terms 'Quaternary' and 'Pleistocene' are often used interchangeably.
3. **Uniformitarian reasoning**, as initially developed by the Scottish geologist James Hutton in the later eighteenth century, emphasises the continuity of geological processes through time. Hence contemporary processes (**modern analogues**) can be used as a basis for interpreting past events. Uniformitarianism is often described by the dictum 'the present is the key to the past'.
4. The **Astronomical Theory of Climate Change** is based on the assumption that surface temperatures of the earth vary in response to regular and predictable changes in the earth's orbit and axis. The three principal components are the **precession of the equinoxes** (apparent movement of the seasons around the sun) with a periodicity of ca. 21 000 years, the **obliquity of the ecliptic** (variations in the tilt of the earth's axis) with a periodicity of ca. 41 000 years, and the **eccentricity of the orbit** (changes in the shape of the earth's orbit) with a periodicity of ca. 96 000 years. Collectively these govern the amount of heat received by the earth and the distribution of this heat around the globe. First developed in its modern form by the Scottish scientist James Croll in

the nineteenth century, the theory was subsequently elaborated by the Serbian geophysicist Milutin Milankovitch in the 1930s. The radiation balance curves that he produced can be calibrated to the orbital parameters and used to provide an **astronomical timescale** for glacial–interglacial cycles (Chapter 5). Further explanation of the astronomical theory can be found in standard Quaternary texts, such as those by Lowe and Walker (1997), Roberts (1998), Williams *et al.* (1998), Wilson *et al.* (2000) and Bell and Walker (2005).

5. A major discovery in physics during the early twentieth century was that sub-atomic particles sometimes behave as if they are waves, a concept that lies at the heart of the science of quantum physics. One consequence is that it is impossible to measure both the *position* of a sub-atomic particle and its *velocity*, an idea that was first proposed by the German physicist Werner Heisenberg in his **Uncertainty Principle**. This indeterminacy in the sub-atomic world can be seen very clearly whenever a single atomic event can be observed, such as in radioactivity. Although quantum physics is a highly complex field, there are a number of accessible texts that deal with this subject or that include sections on it. Ones that I have found particularly informative (and enjoyable!) are by Gribbin (1984), Close *et al.* (1987), Barrow (1988), Gribbin (1995), Penrose (1999), Rees (2000) and, of course, Bryson (2003).
6. During evaporation from the free ocean surface, a **fractionation** (or separation) occurs so that more of the lighter oxygen isotope, ^{16}O, is drawn into the atmosphere than the heavier isotope, ^{18}O. In the cold stages of Quaternary, therefore, large amounts of the lighter isotope would have been transported poleward by moisture-bearing winds and locked into the greatly expanded ice sheets. As a consequence, ocean water would have been relatively 'enriched' in the heavier isotope ^{18}O. The reverse would obtain during interglacial stages for, with reduced land ice cover, more ^{16}O would have been returned to the oceans where water would have become relatively 'depleted' in ^{18}O. Accordingly, the $\delta^{18}O$ trace provides a record of changing volumes of land ice, and hence of glacial/interglacial climatic fluctuations.
7. A **'proxy climatic record'** is one based on an indirect measure of climate. In other words it is based on *inferential* evidence (pollen, plant macrofossils, etc.), as opposed to *direct evidence* obtained using a thermometer or rain gauge.

2

Radiometric Dating 1: Radiocarbon Dating

Life exists in the universe only because the carbon atom possesses certain exceptional properties.

Sir James Jeans

2.1 Introduction

Radiocarbon dating was one of the first radiometric techniques to be developed and, despite the fact that it is applicable to only a relatively short span of Quaternary time (50 000 years or so; see below), it is perhaps the most widely used of all the radiometric techniques. It owes its origin to a remarkable American chemistry professor, Willard Libby, who, in the years immediately following the end of the Second World War, was investigating the possibility that radiocarbon might exist in biological materials. Along with a group of colleagues, Libby was able to demonstrate that radiocarbon could be detected in samples from the Baltimore sewage works, and from these seemingly unprepossessing beginnings, a technique was born that was to revolutionise our view of Late Quaternary time. Not only is the carbon atom the building block of life, therefore (see above), it also provides us with a means of dating life. The first radiocarbon measurements were published in 1949, and since then hundreds of thousands of dates have been produced by more than 100 laboratories all over the world. The story of the development of radiocarbon can be found in Libby's book *Radiocarbon Dating* (1952; 2nd edition 1955), and he was awarded

Quaternary Dating Methods M. Walker

the Nobel Prize for chemistry in 1960. Good overviews of radiocarbon dating can be found in the volumes by Taylor (1987), Bowman (1990) and Taylor *et al.* (1992), and there are shorter accounts in, *inter alia*, Aitken (1990), Lowe and Walker (1997) and Taylor (1997; 2001). The journal *Radiocarbon*, along with its website and associated links, is a valuable source of information on recent developments in radiocarbon dating and on applications of the technique.

2.2 Basic Principles

Radiocarbon (^{14}C) is one of three isotopes of carbon, the others being ^{12}C and ^{13}C. By far the most abundant of these is ^{12}C which comprises around 98.9% of all naturally occurring carbon. ^{13}C forms around 1.1% and ^{14}C one part in 10^{10}%. In other words, only about one in a million million atoms of carbon is ^{14}C. Both ^{12}C and ^{13}C are stable isotopes, but ^{14}C is not and it 'decays' to a stable form of nitrogen, ^{14}N, through the emission of beta (β) particles. One β particle is released from the nucleus for every atom of ^{14}C that decays. It is this instability, or *radioactivity*, which gives us the name 'radiocarbon'.

Atoms of ^{14}C are formed in the upper atmosphere through the interaction between cosmic ray neutrons, which reach the earth's atmosphere from deep space, and nitrogen. This involves neutron capture by the nitrogen (^{14}N) atom, and the loss of a proton, to create ^{14}C. The ^{14}C atoms produced by this process combine with oxygen to form a particular form of carbon dioxide ($^{14}CO_2$) which mixes with the non-radiocarbon containing molecules of CO_2. In this way, ^{14}C becomes part of the global carbon cycle and is assimilated by plants through the photosynthetic process, and by animals through the ingestion of plant tissue. The majority of ^{14}C (more than 95%) is absorbed into the oceans as dissolved carbonate, which means that organisms that live in sea water (corals, molluscs, etc.) will also take up ^{14}C during the course of their life cycle. Although the ^{14}C in the terrestrial biosphere and in the oceans is constantly decaying, it is continually replenished from the atmosphere. Hence the amount of ^{14}C that is stored in plant and animal tissue and in the world's oceans, the *global carbon reservoir*, remains approximately constant through time. In effect, a position has been reached where the carbon that is used to build plant and animal tissue is in *isotopic equilibrium* with the atmosphere; in other words the levels of ^{14}C activity in plants and animals are the same as that in the atmosphere. The ^{14}C reservoir can therefore be likened to a car with a drip-feed to the fuel tank; the car will use fuel as the engine runs, but as the tank is being constantly topped-up, the fuel will remain at more or less the same level.

Once an organism dies, however, it becomes isolated from the ^{14}C source, no further replenishment of ^{14}C can take place, and the 'radiocarbon clock' runs down by radioactive decay, which occurs at a constant rate. Hence, by measuring the amount of ^{14}C that remains in a sample of fossil material (the residual ^{14}C content) and comparing this to modern ^{14}C in standard material, an age can be inferred for the death of the organism. In order to be able to do this, however, we need to know the rate at which ^{14}C decays. Experimental results have shown that the decay rate of ^{14}C is 1% every 83 years. This might imply that after 8300 years, all residual activity in a sample will have ceased. However, as with all radioactive isotopes, the decay curve for ^{14}C is not linear but exponential

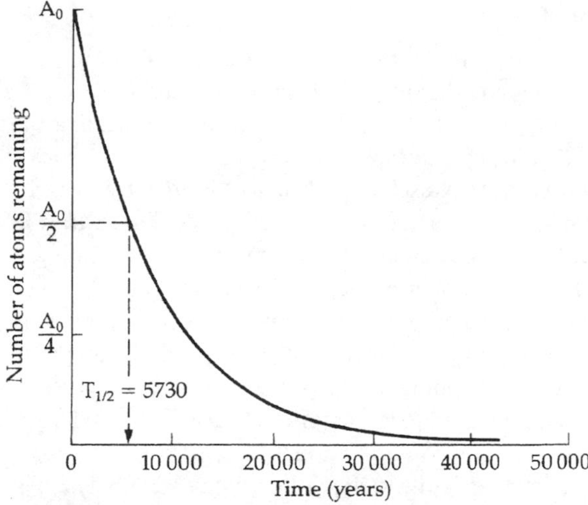

Figure 2.1 *The decay curve for radiocarbon is exponential, not linear. This means that the percentage decrease in number of atoms in a given unit of time is constant. Hence, after each half-life the number of atoms remaining is halved. If there are A_0 atoms of radiocarbon at the beginning of the decay process, then after one half-life there will be $A_0/2$ atoms remaining; after two half-lives, there will be $A_0/4$, after three, $A_0/8$, and so on (after Bowman, 1990). Reproduced by permission of The British Museum Press*

(Figure 2.1), and this means that materials significantly older than this can still be dated. The half-life of a ^{14}C atom is 5730 years,[1] and under normal circumstances, the limit of measurement of ^{14}C activity (i.e. decay rate) is eight half-lives. This translates into an upper age limit of around 45 000 years (but see following sections). Samples older than this are usually described as being of 'infinite age', and are expressed, for example, as >45 000 years.

2.3 Radiocarbon Measurement

Radiocarbon dates can be obtained on a range of biogenic materials. These include, *inter alia*, wood, peat, organic lake sediment, plant remains, charcoal, shell and coral. More problematic are bone and soil, while radiocarbon dates have also been obtained on more unconventional materials, such as cloth, metalwork or fossil pigment. Two approaches are employed to measure residual ^{14}C activity in samples of these materials relative to modern standards: **beta counting**, which involves the detection and counting of β emissions from ^{14}C atoms over a period of time, working on the principle that the rate of emissions will reflect the residual level of ^{14}C activity within the sample, and **accelerator mass spectrometry (AMS)**, which employs particle accelerators as mass spectrometers to count the relative number of ^{14}C atoms in a sample, as opposed to the decay products.

2.3.1 Beta Counting

Beta counting can be carried out in two ways. In **gas proportional counting**, the sample is converted to a gas (carbon dioxide, ethylene or methane) and injected into a counting chamber where each β emission is detected by a charged wire that runs down the centre of the chamber. In **liquid scintillation counting**, the sample is converted to benzene to which a 'scintillant' (usually a phosphoric substance) is added, and each β particle emission stimulates a pulse of light which can be counted photoelectrically. In the early days of radiocarbon dating, laboratories employed gas proportional counting, but the majority of radiocarbon facilities today now use liquid scintillation counting. This is because when converted to benzene, the majority of the sample is carbon, whereas in gas counting, the carbon dioxide is mostly oxygen. Hence significantly more sample material is required for gas proportional than for liquid scintillation counting. In order to ensure comparability between dates, laboratories compare sample activity to a modern reference standard, which is the modern activity of NBS (National Bureau of Standards) oxalic acid held by the American Bureau of Standards. Radiocarbon dates are always measured with respect to this standard and are expressed in years BP (before present), where 'present' is the standard year AD 1950 (van der Plicht, 2002).

In beta counting, it is important to bear in mind that age is not the quantity that is being measured; rather it is the ^{14}C *activity* of the sample that is interpreted as indicating 'age', and there will always be an element of uncertainty in determining this activity partly because of limitations in laboratory procedures, and partly because of the effects of the randomness of radioactive decay on the counting statistics. The former is difficult to estimate, but the latter can be quantified. Because radiocarbon dating is based on repeated measurements, the distribution of results is usually described by a normal (or Gaussian) probability function, and hence the resultant age is always expressed as a mean determination with a plus or minus value of one standard deviation around the mean. The fact that this is a statistical measure is not always appreciated. Hence a radiocarbon date of 5000 ± 50 years (1σ error) does not mean that the true age of the sample lies between 4900 and 5100 ^{14}C years. What it means is that there is a 68% probability (1 standard deviation: 1σ) that the age lies between those two values. There is still a one in three chance that the age could lie outside that range. It is for this reason that users of radiocarbon dates are advised to think in terms of 95% probability (2 standard deviations: 2σ). In the above example, therefore, we would say there is a 95% chance that the true age of the sample lies somewhere between 4800 and 5200 ^{14}C years. Even then, there is a 5% chance that the true age could lie beyond this range.

2.3.2 Accelerator Mass Spectrometry

In the 1980s, a significant breakthrough was made in radiocarbon dating with the employment of particle accelerators as mass spectrometers (Figure 2.2). Mass spectrometers are widely used to detect atoms of particular elements on the basis of differences in atomic weights. Conventional mass spectrometers, however, cannot discriminate between ^{14}C and other molecules of similar weights, such as ^{14}N. However, by accelerating particles to very high speeds, the small ^{14}C signal can be separated from that of other isotopes, hence the term **accelerator mass spectrometry** (often abbreviated to

Figure 2.2 *The 5-Mv National Electrostatics Corporation Accelerator Mass Spectrometer at the Scottish Universities Environmental Research Centre, East Kilbride, Scotland, UK. The accelerator itself is on the left of the photograph, while the ion source (Figure 2.4) is in the screened area to the right. See also Figure 2.3 (photo: Sheng Xu)*

AMS). It is important to emphasise, however, that it is not the *absolute* number of ^{14}C atoms that is being measured; the abundance of ^{14}C atoms is so small that it would be extremely difficult to measure total amounts. Rather, AMS determines the isotope ratio of ^{14}C relative to that of the stable isotopes of carbon (^{13}C or ^{12}C), and the age is determined by comparing this ratio with that of a standard of known ^{14}C content. This is often referred to as **isotope ratio mass spectrometry**. The most commonly used system, which involves two separate phases of acceleration, is a **tandem accelerator** (Figure 2.3). Samples are converted to graphite (or a CO_2 source) and mounted on a metal disc (Figure 2.4). Caesium ions (Cs^+) are then fired at this 'target' and the negatively ionised carbon atoms (C^-) produced are accelerated towards the positive terminal (Figure 2.3). During passage through the 'stripper', four electrons are lost from the C^- ions (five in some AMS systems), and they emerge with a triple positive charge (C^{3+}, or C^{4+} if five electrons are stripped). Repulsion from the positive terminal leads to a second acceleration of the carbon ions through focussing magnets, where deflection occurs according to mass, and the signal of the stable isotope ^{13}C (and sometimes ^{12}C) can be measured using Faraday cups, the ^{14}C signal being collected by ion detectors (Figure 2.3). Radiocarbon ages can be obtained by comparing the ^{14}C to ^{13}C or ^{12}C ratios of the sample material with those for targets in the same set that have been made up from

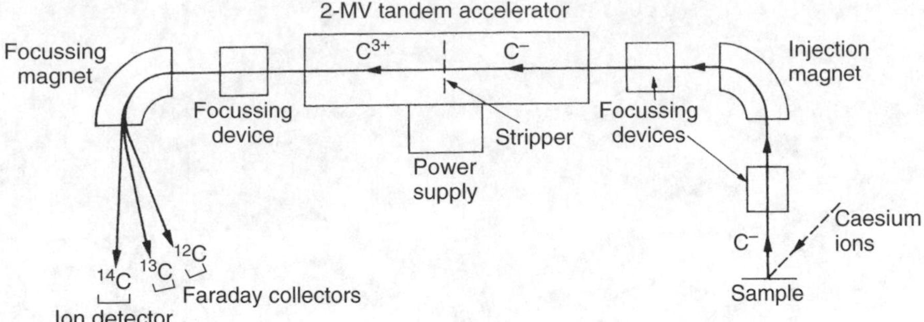

Figure 2.3 *Schematic diagram of a tandem accelerator for the detection and counting of ^{14}C atoms. For explanation, see text (after Bowman, 1990). Reproduced by permission of The British Museum Press*

Figure 2.4 *The exposed sample target wheel (ion source) of the AMS in Figure 2.2. The wheel holds 134 graphite samples, but not all of these will be of fossil material as standards of known age are interspersed at regular intervals (photo: Sheng Xu)*

material of known modern age, usually NBS oxalic acid. The error term that accompanies all AMS dates reflects, as in conventional dating, statistical uncertainties relating to the ^{14}C decay curve, random and systematic errors that occur during the measurement process, as well as uncertainties that arise in measuring the oxalic acid standards and in quantifying the natural background ^{14}C.

The advantages of AMS over beta counting lie in the small sample size required for AMS dating and in the speed and effectiveness with which samples can be processed. Beta counting typically requires several grams of material (typically 1–2 g of carbon) and it can take a number of days (or even weeks) to count a sample. By contrast, AMS laboratories routinely analyse samples containing 1 mg of organic carbon or less. This means, for example, that a radiocarbon date can be obtained from minute quantities of material, such as a single seed or leaf, or even from pollen grains (Mensing and Southon, 1999; Vandergroes and Prior, 2003) or fragments of insects (Hodgins *et al.*, 2001; Walker *et al.*, 2001). The measurement process is relatively rapid and can be completed in a matter of hours rather than days. Moreover, depending on the size of the target wheel (Figure 2.4), upwards of 50 samples can typically be measured during a single run, which is a much more efficient procedure than in beta counting. The disadvantages of AMS are capital cost (the cost of establishing a new AMS facility could be upwards of £3 m at 2004 prices), the cost of each date (75–100% more than a radiometric date) and, until recently, analytical precision. For some time, AMS laboratories were unable to match the levels of precision in beta counting, where the error estimate is typically around 1%, i.e. ±50 years at 5500 and ±80 years at 12 000. In recent years, however, counting statistics have improved and many AMS facilities are now more or less comparable to radiometric laboratories in terms of precision. However, neither type of facility can match the performance of the 'high-precision' radiometric laboratories, such as Belfast, Groningen or Seattle, where dates can be produced with standard errors of 20 years or less, although in order to achieve this level of analytical precision, large amounts of material (5–20 g carbon) and longer counting times are required.

2.3.3 Extending the Radiocarbon Timescale

A number of attempts have been made to extend the radiocarbon timescale beyond the conventional limit of eight half-lives. The problem is that in very old samples, residual ^{14}C activity is so low that impossibly long counting times are needed in order to obtain a meaningful estimate of age. One approach to this problem involves the enhancement of the ratio of ^{14}C to ^{12}C using controlled *isotopic fractionation*[2] (section 2.4.2.) before measurement, a technique known as **isotopic enrichment**. In the radiocarbon laboratory in Seattle, for example, one of the laboratories where this method was pioneered, thermal diffusion has been employed to increase ^{14}C activity in a sample by a factor of 6–7, and this has enabled finite dates of up to 75 000 years to be obtained (Stuiver *et al.*, 1978). However, the process requires very large samples of material and lengthy preparation and counting times (months rather than days or weeks). Another approach to extending the radiocarbon timescale is to use large-volume, high-precision counters such as those at the Groningen laboratory in the Netherlands in which very old samples (up to ten half-lives) can be measured. Finite ^{14}C ages in the 50 000–60 000 year range have been obtained by this facility (Behre and van der Plicht, 1992), but again very large samples and extended counting times are needed. More recently, a technique has been developed for the dating of old charcoal samples which involves more stringent pretreatments (the 'ABOX pretreatment') and innovations in target preparation for AMS dating ('stepped combustion') in order to remove contaminants. Again, the result has been finite ^{14}C ages in the 50 000–60 000 year range (Bird *et al.*, 2003).

2.3.4 Laboratory Intercomparisons

In order to ensure comparability between different laboratories, the International Radiocarbon Community organises inter-laboratory comparisons. These involve the preparation and distribution of typical sample materials (peat, wood, bone, etc.) to different radiocarbon laboratories and which, ideally, will have been dated by independent means, such as dendrochronology. These intercomparison programmes are an essential component of the quality assurance process in radiocarbon dating, for they give confidence to radiocarbon users that laboratories worldwide are producing measurements that are reliable and in accordance with good practice. The most recent of these exercises, FIRI [the fourth international radiocarbon intercomparison (Boaretto *et al.*, 2002; Scott, 2003)], showed an encouragingly broad measure of agreement between radiocarbon measurements made in different laboratories on a wide range of sample materials. Equally importantly, it also demonstrated no statistically significant difference between measurements made by radiometric or AMS techniques.

2.4 Sources of Error in Radiocarbon Dating

In order to translate a measurement of ^{14}C activity in a sample of fossil material into a meaningful calendar age at a reasonable level of precision, five major assumptions have to be made:

1. The $^{14}C/^{12}C$ ratio in each part of the global carbon reservoir (atmosphere, biosphere, freshwater, marine water) has remained essentially constant over time.
2. Complete and rapid mixing of ^{14}C occurs throughout these reservoirs on a global basis.
3. The ratios between the different carbon isotopes (e.g. $^{13}C/^{12}C$) in samples have not been altered, other than by ^{14}C decay prior to or since the death of an organism.
4. The half-life of ^{14}C is accurately known with a reasonable level of precision.
5. Natural levels of ^{14}C can be measured to appropriate levels of accuracy and precision (Taylor, 2001).

Of these, there is a reasonable degree of confidence amongst the radiocarbon community that the half-life of ^{14}C has been accurately determined, and that natural levels of ^{14}C can be established. Questions remain, however, about the remaining three assumptions, and these lead to uncertainties (errors) in radiocarbon dating. Some of the major sources of error are considered in this section.

2.4.1 Contamination

Contamination refers to the addition of younger or older carbon to the sample material and, if unquantified, will result in dates that are aberrant. Contamination can occur prior to field sampling. In a peat or soil profile downward penetration of roots or of younger humic acids can introduce younger materials into older horizons, while in a pond or lake, bioturbation caused by organisms on the lake bed can lead to disturbance of the sediment sequence and the downward movement of younger sediments. The relatively high activity of younger carbon by comparison with older fossil material means that even small amounts of modern contaminant can produce major errors in radiocarbon dates, and the effects will increase with the age of the sample. For example, the addition of 1% modern carbon to a 17 000-year-old sample will reduce the age by 600 years, for a 34 000-year-old sample the same percentage contamination will lead to an error of 4000 years, while if the sample is infinitely old (i.e. beyond the conventional dating range of eight half-lives), the result will be an 'apparent age' of 38 000 years (Aitken, 1990). Contamination can also occur through the incorporation of older carbon residues. This is especially problematical in the dating of lake sediments (section 2.5.1) where older organic carbon washed into the lake from the catchment slopes or carbon from local bedrock can result in an ageing effect in radiocarbon dates from lake muds. The addition of 1% old carbon will increase the age of a modern sample by around 80 years.

Contamination can also occur during field sampling, and may even occur in the laboratory, for example due to fungal growth during sample storage (Wohlfarth *et al.*, 1998a), or through modern contamination by dust, skin, etc. Radiocarbon laboratories employ physical or chemical pretreatment procedures to remove some of these contaminants. Samples are examined for the presence of rootlets and other obvious signs of extraneous material; they are then sieved and treated with hot acid to dissolve older carbonates and with alkali solutions to remove younger humic acids that may coat the samples. However, if the contaminant has the same chemical composition as the sample, selective separation of the contaminant may prove to be impossible. Hence, stringent as modern laboratory procedures are, they cannot guard against all aspects of contamination and this therefore remains a potential source of error in many radiocarbon dates.

2.4.2 Isotopic Fractionation

We have already seen that carbon occurs in three isotopic forms, ^{12}C, ^{13}C and ^{14}C. Although these are chemically indistinguishable, in any biological pathway there will be a tendency for the lighter isotope ^{12}C to be preferentially taken up. By the same token, ^{13}C will be taken up in preference to ^{14}C. In photosynthesis, for example, there is a greater uptake of ^{12}C, and hence there will be an enrichment in ^{12}C and a relative depletion in ^{14}C. This means that a growing plant would be expected to have a lower ^{14}C activity than the contemporary atmosphere. Hence, when fossil plant remains are dated, the radiocarbon dates will appear to be older and, since uptake also varies according to species, different parts of the biosphere will appear to have different radiocarbon ages (Bowman, 1990). The reverse may obtain when marine organisms are being dated. In ocean water there is a preferential absorption of the heavier isotope ^{14}C and a commensurate relative depletion in ^{12}C. Ocean water will, therefore, be relatively enriched in ^{14}C, and organisms that take up ^{14}C from the ocean water during their life cycle (molluscs, corals) will have a higher ^{14}C activity (leading to a younger ^{14}C age), although this may, to some extent, be offset by the marine reservoir effect (section 2.4.3).

Radiocarbon laboratories correct for the effects of isotopic fraction by working on the principle that the heavier isotope ^{14}C is twice as enriched as ^{13}C. Because both ^{12}C and ^{13}C are stable isotopes and can be measured directly, the $^{13}C/^{12}C$ ratio in a sub-sample of fossil material is determined by mass spectrometry, and this ratio is then compared to the $^{13}C/^{12}C$ ratio of a standard PDB limestone (belemnite carbonate from the Cretaceous

Table 2.1 *Approximate $\delta^{13}C$ values for various materials. The ranges on these data are typically ± 2 or 3‰, but greater variability is possible. Each per mil deviation from −25‰ represents ca. 16 years (from Bowman, 1990, with minor modifications by the author)*

Material	$\delta^{13}C$ value (‰)
Wood, peat and many C_3 plants	−26
Bone collagen*	−19
Freshwater plants (very variable)	−16
Arid zone plants (C_4 plants)	−13
Marine plants	−15
Atmospheric CO_2	−8
Marine carbonates	−0

* For direct or indirect C_3 consumers.

Peedee Formation in South Carolina). The deviation of the $^{13}C/^{12}C$ ratio in the fossil sample from that standard (a measure referred to as $\delta^{13}C$) is calculated as follows:

$$\delta^{13}C = \left[\frac{(^{13}C/^{12}C)}{(^{13}C/^{12}C)_{PDB}} - 1\right] \times 1000$$

The $\delta^{13}C$ value, which is expressed in ‰ (meaning parts per thousand or 'permil'), therefore provides an accurate measure of the extent of isotopic fractionation reflected in the fossil material. Most terrestrial samples have negative $\delta^{13}C$ values, but because different photosynthetic pathways exist in plants, different levels of fractionation occur (Table 2.1). It is general practice, however, to 'normalise' these values to a figure of −25‰, which is the mean isotopic composition of wood. Accordingly, if a sub-sample of dated material yields a $\delta^{13}C$ value of −25‰, no correction is made. If a value of −30‰ is obtained, this implies a 5‰ depletion in the $^{13}C/^{12}C$ ratio, which indicates a 10‰ depletion in the $^{14}C/^{12}C$ ratio. In this case, the measured ^{14}C activity would be increased by 10‰, which equates to ca. 83 ^{14}C years (Harkness, 1979).

2.4.3 Marine Reservoir Effects

Exchange reactions between the atmosphere and the ocean surface means that CO_2, and hence ^{14}C, becomes incorporated into sea water in the form of dissolved carbonate. Once the surface water sinks, however, ^{14}C in intermediate and deep water decays without replenishment from the surface and the result is that sea water will have an **apparent age**. The ageing effect varies, from ca. 30 years or less in surface water of parts of the northern Indian Ocean (Dutta *et al.*, 2001) to around 400 years in the North Atlantic (Bard *et al.*, 1991), and almost 600 years in areas of the equatorial East Pacific (Shackleton *et al.*, 1988). In the deep oceans, however, where **residence times** are much longer, the age of sea water may be measurable in thousands of years. Hence, radiocarbon dates on organisms that live in the oceans and which take in dissolved carbonate to build their skeletal structures (such as foraminifers,[3] molluscs, etc.) will have to be corrected for the age of sea water. For example, measurements on present-day molluscs from around the British Isles show that coastal water has an apparent age of 405 ± 40 years BP, and hence radiocarbon dates on fossil marine molluscs from British contexts have tended to be corrected by that value (e.g. Peacock, 1996).

The global variations in the age of sea water referred to above arise largely because of differing patterns of ocean water movement – surface currents, thermohaline circulation,[4] upwelling, etc. However, these changes are not only variable in space; they also vary in time. This poses a serious difficulty because it means that the present-day age of sea water may not necessarily represent an appropriate correction factor for fossil material from a particular locality. For example, data from the North Atlantic indicate that at times during the transition from the last cold stage to the present interglacial the apparent age of surface water may have varied by up to 1900 ^{14}C years (section 2.7.1.5), while in the Norwegian Sea the reservoir age may have been around 1000 (±250) ^{14}C years throughout that time period (Björck *et al.*, 2003). These age values are significantly higher than those indicated by dates on modern marine molluscs, which suggests that finding the appropriate correction factor for marine radiocarbon dates may be a much more difficult proposition than has been imagined hitherto. This is a problem that has yet to be fully resolved.

2.4.4 Long-Term Variations in ^{14}C Production

Long-term variation in ^{14}C production is acknowledged to be one of the major problems in radiocarbon dating, for it is now apparent that one of the fundamental assumptions underlying the technique, namely that atmospheric ^{14}C levels have not varied significantly over time, can no longer be sustained. More than 30 years ago, scientists working on tree-ring chronologies found significant discrepancies between the age of wood based on dendrochronological dating (section 5.1) and that based on radiocarbon, with the radiocarbon ages invariably being younger than the dendrochronological dates (Renfrew, 1973). The cause of this discrepancy was attributed to variations in atmospheric ^{14}C activity. Moreover, comparison between dendrochronological and radiocarbon dating series showed that atmospheric ^{14}C activity appeared to have fluctuated over time (the so-called 'de Vries' effect[5]), not randomly but in a quasi-periodic manner through a series of interlocked cycles ranging in duration from ca. 208 to 2300 years (Sonnett and Finney, 1990). The record of long-term ^{14}C variations is also characterised by 'plateaux' of more or less constant ^{14}C age, for example at ca. 10 400, 10 000 and 9600 ^{14}C years BP (Figure 2.5), each of these plateaux reflecting a period of reduced atmospheric ^{14}C activity. On a longer timescale, extremely large variations in atmospheric ^{14}C concentration have been detected beyond 25 000 years BP (Beck *et al.*, 2001; Hughen *et al.*, 2004), and these are proving particularly problematical for the dating of events in the range of 30 000–50 000 ^{14}C years BP, such as the appearance of modern humans in Europe (Richards and Beck, 2001).

The causes of these long- and short-term variations in atmospheric ^{14}C remain to established, but a number of factors appear to be involved. A major influence may well be modulation of the cosmic ray flux by changes in the earth's geomagnetic field or by variations in the intensity of solar activity (Stuiver *et al.*, 1991). A reduction in magnetic

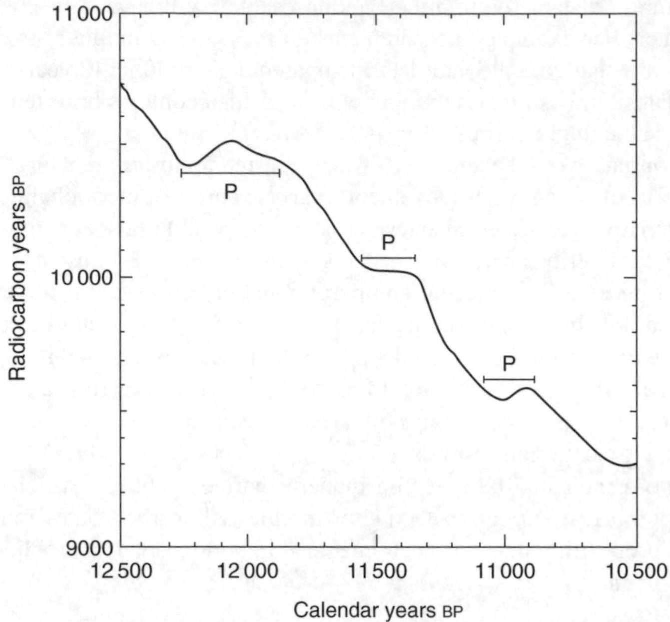

Figure 2.5 *Radiocarbon versus calendar ages for the period 9000–11 000 ^{14}C years BP. The curve is based on radiocarbon-dated samples from lakes, corals, marine sediments and tree-ring series, each of which has also been dated independently in order to obtain a direct comparison between radiocarbon and calendar ages. Note the clearly defined radiocarbon 'plateaux' at ca. 10 400, 10 000 and 9600 ^{14}C years BP. Radiocarbon dates which fall on these plateaux will have calendar age ranges of up to several hundred years (after Goslar* et al., *2000). Reprinted from Nature. Copyright 2000 Macmillan Magazines Limited*

field intensity could weaken the earth's geomagnetic shield and increase the number of cosmic rays entering the atmosphere which would result in a commensurate increase in ^{14}C production (Mazeaud *et al.*, 1992). Similarly, a reduction in solar activity, particularly in the strength of the solar wind,[6] would serve to enable more cosmic rays to enter the atmosphere and again would lead to an increase in ^{14}C production (Van Geel *et al.*, 2003). An alternative hypothesis involves changes in patterns of ocean circulation, particularly changes in ventilation[7] and deep-water formation.[8] These could give rise to an increase (or decrease) in the release of oceanic CO_2 (with relatively low ^{14}C content) into the atmosphere and this would lead to a reduction (or increase) in atmospheric $^{14}CO_2$ levels (Goslar *et al.*, 1995).

In addition to natural variations, atmospheric ^{14}C levels have been affected by human activity. Over the past 250 years or so, the burning of fossil fuels has liberated large quantities of ^{12}C into the atmosphere (Houghton *et al.*, 2001) which have diluted ^{14}C levels, while over the course of the last 50 years, this **industrial effect** has to some extent been offset by the detonation of thermonuclear devices which effectively doubled atmospheric ^{14}C concentrations in the 1960s. Partly for these reasons, and also because of recent natural ^{14}C variations, the young end of the radiocarbon timescale is generally taken to be around 300 years ago (Taylor, 2001).

Irrespective of causal factors, these long- and short-term variations in atmospheric ^{14}C concentration have profound effects for radiocarbon dating, for they mean that unless some sort of correction is made, radiocarbon time will bear little resemblance to calendar time. For example, comparisons between radiocarbon dates and dendrochronological dates suggest that a radiocarbon date of 5000 years corresponds to a calendar age of 5800 years. For older dates, the divergence between radiocarbon and calendar age is much larger. Comparisons between AMS ^{14}C and uranium-series dates (section 3.3) on fossil corals, for example, indicate that radiocarbon dates are around 3000 years younger than 'true' ages at 15 000 ^{14}C years BP, and around 4000 years younger at 20 000 ^{14}C years BP (Stuiver *et al.*, 1998). It is for this reason that it is important to refer to radiocarbon dates in terms of '^{14}C years BP', to make it clear that it is 'radiocarbon time' and not 'actual time' or 'calendar time' that is being referred to. However, there are now ways in which we can begin to convert radiocarbon to calendar years (**radiocarbon calibration**) and these are discussed in section 2.6.

2.5 Some Problematic Dating Materials

As noted above, radiocarbon dating can be undertaken on a range of biogenic materials. Some of these, such as wood, plant macrofossils, charcoal and peat, are relatively straightforward dating media and can, in most instances, be relied upon to generate a meaningful radiocarbon age determination. Other materials, however, are more difficult to deal with, and present both procedural and interpretational problems for the radiocarbon practitioner. Some of these are considered in this section.

2.5.1 Lake Sediments

Organic lake sediments are widely used as a medium for radiocarbon dating, but obtaining a reliable date from these deposits is problematical. Lake sediments are extremely susceptible to contamination by older carbon residues and this leads to a dilution of the ^{14}C concentration and dates that are older than the actual date of sediment deposition. This **mineral carbon error** may arise either through the inwashing of older organic carbon material from slopes around the lake catchment, or from erosion and subsequent incorporation into the lake sediment of even older carbon, such as coal or graphite, from local bedrock. In formerly glaciated areas, such older carbon may have been brought into the area by the passage of glacier ice (Walker *et al.*, 2001). Lake catchments containing geologically old carbonate (e.g. limestone) can be a source of infinitely old, ^{14}C-depleted, carbon for lake or pond water. This dilutes the ^{14}C concentration of the water and is reflected in plants that photosynthesise sub-aquatically, in aquatic algae, and in animals that feed on these plants and also on each other. Hence, where the remains of these organisms occur in lake sediments, there will also be a dilution in ^{14}C levels. This is termed the **hard-water effect** because of the often associated presence of calcium ions, and can add up to 1200 years to the apparent age of limnic material (Peglar *et al.*, 1989). One way around this problem is to use terrestrial material from such sediments (such as leaves and fruits) as these will not be influenced by the hard-water effect.

The often complex biogeochemistry of lake sediments leads to further difficulties in radiocarbon dating. For example, significant age differences have been discovered between different biogeochemical components of lake sediments, such as humic acid fractions, humin (alkali insoluble) fractions, lipids, chlorite treatment residues, etc., from the same stratigraphic horizon (Lowe *et al.*, 1988). Moreover, often significant differences have been found between radiocarbon ages on macrofossil cellulose and the sediments containing those plant macrofossils, with the macrofossils frequently providing younger radiocarbon ages than the host sediments (Peteet *et al.*, 1990; Walker *et al.*, 2001; 2003).

An additional problem that may occur in the dating of lake sediments is that the $^{14}C/^{12}C$ ratio in lake water may be lower than that in the atmosphere because exchange at the lake surface is relatively slow and hence lake water may have a reduced ^{14}C activity by comparison with the atmosphere. Seepage into lakes of groundwater containing dissolved carbonates further complicates this **reservoir effect** (Olsson, 1986). Although standard correction factors have been applied (e.g. 300–400 years for Swedish lakes), as is the case with the marine reservoir discussed earlier (section 2.4.3), it now seems that there may have been significant temporal changes in freshwater reservoir ages, due principally to changes over time in the lake volume/surface area ratio, and also water depth. Indeed, variations of almost 1000 years have been detected in reservoir ages of central and eastern European lakes over the course of the Holocene (Geyh *et al.*, 1998).

2.5.2 Shell

Shells are composed almost entirely of calcium carbonate and interpretation of radiocarbon dates from this material is not straightforward. Terrestrial molluscs (land snails) may ingest older carbonates (limestone or soil carbonates) during their lifetime and this older material can be built into the shell. The result is that shell carbonate is deficient in ^{14}C relative to modern levels, and therefore radiocarbon dates would be anomalously old. Consequently, it is necessary to correct radiocarbon ages of fossil land snails for this anomaly. This can be done by comparing the ^{14}C activity of shell carbonate of modern snails of the same species (preferably from the same region as the fossils) with contemporary ^{14}C activity of the atmosphere, or as recorded in plant carbon (Goodfriend, 1987). A further problem is that fossil land snails often have secondary (pedogenic) carbonates deposited on them and, in some cases, recrystallisation can occur, which would result in anomalously young ages. It is therefore necessary to check for the presence of calcite in the shells using X-ray diffraction (XRD), since the original land snail shell material is always aragonitic (a different form of calcite), whereas secondary carbonate, the contaminant, is usually calcitic (Goodfriend, 1992). Routine laboratory preparation of shells for ^{14}C analysis involves removal of the outer 20% (by weight) of the shell by acid hydrolysis to obtain the inner material for dating.

Similar problems arise in the radiocarbon dating of marine molluscs. Again, the accumulation of solutional carbonate or recrystallisation can occur, but as this exchange usually affects the outer layers more than the inner layers, by comparing dates from the two portions of the shell, it is usually possible to determine whether or not this form of contamination has taken place (Peacock and Harkness, 1990). The development of AMS dating, particularly the ability to date very small fragments of shell, makes this process very much easier. A more intractable problem in the dating of marine molluscs, however, is the determination of the marine reservoir effect which now appears to vary not only spatially but over time (section 2.4.3).

2.5.3 Bone

Bone consists of two elements, an organic fraction that is contained within the shafts of long bones which comprises cell tissues and a fibrous protein called **collagen**, and the bone mineral material, the principal component of which is a phosphate of calcium, **hydroxyapatite**. In theory, both of these elements can be dated, but because of the likelihood of carbon exchange after death between the bone mineral material and the depositional environment, it is usually only the proteinaceous fraction that is used in radiocarbon dating. Where biochemical purification procedures are employed, accurate ^{14}C estimates can be obtained on bone samples retaining significant amounts of bone collagen (Taylor, 1992). The problem with dating collagen, however, is that this material degrades over time and so only very small quantities might be left in older bone materials, and these too may have been contaminated following burial. With the development of AMS dating, it has proved possible to obtain age determinations on individual amino acids (section 6.6) within the collagen to check for consistency in dates and hence for contamination (Bowman, 1990). However, where bones are seriously depleted in protein content, i.e. they contain less than 5% of the original amount – which is often the case with older bone materials – there is a strong likelihood that these will yield radiocarbon ages that are anomalous (Hedges and Law, 1989). In an attempt to overcome this problem, numerous studies have been carried out on other non-collagen organic compounds in bone to see if consistent radiocarbon dates can be obtained, but the results have not so far been promising (Taylor, 2001). More encouraging, however, is recent work on cremated bone which suggests that while collagen does not survive burning, reliable radiocarbon ages might be obtained from the structural carbonate in the mineral fraction of bone (Lantin *et al.*, 2001). Equally promising has been recent dating of human bodies from European peat bogs ('bog bodies') where AMS ^{14}C dating, preceded by careful sample selection and pretreatment, yielded age determinations on bone collagen that were consistent with dates on other materials, such as skin, hair, textile, leather and wood (van der Plicht *et al.*, 2004).

2.5.4 Soil

Although all soils contain carbon and can therefore, in theory, be dated by radiocarbon, the dynamic nature of the soil system means that because they receive organic matter over a protracted time period, any radiocarbon date on a soil will largely be a measure of the **mean residence time** of the various organic fractions within the soil. Because of the continuous input of organic material into soils, measured radiocarbon ages of soil organic matter or its fractions are generally younger than the true ages of soils (Wang *et al.*, 1996). Further complications arise from the circulation of humic acids, root penetration and earthworm and other biological activity. All of this makes soil one of the most difficult of all materials to date using the radiocarbon technique (Matthews, 1985). However, buried soils (**palaeosols**) are important components of the Quaternary stratigraphic record, and a considerable amount of effort has been expended on the development of methods for obtaining coherent radiocarbon dates from such contexts

(Scharpenseel and Becker-Heidemann, 1992). Indeed, meaningful radiocarbon ages have been derived from buried palaeosols by comparing dates from different organic fractions (Matthews, 1991), by using other materials in soils such as charcoal (Goh and Molloy, 1979), by modelling the measured ^{14}C content of soil organic matter (Wang *et al.*, 1996), or by checking radiocarbon dates from soils against results obtained using other dating methods (Dalsgaard and Odgaard, 2001).

2.6 Calibration of the Radiocarbon Timescale

2.6.1 Dendrochronological Calibration

Ever since the discovery of a discrepancy between radiocarbon and dendrochronological ages, the international radiocarbon community has been searching for a reliable basis for converting radiocarbon years into 'actual' or 'calendar' years. Thus far, the major effort has been directed towards dendrochronological records, as these enable a direct comparison to be made between the ages of wood as determined by counting annual tree rings (Chapter 5) and radiocarbon dates obtained from individual wood increments. Very high resolution radiocarbon measurements have been carried out mainly on samples of oak from Germany and Ireland, and on bristlecone pine (Pinus longaeva) (Figure 5.3) and Douglas fir (Pseudotsuga) from the United States. This has been an international collaborative programme with laboratories in Belfast, Groningen, Heidelberg, Pretoria, Seattle and Tucson undertaking high-precision conventional radiocarbon dating (<20 years) and careful cross-checking of tree-ring series, and the result is a continuous calibration curve extending back to 10 279 years BP (Stuiver *et al.*, 1998). Scientists in Germany have linked a chronology of German pine trees to this curve which extends the dendrochronologically based radiocarbon calibration to 12 179 years BP (Kromer and Spurk, 1998). These long dendrochronological series are discussed further in section 5.3.

2.6.2 The INTCAL Calibration

The most recent radiocarbon calibration curve, **INTCAL98** (Stuiver *et al.*, 1998), is based principally on the dendrochronological records described above. Beyond the limit of continuous dendrochronological series, calibration is based on matched uranium-series (U-series: section 3.3) and radiocarbon dates on fossil corals, coupled with radiocarbon-dated organic material from laminated (i.e annually accumulating) marine sediments in the Cariaco Basin, Venezuela (Figure 5.13A). These two forms of evidence have extended the INTCAL98 calibration to 15 585 years BP (Figure 2.6). However, problems arising from past variations in the marine reservoir (section 2.4.3) and also possible errors in the counting of laminated sediments mean that this part of the calibration curve is less secure than that based on tree-ring records. Although the INTCAL98 calibration officially extends to 24 000 calendar years BP, the resolution of the last 8400 years of the curve is so low (only eight pairs of U-series and radiocarbon-dated samples) that it cannot, at present, be considered as a true 'calibration curve' (van der Plicht, 2002).

Despite the uncertainties associated with the older part of the age range (ca. 15 600–24 000 cal. years BP), the international radiocarbon community has recommended that INTCAL98

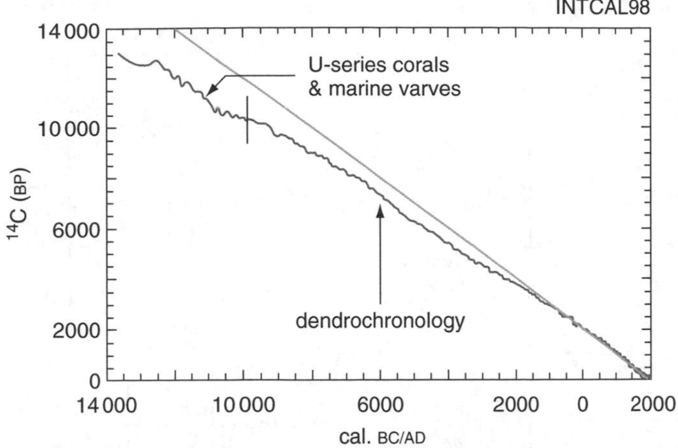

Figure 2.6 *The INTCAL calibration curve back to ca. 15 600 cal. BP. The right-hand section of the curve is based on a comparison between dendrochronological and radiocarbon ages on wood samples, while the left-hand part is based on paired radiocarbon and U-series dates on corals, and radiocarbon-dated marine varves (after van der Plicht, 2002). Reproduced by permission of Stichting Netherlands Journal of Geosciences*

be used as the basis for calibration for the time being (Stuiver and van der Plicht, 1998). The curve is decadal, in other words it has a resolution of 10 cal. years. The most recent (2004) calibration program based on INTCAL98 (CALIB 4.4) can be downloaded from the Internet, and is available in both PC and Macintosh formats. Ages obtained using CALIB 4.4 are always expressed in terms of 'cal. years BP' (or AD/BC). However, in applying calibration to radiocarbon dates, two points must be borne in mind. First, we have already seen that radiocarbon dates are expressed in terms of probabilities, and that it is important to think not in terms of 1σ (68% probability), but rather of 2σ (95.4% probability) when making an interpretation based on a particular age estimate. There is, however, also an element of uncertainty in the calibration program arising, for example, from errors in tree-ring counting or in the measurement of U-series in coral samples. This means that when dates are generated using CALIB 4.4, they are invariably expressed as a *range*, rather than as a single age value in order to reflect these uncertainties. Second, the perturbations in the calibration curve, which reflect the natural variations in long-term ^{14}C activity (see above), can result in radiocarbon dates having more than one calendar (or calibrated) age. In Figure 2.7, for example, which shows part of the calibration curve for the Mid-Holocene, a ^{14}C date of 4500 ± 50 years BP corresponds to seven separate points on the calibration curve with a calibrated age range of 5050–5275 years BP. When the 2σ confidence interval on the radiocarbon date is taken into account, the calibrated age range could lie anywhere between ca. 4950 and 5350 years BP. This is before any of the errors in the calibration program are taken into account! Calibration, therefore, although valuable in giving an approximation of 'true' age, is still a relatively imprecise tool and must be treated accordingly.

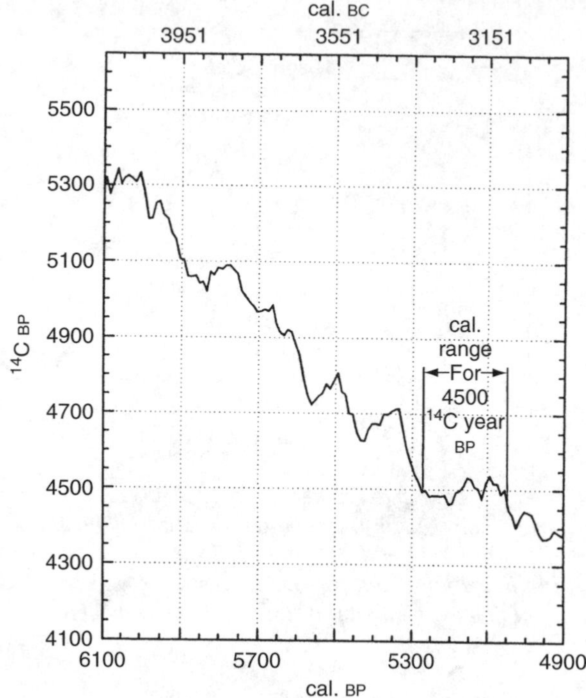

Figure 2.7 Part of the INTCAL98 calibration curve for the Mid-Holocene. Note that a ^{14}C age of 4500 years BP corresponds to a calibrated age range of 5050–5275 years (Stuiver et al., *1998, Figure A14). Reprinted by permission of Radiocarbon*

2.6.3 Extending the Radiocarbon Calibration Curve

Extending the radiocarbon calibration curve beyond the current limits of INCAL98 is a major aim of the radiocarbon community. Variations in atmospheric ^{14}C (expressed as $\Delta^{14}C$) back to 40 000 years and beyond have been demonstrated, *inter alia*, by comparisons between radiocarbon dates and independently dated laminated (annual) sediment sequences (section 5.3.3.4), and by paired radiocarbon and U-series dates on speleothem carbonate[9] (Beck *et al.*, 2001). Figure 2.8 shows the relationship between U-series and ^{14}C ages from a speleothem from the Bahamas over the period 11 000–45 000 years. It should be emphasised, however, that this is not a 'calibration curve' in the strict sense as it cannot be used to *correct* radiocarbon dates. Nevertheless, it can be regarded as a **comparison** curve, in other words it can be used to compare radiocarbon and U-series dates in order to gain an impression of the extent of divergence between the two dating series (Richards and Beck, 2001).

A different approach to this problem has recently been described by Hughen *et al.* (2004). They generated a record of long-term atmospheric ^{14}C activity back to 50 000 years ago, based on the correlation between ^{14}C dates on marine sediments from the Cariaco Basin, Venezuela, and the annual-layer timescale of the Greenland GISP2 ice core

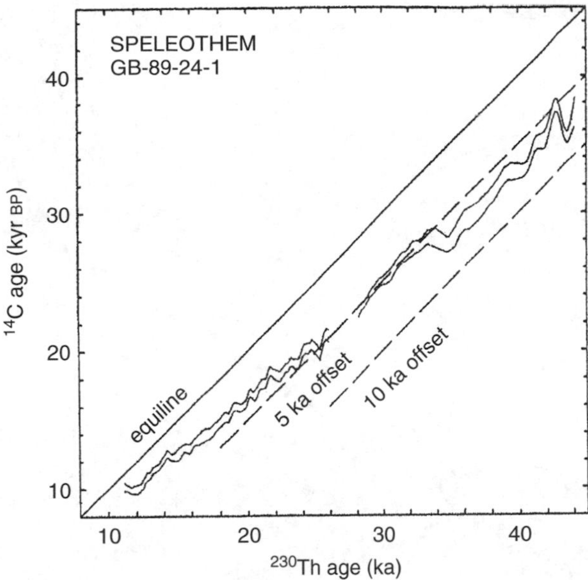

Figure 2.8 *Radiocarbon 'comparison' curve showing replicate ^{14}C and ^{230}Th ages on a Bahamas speleothem for the period 10 000–45 000 ^{14}C years BP. This shows an offset between ^{14}C and U-series dates of ca. 5000 years around 20 000 ^{14}C years BP. Beyond 30 000 ^{14}C years BP, the offset increases to a maximum at ca. 36 000 ^{14}C years BP where ^{14}C years may be 8000 years younger than U-series years (after Richards and Beck, 2001, Figure 1). Reproduced by permission of David Richards and Antiquity Publications Ltd*

(Chapter 5). This correlation is underpinned by proxy climatic data, and by independent radiometric dating of climatic events linked to GISP2. The curve (Figure 2.9) shows prominent ^{14}C plateaux at ca. 33 000, 28 000, 24 000 and 13 300 years ^{14}C years BP, and there is an abrupt shift at 42 000–40 000 cal. years BP in which nearly 7000 ^{14}C years elapsed in only 2000 cal. years. Prior to that there is a general trend of decreasing calendar-^{14}C age offset from 40 000 to 15 000 cal. years BP. Although there are slight age offsets between this curve and those based on lake sediments and carbonates (Figure 2.8), the overall ^{14}C structures are remarkably similar, suggesting that an accurate reconstruction of long-term atmospheric radiocarbon activity, and hence a secure basis for long-term radiocarbon calibration, may eventually be achievable.

2.6.4 Bayesian Analysis and Radiocarbon Calibration

One way in which radiocarbon calibration might be improved is to use a form of statistics known as **Bayesian analysis**.[10] Bayesian probability differs from classical statistics in that concepts of uncertainty, probability and subjectivity are all interrelated. In particular, prior knowledge and assumptions are expressed explicitly in the calculation of probability, as opposed to being applied before statistical manipulation. The use of Bayesian analysis allows such prior information to be incorporated into the radiocarbon calibration process.

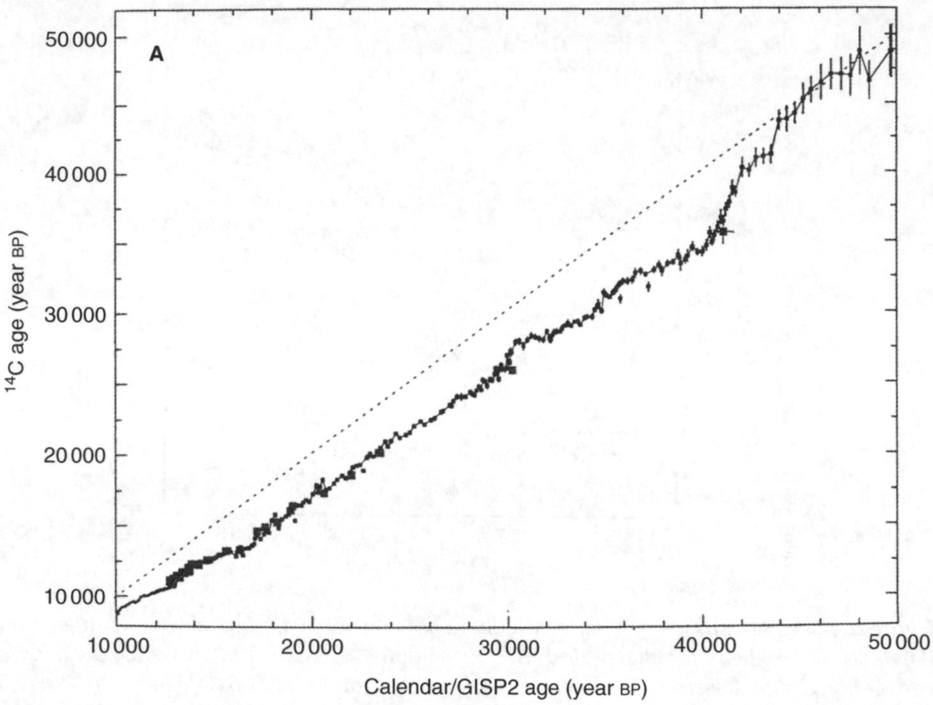

Figure 2.9 *Radiocarbon dates from the Cariaco Basin, Venezuela, plotted against the Greenland GISP2 timescale for the time period 12 000–50 000 cal. years* BP. *The initial part of the curve (up to 12 000 cal. years* BP*) is the tree-ring record, while the squares show paired* ^{14}*C/U-series dates. Note the abrupt shift in* ^{14}*C age at 40 000–42 000* ^{14}*C years* BP *(after Hughen* et al.*, 2004). Reprinted from Science*

This approach can be most usefully applied to the dating of sedimentary sequences where the aim is to generate an age–depth model for the profile. Such sequences tend to be marked by a degree of scatter in the radiocarbon dates, often with inversions in age–depth relationships, and developing a coherent timescale may be difficult (e.g. Walker *et al.*, 2003). Here the prior information that is incorporated into the analysis is (i) stratigraphic context and (ii) succession, working on the principle that in a body of stratified sediment the age should increase with depth. Radiocarbon calibration programmes such as OxCal (Bronk Ramsay, 1995) employ a Bayesian statistical framework, and these enable probability distributions to be calculated for each date in a sequence. At each level in the profile, two probability calibrated age distributions can be generated, one constrained by knowledge on age–depth relationships (posterior distribution) and one unconstrained by such knowledge (prior distribution). A strong compatibility between the two distributions will give greater credence to the calibrated age obtained. Where the two diverge significantly, the dates may be rejected as being aberrant, in other words they are out of stratigraphic order. In this way, a chronology can be constructed for a sedimentary sequence that has a more secure statistical basis than one in which radiocarbon dates alone, either in calibrated or in uncalibrated form, have been employed.

Bayesian statistical methods are being increasingly widely used in the analysis of Quaternary dating series, and further details can be found in Bronk Ramsay (1998), Buck *et al.* (1996; 2003), Buck (2001) and Blockley *et al.* (2004).

2.6.5 Wiggle-Match Dating

One way to circumvent the problems caused by fluctuations in the radiocarbon curve is to use **wiggle-match dating**. This is a technique which, paradoxically perhaps, takes advantage of the short-term fluctuations in the radiocarbon calibration curve and which can, in certain circumstances, provide more accurate chronologies than can be obtained through the calibration of individual radiocarbon dates. The simplest example involves the use of tree rings. Here, high-precision radiocarbon dating is carried out on individual rings whose age differences are known precisely and the resulting curve of dendro-age plotted against radiocarbon age can be fitted (wiggle-matched) to the shape of the radiocarbon calibration curve. A good example of this approach is the radiocarbon dating of the Kaharoa Tephra[11] in New Zealand (Figure 7.4). Wiggle-matching dating has also been applied to peat sequences, although here only the radiocarbon age of individual horizons is known, and the independent (i.e. non-radiocarbon-based) timescale has to be derived by making assumptions about the rate of peat accumulation (Blaauw *et al.*, 2003; 2004). The technique of wiggle-matching is best used over a short section of the calibration curve where, in some cases, it may prove possible to match the dated section to within a few years on the calendar timescale (e.g. Hogg *et al.*, 2003). Various statistical approaches to wiggle-matching have been developed, some again including Bayesian analysis (Christen and Litton, 1995), and these can generate dating ranges that are even tighter than the combined errors in the original radiocarbon measurements. The result is a higher level of overall dating precision (Bronk Ramsay *et al.*, 2001). Computer programs are now available for curve fitting, and these are available via the Internet. They include Cal25 (van der Plicht, 1993) and OxCal which can perform both radiocarbon calibration and automatic wiggle-matches (Bronk Ramsay, 1995; 1998).

2.7 Applications of Radiocarbon Dating

2.7.1 Radiocarbon Dating: Some Routine Applications

Although radiocarbon dating is really only applicable to the last 45 000 years or so of Quaternary time, the fact that ^{14}C occurs in a number of different organic materials means that the technique has applications across a wide spectrum of the earth, environmental and archaeological sciences. The majority of samples that are dated on a day-to-day basis in radiocarbon laboratories are of plant macrofossils (including charred remains), charcoal, peat, organic lake mud, marine microfauna and microflora and, perhaps to a lesser extent, shell and bone. Here we discuss examples of the use of these different dating media to show how radiocarbon chronologies can be developed in order to provide answers to a range of archaeological, palaeoenvironmental or palaeoecological questions.

2.7.1.1 Dating of plant macrofossils: Lateglacial cereal cultivation in the valley of the Euphrates. Plant macrofossils have long been used in radiocarbon dating. This is because they are well preserved in a range of depositional contexts (peats, lake sediments, buried soils, etc.) and although it is possible for them to be reworked, for example, by fluvial or colluvial activity, they tend to retain their sample integrity and, unlike many other organic materials, are seldom affected by contamination. With the advent of AMS dating, it has proved possible to date individual fruits or seeds, or very small fragments of leaf. Plant macrofossils provide valuable data on local vegetation cover, on land-use change and farming activity, and on patterns of climatic and environmental change (Lowe and Walker, 1997). Moreover, they are frequently preserved as charred or carbonised remains on archaeological sites where they can provide evidence on plant domestication and cultivation (Jones and Colledge, 2001).

In this context, Hillman *et al.* (2001) describe the results of an AMS dating programme on charred cereal grains from the prehistoric (Epipalaeolithic) settlement site of Abu Hureyra in the Euphrates Valley, Syria, about 130 km east of the present-day city of Aleppo (Figure 2.10A). The first settlement there, Abu Hureyra 1, was occupied from ca. 11 500–10 000 ^{14}C years BP, and may have had a population of 100–200 people. Abundant plant remains recovered from the site show that the inhabitants were initially hunter-gatherers who lived on a range of wild food plants. From shortly before ca. 11 000 ^{14}C years BP, however, there are signs of cultivation, first of wild cereals, and subsequently of domestic types. The evidence for domesticates is in the form of charred grains of rye (*Secale cereale*), which are readily distinguishable from their wild counterparts on the basis of size (Figure 2.10B). Of the five domestic-type rye grains that were AMS-dated, three gave Epipalaeolithic dates that matched other dates from the levels in which they were found, namely 11 140±100, 10 930±120 and 10 610±100 ^{14}C years BP. These findings are of considerable significance for our understanding of the origins of agriculture for, hitherto, the earliest archaeological finds of domestic cereals in southwestern Asia have involved wheat and barley and date from around the beginning of the Holocene, i.e. 11 000–12 000 cal. years BP. The Abu Hureyra evidence suggests, however, that systematic cultivation of cereals began before the end of the Pleistocene, by at least 13 000 cal. years BP, and that rye was amongst the first crops. The evidence also suggests that the settlers at Abu Hureyra began cultivating crops in response to a steep decline in wild plants that had served as staple foods for the previous four centuries, and which may have resulted from an increase in aridity. If this is the case, then the primary trigger for plant domestication in the Near East may have been climate change.

2.7.1.2 Dating of charcoal: a Holocene palaeoenvironmental record from western Germany. Charcoal is preserved in a wide range of deposits, including lake sediments, peats, palaeosols and cave sediments, and it is often found in archaeological contexts where it can provide useful ecological and cultural information (Mellars and Dark, 1998). In lake sediments and peats, it has proved to be especially valuable as an indicator of past fire regimes, reflecting the influence of both climate (increased incidence of fire under drought conditions: Brunelle and Anderson, 2003) and human activities (Huber and Markgraf, 2003).

Charcoal is also valuable as a dating medium, however, and this is especially the case where it is found in contexts that otherwise would be difficult to date by radiocarbon. One

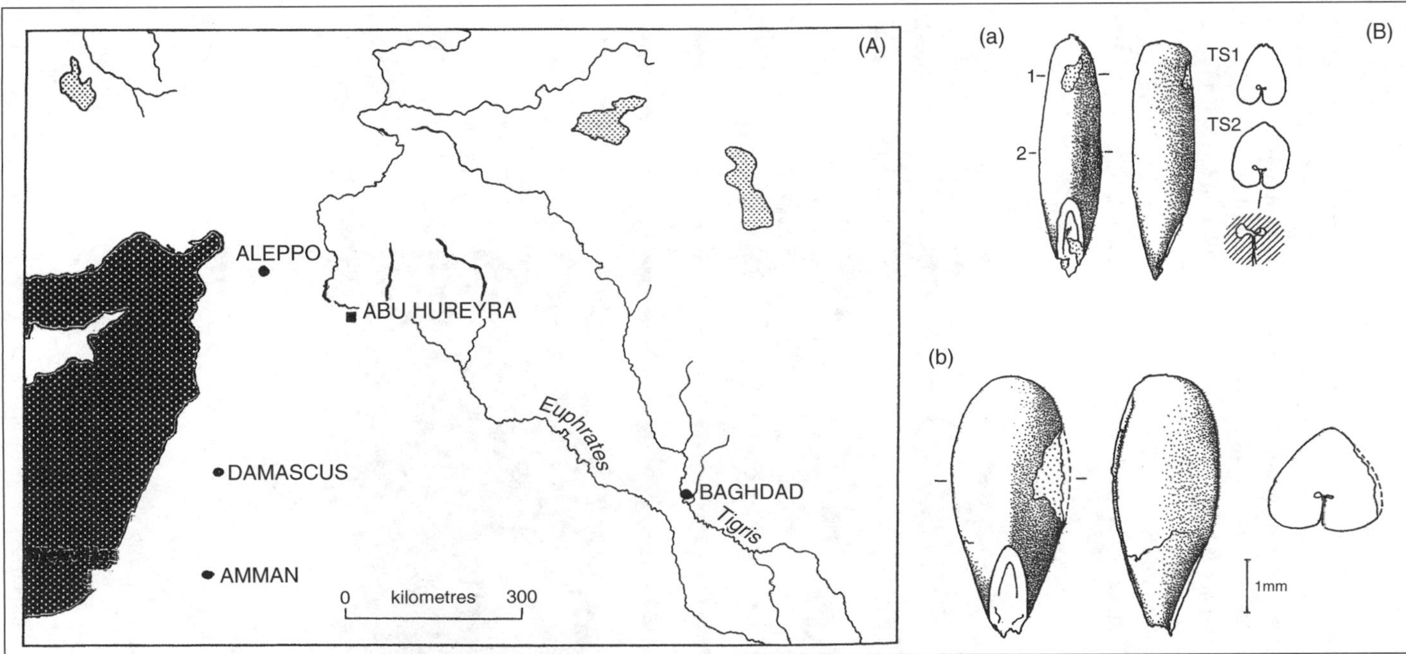

Figure 2.10 *A. The location of Abu Hureyra in the Euphrates Valley, Syria; B. Examples of charred remains of rye grains from the Epipalaeolithic levels at the site: (a) a typical grain of wild rye which predates the earliest level of cultivation; (b) a typical grain of domestic-type rye which is significantly larger than its wild counterpart. The latter comes from a sightly higher level than the wild specimen (a) and has been AMS radiocarbon dated to 10 930± 120* BP *(after Hillman* et al.*, 2001). Reproduced by permission of Arnold Publishers*

example is the Holocene palaeoenvironmental record obtained from fossil molluscs at Kloster Mühle, western Germany (Meyrick, 2003). The difficulty at this site is that the molluscs are contained within a tufa,[12] and any attempt to date either the molluscs themselves or the carbonaceous tufa would be confounded by the likelihood of errors in the dates arising from the incorporation of older carbon residues (sections 2.4.1 and 2.5.1). Fortunately, however, charcoal fragments have become incorporated into the tufa in sufficient quantities for AMS radiocarbon dating. Six age determinations were obtained from the sequence which provided a chronology for the environmental reconstruction based on the molluscan faunas (Figure 2.11). The base of the sequence, dated at 9720±75 ^{14}C years BP, shows that the Early Holocene was characterised by moist, open ground fauna, which was followed by an aquatic phase bracketed by statistically indistinguishable dates of 9530±60 and

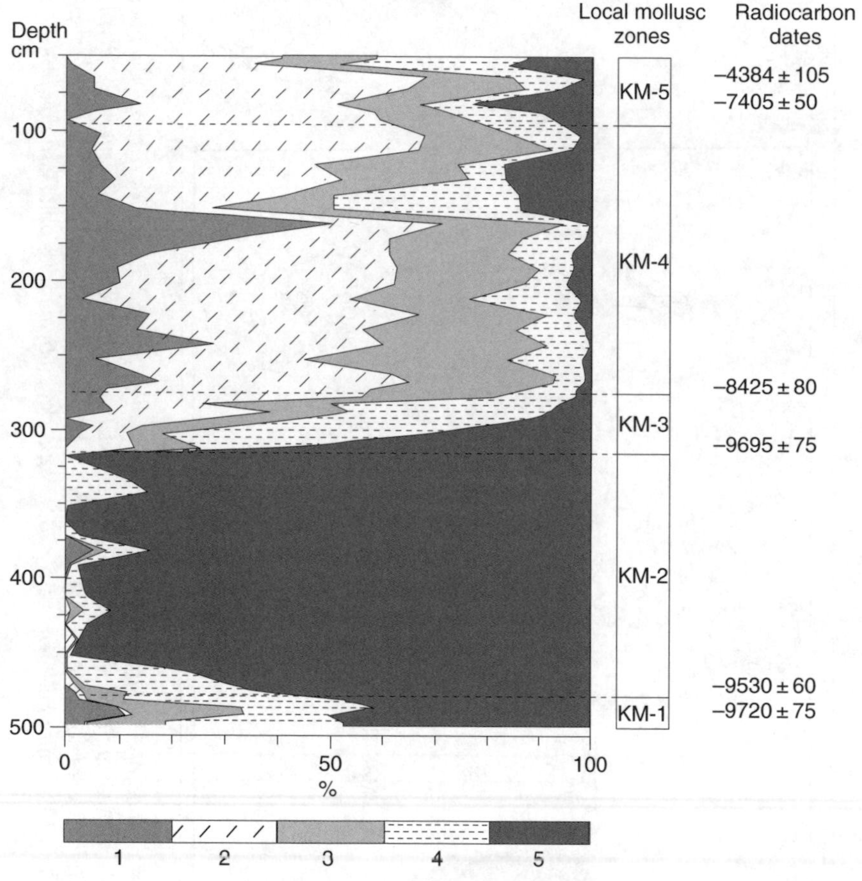

Figure 2.11 *Summary diagram of the ecological classifications of the molluscan fauna from Kloster Mühle, Germany: 1. open ground; 2. shade demanding; 3. catholic (no clear environmental affinities); 4. marsh; 5. aquatic. Note how the aquatic taxa are dominant in zones KM-1 to KM-3, while the upper parts of the record are increasingly dominated by shade-demanding taxa, reflecting the spread of deciduous woodland (after Meyrick, 2003). Reproduced by permission of John Wiley & Sons, Ltd*

9695 ± 75 ^{14}C years BP. This, in turn, was succeeded by a woodland phase, with optimum deciduous forest conditions established by 7405 ± 50 ^{14}C years BP. The radiocarbon date from the uppermost horizon (4384 ± 105 ^{14}C years BP) is almost certainly too young. This is probably because only a small amount of charcoal was available from this horizon (0.1 mg of carbon compared with 1.0–4.3 mg in the other five samples), and hence background contamination could not be accurately quantified for this very small sample.

2.7.1.3 Dating of peat: a Holocene palaeoclimatic record from northern England. Peat bogs are rich archives of palaeoenvironmental information. Fossil pollen preserved within the peat provides a record of local and regional vegetational change, while plant macrofossils (wood fragments, leaves, fruits, seeds, etc.) indicate the nature of former vegetation communities growing on the mire surface and in the immediate hinterland of the site. By using this type of evidence, inferences can be made about long-term ecological processes within the mire and its hinterland (Charman, 2003). Some of these will reflect the activities of human groups (woodland clearance, agricultural practices), while others will have resulted from climate change. The latter may be reflected in changes in the abundance of pollen of different plants, but climate is also reflected in other lines of fossil evidence, such as plant macrofossils or testate amoebae,[13] which indicate past variations in mire surface wetness. Where such changes can be detected in ombrotrophic mires (mires that are entirely rain-fed), it is possible to reconstruct a climatic record of past precipitation (Barber *et al.*, 1994; Chambers *et al.*, 1997; Langdon *et al.*, 2003). In the majority of cases, such records typically rely on radiocarbon dating as the basis for a chronology.

At Walton Moss, an ombrotrophic peatland complex in Cumbria, northern England, Hughes *et al.* (2000) developed a high-resolution record of Holocene palaeoprecipitation changes based on variations in plant macrofossil content within the peat profile. Calibrated radiocarbon dates from the Walton Moss profile, and also from the adjacent site of Bolton Fell Moss (Figure 2.12A), provided the basis for a temporal reconstruction of mire surface wetness changes from the Early Holocene to the present day. This revealed a series of clearly defined wet-shifts, each reflecting episodes of higher precipitation, from ca. 7800 onwards (Figure 2.12B). These climatic changes do not appear to be random; rather there is a cyclic trend with a periodicity of around 1100 years. Interestingly, a similar cyclicity has been detected in North Atlantic and in Greenland ice-core records, suggesting a powerful climatic forcing mechanism during the course of the Holocene that was affecting not just the British Isles but the North Atlantic region as a whole.

2.7.1.4 Dating of organic lake mud: a multi-proxy palaeoenvironmental record from Lake Rutundu, East Africa. Lakes are natural sediment traps and analysis of the sedimentary record in lake basins allows inferences to be made about environmental changes that have taken place around the lake catchment. Perhaps the most widely used palaeoecological indicator in lake sediment sequences is fossil pollen, and many of the pollen diagrams that have been published over the years have been based on samples from lake sites. More recently, however, pollen records have been integrated with other data from lake sediment cores, these 'multi-proxy' studies providing a more detailed palaeoenvironmental reconstruction than would be possible using pollen evidence alone (Jones *et al.*, 2002; Walker *et al.*, 2003).

(A)

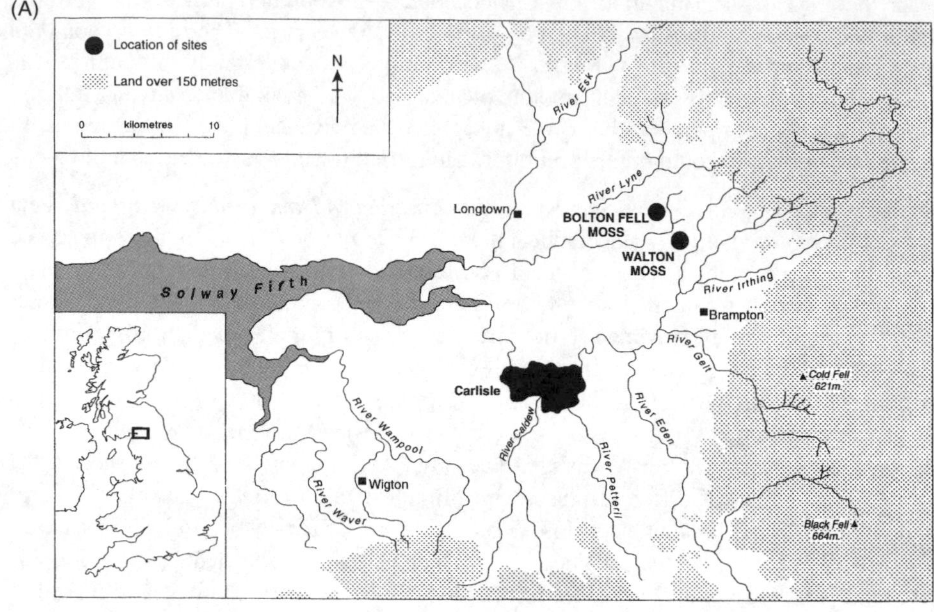

(B)

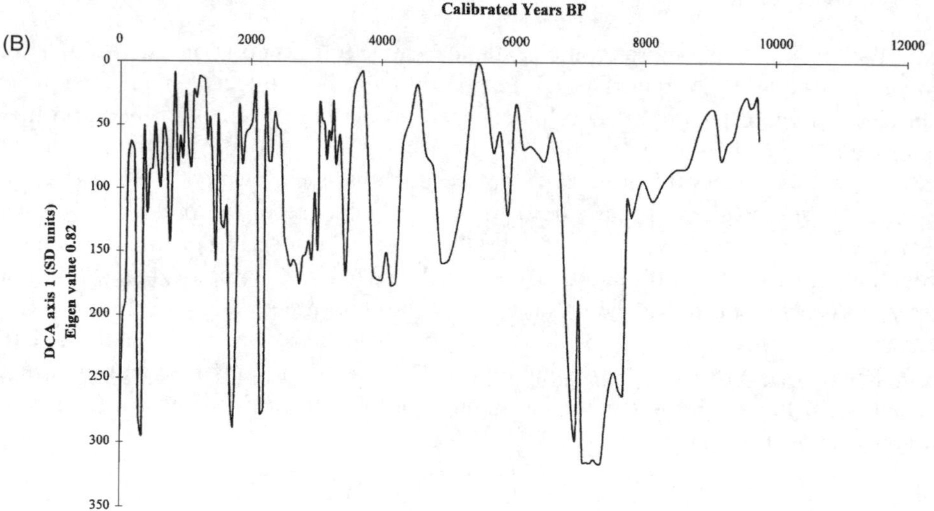

Figure 2.12 *(A) Location of Walton Moss and Bolton Fell Moss, northern England. (B) Reconstructed mire surface wetness changes at Walton Moss. Increased wetness (and, by implication, higher levels of precipitation) is reflected in the higher values of DCA (detrended correspondence analysis) shown on the y-axis. DCA is a standard mulitivariate statistical technique which is used here to analyse the variation in plant macrofossil components throughout the peat profile and to relate them to particular controlling factors (such as surface wetness). Note the marked wet-shifts at ca. 7800, ca. 5300, 4410–3990, ca. 3500, 3170–2860, 2320–2040, ca. 1750, ca. 1450, ca. 300 and ca. 100 cal. years* BP *(modified after Hughes* et al., *2000). Reproduced by permission of Arnold Publishers*

An example of such an approach is the study by Wooller *et al.* (2003) of Lake Rutundu which is located at around 3078 m on the northeast slope of Mount Kenya, East Africa. Thirteen AMS radiocarbon dates were obtained from the lake sediments, the oldest of which is 22 300±250 BP (ca. 26 282 cal. years BP). These directly dated records therefore cover the last glacial–interglacial transition and the whole of the Holocene. The extrapolated basal age is 38 300 cal. years BP, indicating that the lower part of the sequence predates the Last Glacial Maximum in the northern hemisphere. A pollen sequence was obtained from the core (Figure 2.13) along with data on grass cuticles,[14] stable carbon isotopes and other plant biomarkers.[15] The evidence indicates that prior to ca. 25 000 cal. years BP, an open vegetation subject to disturbance by fire existed in the area, although bushed grassland and dry montane forest were also present. The most fire-disturbed and open vegetation surrounding Lake Rutundu developed between 25 000 and 13 400 cal. years BP, however, when the local vegetation was one of either bushed grassland or open grassland, with a low canopy cover. From 13 400 cal. years BP there has been a gradual trend towards woody shrubland and open woodland culminating in the landscape of woody, sub-alpine shrubs and grasses that surrounds Lake Rutundu at the present day. The data also show that while grasses have been the major constituent of the vegetation over the past 38 000 years, the proportion of plants using the C_4 photosynthetic pathway[16] was greater during the Late Pleistocene than in the Holocene.

2.7.1.5 Dating of marine micropalaeontological records: an example of a problem from the North Atlantic. As noted in Chapter 1, one of the major advances in Quaternary science during the course of the twentieth century was the development of a technology which enabled long and complete cores of sediment to be recovered from the deep ocean floors. With the advent of AMS radiocarbon dating, which required only very small amounts of sample material, a chronology for the upper parts of these core sequences could be obtained on the basis of AMS dates from marine microfauna, most notably foraminifers. Over the past 20 years or so, a very large number of research papers have been published on palaeoceanography (changes in surface water temperatures and salinity, changes in deepwater formation,[8] changes in both surface and deepwater circulation, etc.) that employ timescales based on AMS radiocarbon dates (e.g. Andrews *et al.*, 1996).

However, while radiocarbon dating has been widely used to derive calendar ages for marine sediments, it rests on the assumption that the 'apparent age' of surface water (in other words, the age of surface water relative to the atmosphere) has remained constant over time. There is now a growing body of evidence to show that this is not the case (Bard *et al.*, 1994; Austin *et al.*, 1995) and that, often during periods of rapid climatic and oceanographical change, significant variations can occur in the extent of the marine reservoir effect (section 2.4.3). An elegant demonstration of the spatial and temporal variability in the marine reservoir during the last glacial–interglacial transition is provided by Waelbroeck *et al.* (2001). A record of changes in sea-surface temperature was obtained from planktonic foraminifers in three cores from the North Atlantic between 38°N and 55°N, and these fossils also provided material for AMS radiocarbon dating. In Figure 2.14, the sea-surface temperature curves are compared with the oxygen isotope signal from the GRIP Greenland ice core, in which the isotopically more positive values ('peaks') in the $\delta^{18}O$ trace reflect warmer conditions and more negative values ('troughs') indicate colder episodes. It is apparent in Figure 2.14 that the first sharp increase in sea-surface temperature leads the first temperature rise in Greenland by 1940±750 cal. years at 55°N

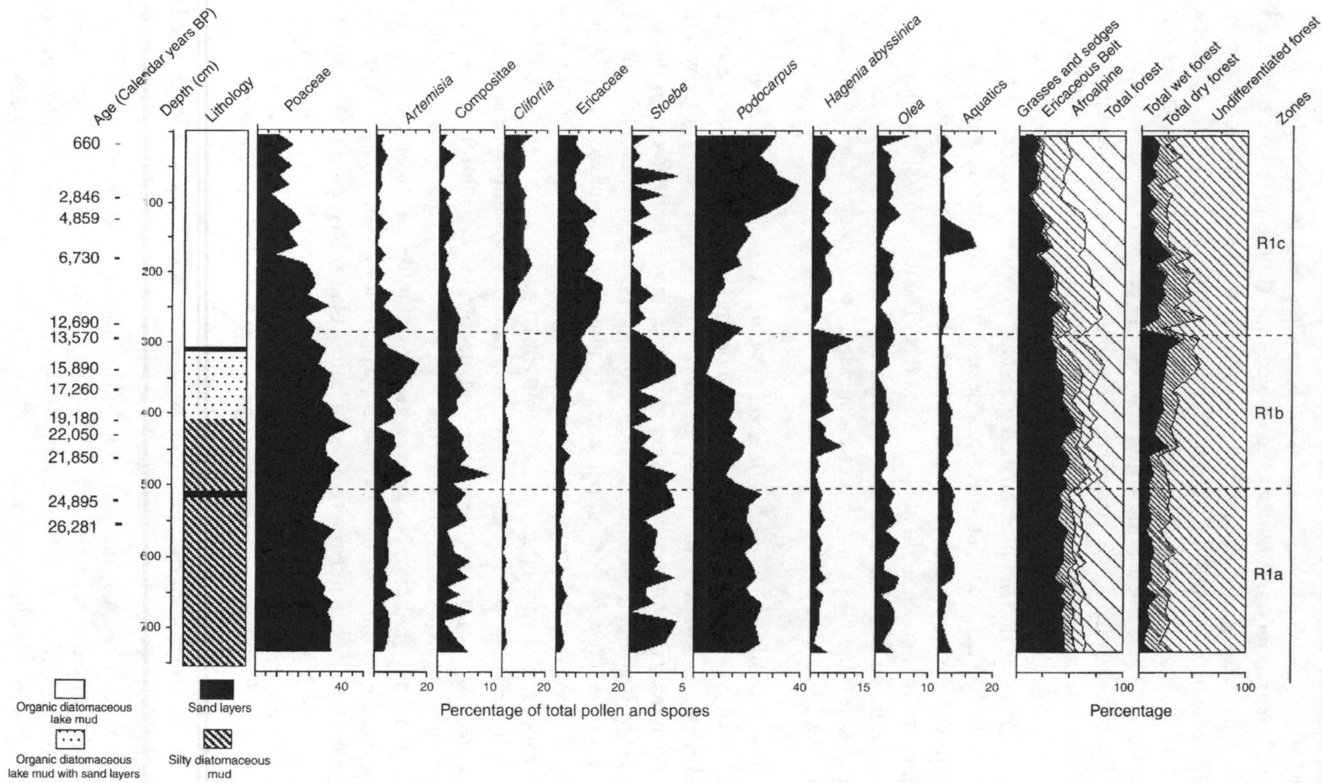

Figure 2.13 *Radiocarbon dates, lithostratigraphic record and percentage pollen diagram (selected taxa only) from Lake Rutundu, Mt Kenya. Note the dominance of grass pollen (Poaceae) throughout the record, and the increase in pollen of shrubby taxa (*Clifortia*, Ericaceae) in zone R1c (post ca. 13 570 cal. years BP). Note also how tree pollen (notably* Podocarpus*) is initially abundant (R1a), declines in R1b, and then increases steadily towards the present. The pollen record for open-habitat herbs (*Artemisia*, Compositae), however, shows the opposite trend (after Wooller* et al.*, 2003). Reprinted by permission of John Wiley & Sons, Ltd*

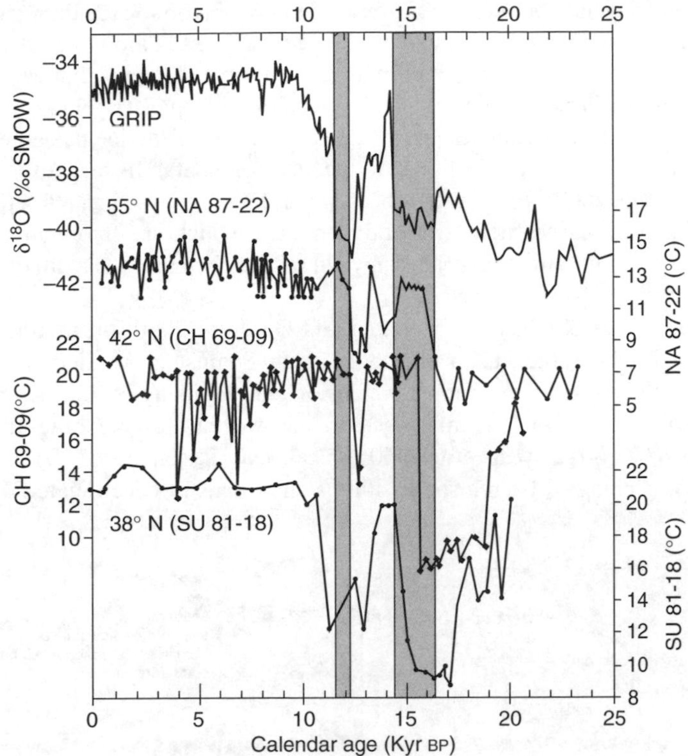

Figure 2.14 *Surface temperature records from three cores from the North Atlantic, NA 87-22 (55°N), CH 69-09 (42°N), SU 81-18 (38°N), compared with the GRIP $\delta^{18}O$ record. The y-axes for the ocean cores show sea-surface temperatures (°C) and the x-axis is in cal. years BP. The grey stripes indicate the large lead times between NA 87-22 (the most northerly of the three cores) sea-surface temperature and inferred GRIP air temperatures at around 15 000 and 11 500 cal. years BP (after Waelbroeck et al., 2001). Reprinted from Nature. Copyright 2001 MacMillan Magazines Limited*

(core NA 87-22) and by 1230±600 cal. years at 42°N (core CH 69-09). The second abrupt increase in sea-surface temperature leads the temperature rise reflected in the ice by 820±430 cal. years at 55°N and by 1001±340 cal. years at 42°N. By contrast, in the most southerly record from 38°N (core SU 81-18), the sea-surface temperature signal coincides very closely with air temperature variations in Greenland. As it is inconceivable that the northern parts of the Atlantic would experience deglacial warming *before* more southerly areas, the only explanation can be that a marine reservoir factor has affected the dates from the central and northern cores, resulting in an ageing effect of around 1900 ^{14}C years at around 15 000 BP, and ca. 800 ^{14}C years at ca. 11 500 BP. Clearly, these findings have far-reaching implications for the radiocarbon dating of marine records, particularly during periods of rapid oceanographical change.

2.7.1.6 Dating of marine shell: a Holocene aeolianite from Mexico. Although shells can be problematic dating media (section 2.5.2), careful selection of shell material and the application of stringent laboratory procedures, typically involving the dating of both inner and outer parts of shell fragments, can produce reliable radiocarbon age determinations. This is important because marine shell material can provide dates that are important in a range of contexts, including glacial and deglacial chronologies (England, 1999), relative land- and sea-level change (Ingólfsson *et al.*, 1995), studies of marine palaeo-environments (Peacock, 1999), and reconstructions of the environmental history of coastal zones (Christiansen *et al.*, 2002). In addition, coastal areas have always been attractive to human groups, and hence marine shells from midden sites, which are frequently encountered around the coasts of northwest Europe, can yield radiocarbon dates that are of considerable interest to archaeologists (Miracle, 2002).

Radiocarbon dates obtained by McLaren and Gardner (2000) on marine shells in an aeolianite (cemented sand dunes) on the Yucutan Peninsula, Mexico (Figure 2.15A), illustrate both the physical and human dimensions in the dating of marine molluscs. The shells were obtained from four different horizons within the aeolianite and gave ages ranging from 4082±50 cal. years BP to 2460±45 cal. years BP (Figure 2.15B). The lowermost date, which was obtained from a shell within a small lens of beach material cut into the

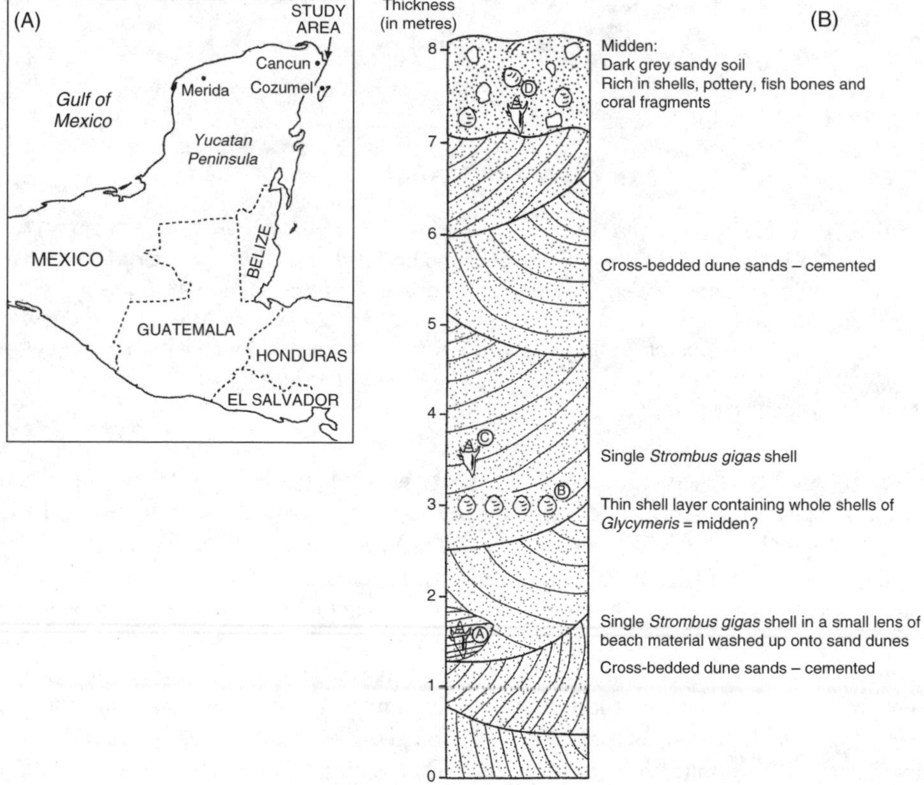

Figure 2.15 *(A) The Yucutan Peninsula, Mexico. (B) Generalised lithostratigraphic section through the aeolianite showing the relative positions of the dated shells (after McLaren and Gardner, 2000). Reproduced by permission of Arnold Publishers*

sand dune probably under storm conditions, suggests that sea level was close to its present height around 4000 cal years BP. This, in turn, implies that the sand dunes began to form during the later stages of the Holocene marine transgression. The layer of shells at around 2.9 m within the aeolianite, and dated to 4082 ± 50 cal. years BP, is believed to be a midden, and suggests human occupation of the northeastern Yucutan coastal plain as long ago as 4000 years BP. Sample C, dated to 3571 ± 45 cal. years BP (inner date), may also have been discarded by humans. The youngest dates (D) come from a clearly defined midden, rich in shells, pottery fish bones and coral fragments, on top of the dune sands. The dates, of around 2500 cal. years BP, provide a minimum age for completion of sand-dune formation and again further evidence of human occupation at that time.

2.7.1.7 Dating of bone: the earliest humans in the Americas. One of the great problems in New World archaeology concerns the timing of the peopling of the Americas. The general view is that this event took place late in the Pleistocene following the wastage of the great Laurentide and Cordilleran ice sheets which, hitherto, had acted as a barrier to migrating peoples who had crossed the Bering Straits from eastern Asia. Retreat of glacier ice from the western prairies and from the coastal regions of western Canada enabled human groups to move southwards for the first time. The chronology of this southward migration rests largely on radiocarbon dates, but these have often proved to be contentious, and a vigorous debate has ensued between those arguing for an early human colonisation of the Americas (pre-12 000 ^{14}C years BP) and those who would see the well-documented terminal Pleistocene 'Clovis culture' (ca. 11 500 ^{14}C years BP) as representing the first securely attested human presence (Taylor, 2000; Marshall, 2001).

A curious aspect of this debate is that, despite the very large number of radiocarbon dates that have been obtained on contexts associated with human activity, there are relatively few *directly dated* Palaeo-Americans. Moreover, of the skeletal material that has been dated, none has yielded a radiocarbon age in excess of the known age of the Clovis period (Taylor, 2001). The radiocarbon dating of four early human specimens from Mexico is therefore of significance in the context of these previous findings. Gonzalez *et al.* (2003) obtained AMS radiocarbon dates on five specimens that had been collected from palaeoindian localities around the Basin of Mexico (Figure 2.16). Four of the specimens yielded sufficient bone collagen (>10 mg g^{-1}) for dating purposes, the two oldest dates being 10 755 ± 75 ^{14}C years BP (Peñon III) and 10 200 ± 65 ^{14}C years BP (Tlapacoya 1). These dates are highly significant in the discussion of the peopling of the New World, partly because they increase the number of directly (and apparently reliably) dated individuals, but also because they fill a geographical gap (central America) in terms of human occupation of the Americas at the end of the Pleistocene. They also pose a question, for the skulls indicate that a long- and narrow-headed (dolicocephalic) people of non-Mongolian affinity were present in Mexico around 12 000 cal. years BP. Where these people came from and how they arrived in the Basin of Mexico remain to be established.

2.7.2 Radiocarbon Dating of Other Materials

In the previous section we have considered radiocarbon dating of what might be termed the more 'conventional' materials. In other words, we have dealt with some of the more typical dating applications that are routinely carried out in radiocarbon laboratories

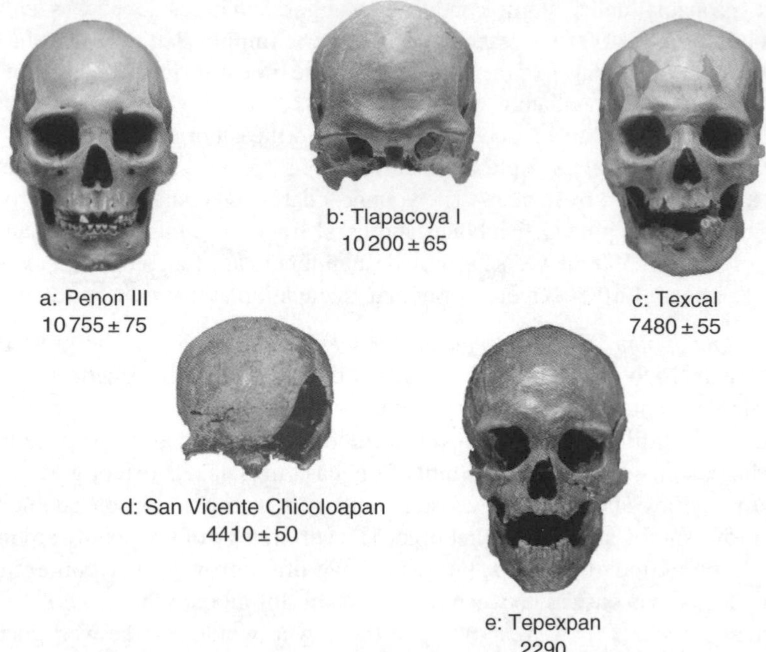

Figure 2.16 *Radiocarbon-dated Mexican Palaeo-Americans. The age determination for Tepexpan is shown without confidence limits, for chemical analyses suggest that molecular contamination with preservatives has taken place which has resulted in an anomalously young age (Gonzalez* et al.*, 2003). Reproduced with permission of Elsevier*

throughout the world. However, radiocarbon is proving to be a much more versatile technique than has traditionally been envisaged, for with the advent of AMS, it is proving possible to obtain radiocarbon dates on materials, some of which only 20 years or so ago would have been considered unsuitable for radiocarbon dating. Examples of these more specialised and, in some cases, experimental approaches to the use of radiocarbon are considered in the following sections.

2.7.2.1 Dating of textiles: the 'Shroud of Turin'. This was perhaps one of the most striking demonstrations of the value of AMS dating, and was one of the earliest archaeological applications of the technique. The Shroud of Turin is a ca. 4-m piece of linen cloth which is housed in the Cathedral of St John the Baptist in Turin, Italy. It bears the shadowy image of the front and back of a man who appears to have been scourged and crucified, and is alleged to have been the 'True Burial Sheet of Christ'. In October 1987, offers to date the Shroud were made by three AMS laboratories (Arizona, Oxford and Zurich) and were accepted by the Archbishop of Turin, acting on instructions from the Holy See, the owner of the Shroud. The British Museum was invited to help in the certification of the samples and in the statistical analysis of the results. A thin strip ($10 \times 70\,mm^2$) was cut from the Shroud and three samples, each 50 mg in weight, were prepared from the strip and were dated alongside three control samples. The results (Table 2.2) were remarkably consistent and produced a weighted mean age of 689 ± 16 ^{14}C years BP and a calibrated age of AD

Table 2.2 *Radiocarbon dates on the Shroud of Turin. Each laboratory split the samples and subjected these to different pretreatment procedures. The calibrated age range for the sample, at 95% level of confidence, is* AD *1262–1384 (after Damon* et al.*, 1989)*

Laboratory	Sample dates	Mean age	Overall mean age	Calibrated age range (AD)
Arizona	591±30			
	690±35	646±31		
	606±41			
	701±33			
Oxford	795±65			68%, 1273–1288
	730±45	750±30	**691±31**	95%, 1262–1312
	745±55			1353–1384
Zurich	733±61			
	722±56			
	635±57	676±24		
	639±45			
	679±51			

1260–1390 (95% confidence). This suggests that the flax from which the linen was woven was probably growing during the late thirteenth or fourteenth centuries, which matches the date (AD 1353) when the Shroud was first historically documented (Damon *et al.*, 1989). Since the publication of these dates, questions have been raised about their authenticity because of possible contamination of the Shroud by, for example, bacteria (Gove *et al.*, 1997), or as a consequence of fire-induced chemical modifications (scorching) which could have resulted in a change in ^{14}C activity in portions of the Shroud leading to an aberrantly young radiocarbon age (Kouznetzov *et al.*, 1996). However, experiments designed to test the ways in which the latter, in particular, could have operated indicate that if scorching had indeed occurred, it would have been unlikely to have influenced the ^{14}C activity of cellulose-like materials such as the linen of the Shroud of Turin (Jull *et al.*, 1996; Long, 1998; Hedges *et al.*, 1998a).

2.7.2.2 Dating of old documents: the Vinland Map. One of the great advantages of AMS, as exemplified by the dating of the Shroud of Turin, is that radiocarbon dates can be obtained on culturally valuable objects using only a very small amount (1 mg or less) of material. The damage to the object or artefact is therefore minimal. Another example of the dating of such materials is the application of AMS to ancient documents, an approach that has yielded historically coherent ages from, *inter alia*, the famous Dead Sea Scrolls (Jull *et al.*, 1995) and medieval Japanese documents (Oda and Nakamura, 1998).

A further instance is the radiocarbon dating of the Vinland Map (Figure 2.17). This apparently old map, drawn on parchment, is housed in the Rare Book and Manuscript Library at Yale University. In the northwest Atlantic Ocean it shows 'the Island of Vinland', discovered by the Norse seafarers Bjarni and Lief and colleagues. If authentic, it is the first known cartographic representation of North America and its date is important in establishing the history of European knowledge of the lands bordering the western North Atlantic. Some scholars have strongly argued the map's authenticity and have associated it with the Council of Basle (AD 1431–1449), in other words dating the map to at least half a century before Columbus's voyage. Others, however, are convinced that the map is

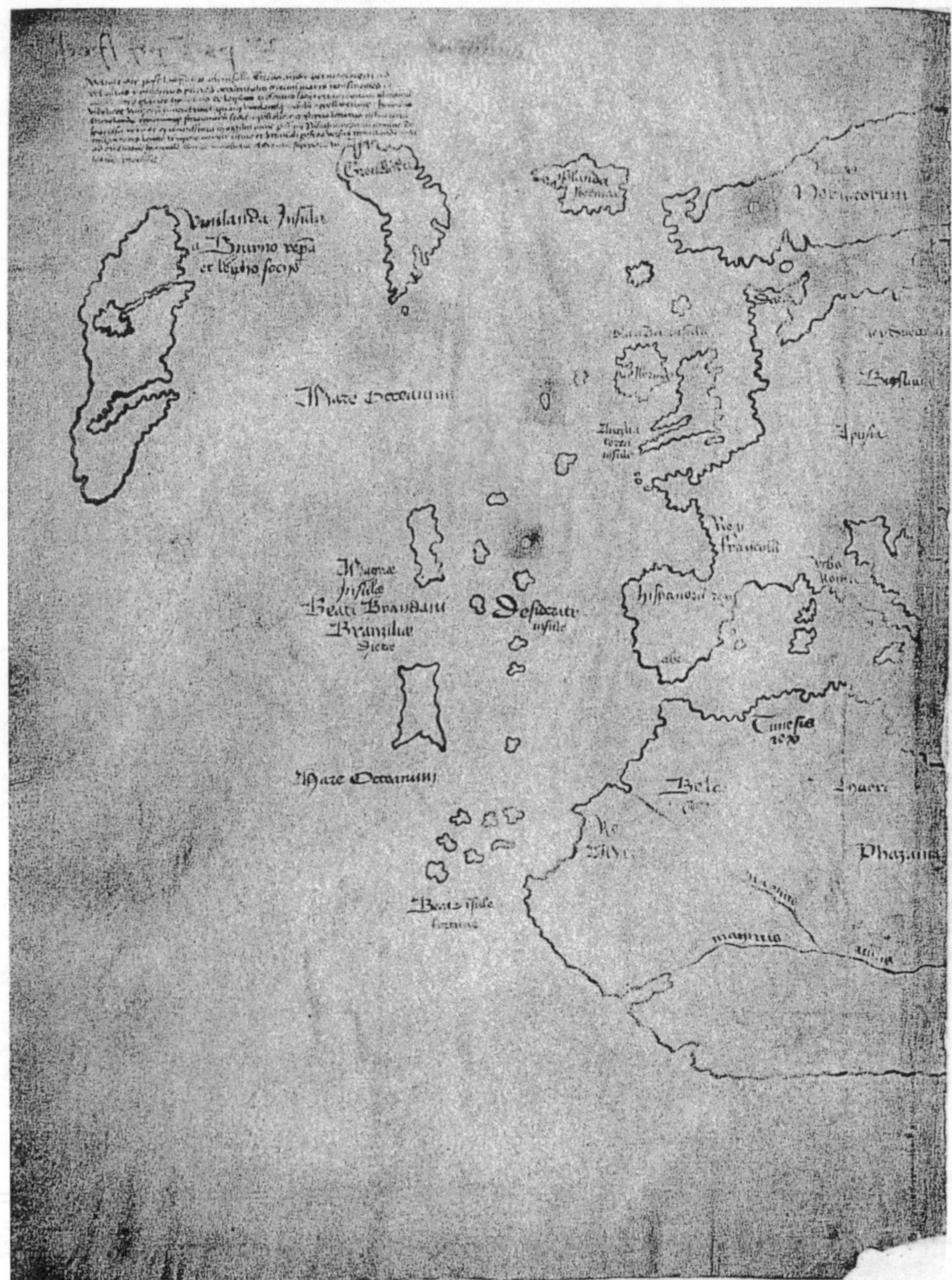

Figure 2.17 *The Vinland Map (after Donahue* et al.*, 2002). Reproduced by permission of Radiocarbon*

a clever forgery and may even date from the twentieth century. A fragment of parchment from the map was radiocarbon dated in the University of Arizona NSF AMS Facility (Donahue *et al.*, 2002). Five individual measurements were made which yielded a weighted average of 467 ± 27 ^{14}C years BP. This gives a calibrated age range (at 95% probability) of AD 1411–1468. The one-sigma calibrated calendrical date range is 1434 ± 11 years. Clearly, the AMS date only determines the age of the parchment of the Vinland map and it cannot provide a conclusive date for the production of the map itself (Ambers and Bowman, 2002), although it is an important contribution to the ongoing debate on the authenticity of the map. In this context, it may be significant that the AMS ^{14}C age determination coincides very well indeed with the dates for the Council of Basle, with which the map has been associated.

2.7.2.3 Dating of lime mortar: medieval churches in Finland. Dating of buildings has traditionally relied on the dating of wood, either by radiocarbon dating or by dendrochronology. Such evidence, however, will not necessarily date the initial construction of an old building, for repeated rebuilding and replacement of timber because of rot or fire will lead to younger ages, while in other cases the use of older timbers will often result in ages that predate building construction. For this reason, attempts have been made to date lime mortar. During the production of building lime, limestone is heated to liberate CO_2 and produce quicklime (calcium oxide: CaO). This is then slaked with water to form calcium hydroxide ($Ca(OH)_2$) or building lime which, in turn, is mixed with sand and water to form mortar. On exposure to air, the lime mortar hardens by taking up atmospheric CO_2 (and hence also ^{14}C) to form carbonate ($CaCO_3$). Accordingly, the ^{14}C content of a mortar sample can give an indication of the time that has elapsed since the mortar hardened, and hence the age of the construction of the building. Dating of lime mortar was shown to be feasible more than 20 years ago (Folk and Valastro, 1976; Van Strydonck *et al.*, 1983), but the advent of AMS radiocarbon dating, especially the requirement for very small quantities of sample material, along with more refined laboratory protocols and pretreatment procedures (Heinemeier *et al.*, 1997; Sonninen and Jungner, 2001), has resulted in significant developments in this area of radiocarbon dating.

A successful application of this approach has been in the dating of medieval stone churches on the Åland Islands, Finland (Heinemeier *et al.*, 1997). In this area, there is a particular need for mortar dating because of the lack of written sources or of suitable material for other scientific dating techniques. Multiple dates were obtained on lime mortar from three different church buildings: Hammarlund (18 samples), Eckerö (8 samples) and Saltvik (4 samples). The weighted mean radiocarbon ages from the three churches were 686 ± 20, 718 ± 25 and 710 ± 30 BP, respectively. Calibration of the weighted mean ages suggests that the churches were constructed either late in the thirteenth or during the course of the fourteenth centuries, with the most probable date being just before AD 1300.

2.7.2.4 Dating of hair: radiocarbon dates and DNA from individual animal hairs. Both human and animal hair have the potential to become an important source of information about human and animal palaeobiology, palaeoecology and palaeoanthropology. Biochemically, hair is characterised by the protein keratin. This material is extremely durable in the natural environment and it survives over very long periods of time in archaeological, palaeontological and palaeoecological sites (Bonnichsen *et al.*, 2001). Moreover, its chemical properties make it more resistant to the various processes that complicate the radiocarbon dating of bone (section 2.5.3). Although samples of human hair (from an Egyptian predynastic site) were dated at an early stage in the development of radiocarbon (Libby, 1955), the limitations imposed by sample size meant that it was only with the introduction of AMS that the dating of human and animal hair became a realistic proposition (Taylor *et al.*, 1995).

One area where this advance in dating may be especially significant is in the study of ancient DNA (Brown, 2001). Genetic data from archaeological specimens are having a growing influence on archaeology and there is an increasing collaboration between archaeologists, palaeoecologists and molecular biologists. In this respect, hair from archaeological sites may be an important but hitherto overlooked data source which may be amenable to molecular analysis. Thus far, DNA identification from hair has largely been restricted to modern samples, but Bonnichsen *et al.* (2001) have described a study of a Palaeo-American occupation level in a cave site in Nevada, USA, in which mitochondrial DNA (mtDNA)[17] could be extracted from animal hair from a fossil specimen of bighorn sheep (*Ovis canadensis*). Three samples of hair from the skin were AMS-dated and these produced remarkably consistent ages of 9830 ± 210, 9820 ± 300 and 9900 ± 90 BP. Moreover, the dates are in excellent agreement with five radiocarbon dates on hearths associated with the occupation layer, which range from 9940 ± 160 to 11 140 ± 200 BP. This is the first time that ancient DNA has been obtained from hair of this antiquity, and it opens up new possibilities for addressing a range of biological questions in prehistory. Bonnichsen *et al.*'s conclusion is that '... hair could become a "golden fleece" for prehistoric archaeologists and evolutionary biologists, allowing us to address old questions in new ways'.

2.7.2.5 Dating of iron artefacts: the Himeji nail and the Damascus sword. Radiocarbon dates can be obtained from iron artefacts provided that the source of carbon in the steel-making process was charcoal from freshly cut young wood. Other carbon sources, such as coal, coke or old wood, are depleted in ^{14}C and will cause artefacts to appear too old (Cook *et al.*, 2001). The possibility of using iron as a medium for radiocarbon dating has been known for some time (van der Merwe, 1969), but again it has been the development of AMS that has given added impetus to research into the radiocarbon dating of iron (Cresswell, 1991; 1992). Indeed, recent experimental work suggests that reliable dates may also be obtained from rusty iron, opening up the possibility of dating iron artefacts from waterlogged terrestrial sites or from shipwrecks (Cook *et al.*, 2003).

A considerable amount of work has been undertaken to improve laboratory methods for extracting carbon from ancient iron, and Cook *et al.* (2001) describe the results of such research in relation to the dating of two iron artefacts, a nail from the medieval Japanese castle of Himeji believed to date to around AD 1600, and a sword made originally in India but exported to Damascus and considered to date from ca. AD 1650. The results, using a new carbon extraction method, show that the nail was made between AD 1550 and 1640 (39.7% confidence) while the sword was manufactured between AD 1640 and 1670 (71% confidence). These experimental results are encouraging and indicate that as laboratory protocols become more refined, increasingly reliable dates may be obtained from iron artefacts.

2.7.2.6 Dating of pottery: the earliest pottery in Japan. Ancient pottery will contain a number of carbonaceous residues. Some of these will be from older carbon incorporated into the clay prior to firing, while others will be from younger materials that have been absorbed onto the surface of the pottery following burial. However, where there is an organic-rich coating on old pottery fragments (potsherds), for example from fuel or from food residues, this can also provide a reasonably reliable sample for AMS radiocarbon dating (Hedges *et al.*, 1992; Stott *et al.*, 2001).

Nakamura *et al.* (2001) report the results of a dating programme on charred adhesions on five potsherds from the Odai Yamamoto 1 site at the northern end of the main island, Honshu, in Japan. The site was first discovered in the 1970s and has since yielded stone artefacts, arrowheads and a number of potsherds characteristic of the Jomon culture of the Late Palaeolithic. Many of the potsherds were coated in carbonaceous material believed to derive from boiling food. Samples from five of the potsherds were extracted for AMS dating, and the ages ranged from 12 680 to 13 780 ^{14}C years BP. The average age was 13 070 ± 440 BP, and the calibrated age range was 15 710–16 540 cal. years BP. These dates are consistent with dates on other charcoal from the site, and also with dates from sites with ancient pottery in the Russian Far East. The Jomon culture, with the first use of pottery in Japan, was previously considered to have begun in the Holocene. If these dates are correct, they suggest that this cultural phase may have begun several thousand years earlier.

2.7.2.7 Dating of rock art: Palaeolithic cave paintings in Spain and France. Since the advent of AMS, a considerable amount of attention has been directed towards the dating of rock art, including cave paintings, paintings and engravings in rockshelter sites, and pictographs and petroglyphs on exposed rock surfaces, particularly in desert regions. In the case of the last named, the aim has been to obtain a minimum age for the rock engravings by isolating and dating organic carbon residues associated with desert varnish, a naturally occurring coating that develops on rocks in arid regions. However,

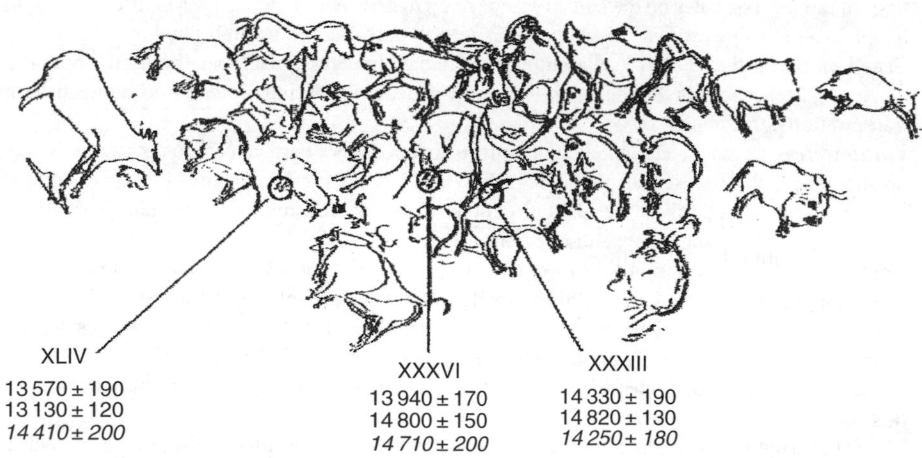

Figure 2.18 *Dated images from the Altamira painted ceiling (after Valladas* et al.*, 2001). The locations of the three dated bison are indicated along with the dates obtained on purified charcoals and humic fraction (in italics). Reproduced by permission of Radiocarbon*

the complex processes involved in desert varnish and organic carbon accumulation mean that the dating results have often proved to be controversial (Beck *et al.*, 1998; Dragovich, 2000). In other contexts, by contrast, realistic radiocarbon age determinations have been obtained through careful chemical separation of carbon residues from patinas and crusts overlying the rock paintings (Hedges *et al.*, 1998b; Watchman *et al.*, 2001).

In some cave sites attention has focussed on charcoal drawings and Valladas *et al.* (2001) have obtained radiocarbon dates on nine bison drawings from three caves (Covaciella, Altamira and El Castillo) in the Cantabrian region of northern Spain (Figure 2.18). In all, 27 dates were obtained, all of which fell between 13 000 and 14 500 ^{14}C years BP. Charcoal residues from rock paintings were also dated at two caves in France, Cosquer near Marseilles and Chauvet in the Ardeche region of the southeastern Massif Central. The 24 dates for the paintings in the Cosquet cave indicate 2 distinct phases of cave painting activity, one around 28 000 ^{14}C years BP and a later one around 19 000 ^{14}C years BP. The Chauvet paintings, of bison and woolly rhinoceros, yielded five dates between 32 000 and 31 000 ^{14}C years BP, which are the oldest radiocarbon age determinations so far obtained for European cave paintings. Although they lie beyond the limit of the INTCAL98 calibration, an estimated calendar age based on the 'comparison curve' provided by the Bahamas speleothem record (Figure 2.18) would date the paintings to around 38 000 years BP (Bard, 2001).

Notes

1. Libby's early work indicated that the half-life of ^{14}C was 5568±30 years and the early radiocarbon dates were all calculated on the basis of this estimate of half-life. Subsequently, however, the half-life of ^{14}C has been recalculated at 5730±40 years (Godwin, 1962). However, many dates were published using the old half-life, and hence in order to avoid confusion, it is convention to base all radiocarbon dates on the half-life value of 5570±30 years (Mook, 1986). Strictly speaking, to obtain an accurate estimate of radiocarbon age, all dates should be multiplied by 1.03.
2. **Fractionation** is the selective separation of chemical elements or isotopes during the operation of physical, chemical or biological processes, for example during evaporation, condensation, transpiration or metabolism.
3. **Foraminifers** are small, single-celled organisms that possess a hard calcareous shell that is often distinctively coiled to resemble that of a gastropod. They are found in a range of marine contexts from estuarine situations to the abyssal depths of the oceans, and are especially valuable as palaeoceanographic and palaeoclimatic indicators.
4. **Thermohaline circulation** is the movement of ocean water that occur because of differences in temperature and salinity (density). Such circulation is a major factor in the transfer of heat from low to high latitudes, and past changes in patterns of thermohaline circulation are considered to be major forcing factors in Quaternary climate change (Lowe and Walker, 1997).
5. Named after **Henrik de Vries** whose important paper published in 1958 first drew attention to this phenomenon.
6. The **solar wind** is not a 'wind' in the strict sense of the word. Rather it is a stream of charged particles (protons and electrons) emitted by the sun.
7. **Ventilation** is part of the thermohaline circulation in the oceans (Footnote 4) and refers to that component of the circulatory system by which oxygenated surface water are moved downwards into the deeper parts of the oceans.

8. **Deepwater**, as the name implies, is a water mass with specific temperature and salinity characteristics, which forms in the depths of the oceans.
9. **Speleothem** is the collective term for the reprecipitated carbonates that are found in caves in limestone regions. These can occur in a variety of forms of dripstone, the best known of which are stalagmites and stalactites, or flowstone.
10. **Bayesian analysis** derives from the work of the eighteenth-century mathematician, Thomas Bayes. His 'Theorem' has three components: the *likelihood*, which is the probability of observing a particular data value, given the values of a set of unknown parameters (i.e. the relationship between what we want to learn about and the data we collect); the *prior*, which is the probability of observing specified values of the unknown parameters before we observe the data (i.e. what we knew before the latest data were collected); and the *posterior*, which is a combination of the information contained in the data, the likelihood and the prior (i.e. the probability we attach to specific values of the unknown parameters after observing the data). In Bayes Theorem, the posterior is then proportional to the likelihood multiplied by the prior, and this provides us with a probabilistic framework within which to make interpretations about the posterior (Buck, 2001). In other words, Bayes Theorem aims to ascribe probabilities to hypotheses or theories in the light of particular specified lines of evidence (see also Chalmers, 1999).
11. **Tephra** is volcanic ash (Chapter 7).
12. **Tufa** is calcium carbonate precipitated from calcium carbonate-saturated spring water in limestone regions. It often forms a calcareous crust around lake margins, stream edges or springs, and frequently contains inclusions of plant macrofossils.
13. **Testate amoebae** (or **rhizopods**) are microscopic unicellular protozoans that live in freshwater. Fossils are well preserved in *Sphagnum*-rich peats (Charman *et al*., 2000).
14. A **cuticle** is the thin wax layer that forms on the outer walls of plant cells that are exposed to the air. Being waxy, these may preserve well in certain depositional contexts, such as lake sediments. Grass cuticle analysis can be an important adjunct to pollen analysis, for whereas grass pollen is very difficult to identify below species level, particular grass types may be identifiable from their cuticles.
15. **Plant biomarkers** that can be obtained from the **lipids** (fats or fat-like compounds) found in lake sediments include *n*-alkanes, *n*-alkanols and *n*-alkanoic acids.
16. In plants, there are two photosynthetic pathways: C_3 and C_4. In general, C_4 plants photosynthesise more efficiently than C_3 plants and are capable of more rapid growth; indeed, they use only around half the water per unit increase in dry weight. C_4 plants are therefore well adapted to arid climates where they are typically found at the present day.
17. **Mitochondria** are small cytoplasmic particles within living cells and are intimately associated with processes of respiration.

3

Radiometric Dating 2: Dating Using Long-Lived and Short-Lived Radioactive Isotopes

They (Ernest Rutherford and Frederick Soddy) *discovered that radioactive elements decayed into other elements – that one day you had an atom of uranium, say, and the next day you had an atom of lead. This was truly extraordinary. It was alchemy pure and simple; no-one had ever imagined that such a thing could happen naturally and spontaneously.*

Bill Bryson

3.1 Introduction

In this chapter we move on to consider other radioactive isotopes that form a basis for dating. Some of these have long half-lives, enabling episodes in deep Quaternary time to be dated; others, possessing much shorter half-lives, can be used for dating only very recent events. **Argon-isotope dating** is a technique that can provide ages of volcanic rocks across the entire Quaternary time range. **Uranium-series dating**, which is mostly applied to carbonate materials such as speleothems and corals, has a shorter time span of a few hundred thousand years. Also applicable over time ranges of many thousands of years is **cosmogenic nuclide dating**, a method that is based on the progressive accumulation of cosmic-ray radionuclides on exposed rock surfaces. All of these techniques involve radioactive nuclides with relatively long half-lives. By contrast, **lead-210 (^{210}Pb)**,

Quaternary Dating Methods M. Walker

caesium-137 (**^{137}Cs**) and **silicon-32** (**^{32}Si**) are nuclides with very much shorter half-lives, and these can be used for dating sediments over timescales that are measurable in tens of years.

3.2 Argon-Isotope Dating

Two approaches are employed in argon-isotope dating. **Potassium–argon dating** is based on the decay of the radioactive isotope of potassium, potassium-40 (^{40}K), to the relatively unreactive argon isotope ^{40}Ar which is a gas. The technique was developed in the 1960s and has been used largely to date igneous rocks. Indeed, because of the very long half-life of ^{40}K (1250 million years), it is possible to obtain ages for some of the oldest rocks on earth. In terms of Quaternary applications, however, the ^{40}K/^{40}Ar technique is appropriate only for the dating of volcanic rocks that are older than ca. 100 000 years. This is because the lack of precision in younger dates, where standard errors may be of the order of ±100%, means that such age determinations have little practical value. A variant of the method, developed in the late 1960s and early 1970s, involves the measurement of the ratios between two argon isotopes, ^{40}Ar and ^{39}Ar, the latter being generated in the rock sample by neutron irradiation in the laboratory. **Argon–argon dating** has been particularly significant for Quaternary science, as the greater level of analytical precision that is associated with the ^{40}Ar/^{39}Ar method means that much younger samples can be dated, and meaningful ages of 10 000 years or less can be routinely obtained. Useful reviews of potassium–argon and argon–argon dating can be found in Aitken (1990) and Richards and Smart (1991), while a more in-depth account of the ^{40}Ar/^{39}Ar is provided by McDougall and Harrison (1999).

3.2.1 Principles of Potassium–Argon Dating

Potassium is one of the most abundant elements on earth and is found in a range of minerals, including micas, clay minerals and evaporites. It is a major element in minerals of volcanic origin, most notably the feldspars such as sanidine.[1] It exists in three isotopic forms, the most common of which, ^{39}K, comprises 92.23% of all naturally occurring potassium, while ^{41}K comprises 6.73%. Both of these are stable isotopes. The rarest of the three isotopes, ^{40}K (0.00118%), is radioactive and decays via a branched decay scheme to form two daughter nuclides, ^{40}Ca and the inert gas ^{40}Ar, both of which are stable. Despite the fact that most ^{40}K (almost 90%) decays to ^{40}Ca, the $^{40}K \rightarrow {}^{40}Ca$ decay pathway cannot be used as a means of dating. This is because ^{40}Ca is the most commonly occurring stable isotope of calcium and hence the very small amounts that would be generated through the decay of ^{40}K will result in only a minimal, and therefore effectively unmeasurable, increase in its abundance (Richards and Smart, 1991). The ^{40}Ar content of a sample can be determined, however, and it is this decay pathway that constitutes the basis for dating.

The principles underlying ^{40}K/^{40}Ar dating are relatively straightforward. When in a molten state, volcanic rocks such as lavas or the crystal components of volcanic ashes will release any ^{40}Ar that is generated by the decay of ^{40}K. Once the rocks or minerals begin to cool, however, argon can no longer escape and is trapped within the mineral crystal lattices where its abundance increases over time. In the laboratory, the ^{40}Ar concentrations in rock samples can be determined by heating them to the melting point in a furnace. This releases the gas which can be measured in a mass spectrometer. A second sample is used to determine ^{40}K content using an atomic absorption spectrophotometer or a flame photometer. The ratio between ^{40}K and radiogenic ^{40}Ar then gives an indication of the time that has elapsed since the volcanic event occurred.

3.2.2 Principles of Argon–Argon Dating

One of the problems with conventional potassium–argon dating is that measurements have to be made on separate *aliquots* (equal proportions of the sample) and this can result in erroneous ages where the sample material is heterogeneous (in basalts, for example). A way around this difficulty was the development of the argon–argon dating method wherein measurements are made on a single sample. Here, the ^{40}Ar content is measured directly as before, but the ^{40}K concentration is measured indirectly by using the known proportions between the potassium and argon isotopes. Sample grains are placed in a nuclear reactor and irradiated. This converts a portion of the stable ^{39}K isotopes to another form of argon, ^{39}Ar. As ^{39}Ar abundance is proportional to that of the stable isotope ^{39}K which is, in turn, proportional to ^{40}K (see above), the ^{40}Ar/^{40}K ratio in the samples can be inferred from a single mass spectrometric measurement (McDougall and Harrison, 1999).

In order to determine the argon content accurately, the rock samples are heated in a stepwise manner from room temperature to the melting point of the crystal (>1500 °C for sanidine). Both of the argon isotopes, ^{40}Ar and ^{39}Ar, are driven off and an age determination is made for each step. For a well-behaved sample, the age stays the same for all steps of gas release and a plateau is obtained which gives the date of the sample (Figure 3.1a). The absence of a well-defined plateau provides a warning that the sample may be unreliable (see below). One reason why this might occur is that older feldspar grains (xenocrysts[2]) with a different K–Ar concentration may have become incorporated into the dated sample. This difficulty can be circumvented by hand-picking the grains, but this is a laborious and not always reliable approach. A more recently developed alternative is to target single mineral crystals of the order of 1 mg or less and to use a high-powered laser to drive off the argon for measurement using a super-sensitive mass spectrometer. This is known as **single-crystal laser fusion (SLCF) ^{40}Ar/^{39}Ar dating** and has led to a significant improvement in both the accuracy and the precision of argon–argon age estimates (Wintle, 1996).

3.2.3 Some Assumptions and Problems Associated with Potassium–Argon and Argon–Argon Dating

There are two fundamental assumptions that underlie potassium–argon dating. The first is one that is common to all radiometric dating techniques, namely that the system remains closed following the volcanic event. In other words, there has been no loss of argon following crystallisation. Loss of argon could occur if the mineral sample has been weathered or if there has been a further episode of reheating; in both cases an underestimate of age will be obtained. However, petrographic examination may reveal evidence of weathering, while the presence of secondary minerals, such as xenocrysts, will tend to indicate that recrystallisation has occurred. A further way in which possible argon loss can be detected is to compare age estimates derived from the whole rock and mineral fractions to determine whether or not they are concordant. This approach can also be used to establish whether older mineral fractions have been incorporated into the rocks, a relatively common phenomenon in volcanic material and which, if undetected, can lead to an overestimate of age (Richards and Smart, 1991). Overestimate of age can also arise if there is excess argon present in the sample, in other words previously accumulated gas that has failed to escape while the rock was molten. Figure 3.1b shows the age spectra that could result from these various effects. One possible cause for the lower age estimate indicated by the initial plateau at around 290 000 years is that mineral alteration during weathering or secondary reheating occurred during burial thereby allowing the release of less firmly held argon, which has resulted in an aberrantly low age estimate. The true age is likely to be given by the final hearing step, in other words ca. 800 000 years. Alternatively, it could be that the younger date is correct and that the higher ages obtained during the later heating steps are due to excess argon. In that case, the date of 290 000 years is probably correct (Aitken, 1990).

A second important assumption is that all the ^{40}Ar in the sample has been derived from the decay of ^{40}K. However, ^{40}Ar is also a constituent of the atmosphere and hence a proportion of this **atmospheric ^{40}Ar** (as opposed to **radiogenic ^{40}Ar**) will be present in all minerals. It will also contaminate samples in the laboratory via the measuring apparatus. However, atmospheric ^{40}Ar is always accompanied by ^{36}Ar and as the latter will be

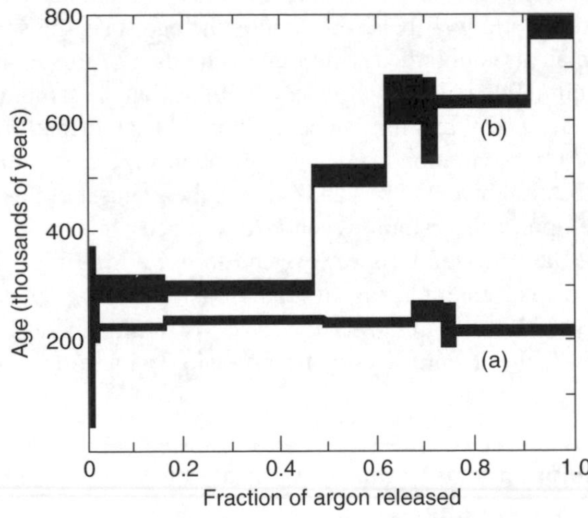

Figure 3.1 *Age spectra obtained using the argon–argon dating technique. The argon is released in steps by heating the sample to successively higher temperatures, and an age determination is made on each portion released. The less firmly held argon is released in the early steps. In (a) a clear plateau is evident which gives confidence in the age estimate (ca. 220 000 years* BP*) of the sample. Sample (b), however, is more problematical (see text for discussion). The samples are from two basalt flows in Israel between which there was a palaeosol containing Upper Acheulean artefacts (after Aitken, 1990). Reproduced by permission of Pearson Education Ltd*

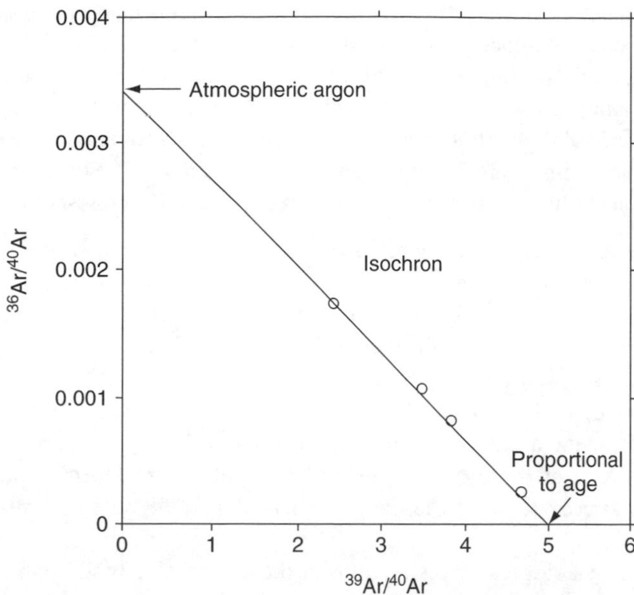

Figure 3.2 *Schematic illustration of an isochron plot. The two intercepts correspond to the pure trapped (y-axis) and pure radiogenic (x-axis) components of the sample (after McDougall and Harrison, 1999). Reproduced by permission of Oxford University Press*

present in a sample only as a result of atmospheric contamination, measurement of ^{36}Ar will enable a correction to be made for the presence of atmospheric ^{40}Ar, using the known ratio of ^{40}Ar/^{36}Ar in the contemporary atmosphere (296:1). Matters are more complicated in argon–argon dating because ^{36}Ar is produced from calcium when the sample is exposed in the nuclear reactor. In that case, the correction for the presence of atmospheric argon involves the measurement of another non-naturally occurring argon isotope, ^{37}Ar, which is also produced from calcium (Aitken, 1990).

The measurement of ^{36}Ar in argon–argon dating also enables the **isochron technique** to be employed to determine the age of a sample using ^{40}Ar as the reference isotope (McDougall and Harrison, 1999). The latter is usually the most abundant of the stable argon isotopes and can therefore be measured very precisely. During successive heating steps, the ratios between ^{36}Ar/^{40}Ar and ^{39}Ar/^{40}Ar are determined, and these are plotted on a graph (Figure 3.2). The result will be a series of data points that range from pure atmospheric argon to pure radiogenic argon. A regression line through these data points forms an **isochron**, and the point at which the isochron intercepts with the *x*-axis (where the ^{36}Ar/^{40}Ar ratio is equal to 0) enables the age of the sample to be determined from the ^{39}Ar/^{40}Ar ratio at the intercept point (Figure 3.2).

3.2.4 Some Applications of Potassium–Argon and Argon–Argon Dating

The principal contributions of the potassium–argon and argon–argon techniques in Quaternary research have been in the dating of lavas and other volcanic materials associated with early hominid finds, especially in Africa (Deacon and Deacon, 1999); in the development of glacial chronologies by dating tephras interbedded with glacial deposits, for example in the Yukon (Westgate *et al.*, 2001) and Patagonia (Singer *et al.*, 2004); in the dating of volcanic events, such as the great 'super-eruption' of Mt Toba in Sumatra around 74 000 years (section 4.3.3.4); and in the provision of a chronology for the palaeomagnetic timescale, which is discussed in Chapter 7. Some of these, and also other applications of the technique, are considered in the following section.

3.2.4.1 Potassium–argon and argon–argon dating of the dispersal of Early Pleistocene hominids. There is now a general acceptance that the earliest hominids evolved in Africa (Strauss and Bar-Yosef, 2001). However, the identity of the first hominid species to disperse out of Africa and the timing of this dispersal remain controversial. The long-held view, which was the received wisdom until at least the early 1990s, that hominids remained in Africa until around 1 million years ago has been challenged by a series of recent archaeological discoveries (Mithen and Reed, 2002). Perhaps the most significant of these have been in the Republic of Georgia and on the island of Java which appear to shift the date of early dispersal back by more than half a million years.

In Georgia, archaeological investigations at the site of Dmanisi, approximately 85 km southwest of Tbilisi, revealed the presence of two partial Early Pleistocene hominid crania (Gabunia *et al.*, 2000). These fossils are comparable in size and morphology with *Homo ergaster* (also known as *Homo erectus*) from Kenya. Potassium–argon and argon–argon dating of basalt immediately underlying the fossil-bearing sediment produced maximum ages for the fossil assemblages of between 1.8 and 2.0 million years, results that are corroborated by palaeomagnetic dating (Chapter 7) which suggests an age of around 1.7 million years for the hominid fossils. In Java, potassium–argon and argon–argon ages of between 1.8 and 1.9 million years have been obtained on hominid fossils (including skull and mandible) believed to be early *Homo erectus* (Swisher *et al.*, 1994). Although there has been some discussion about the provenance and reliability of the dating of these specimens (Langbroek and Roebroeks, 2000; Huffman, 2001), subsequent $^{40}Ar/^{39}Ar$ dating, again supported by palaeomagnetic evidence, appears to confirm a maximum age for early hominids in Java of around 1.7 million years (Sémah *et al.*, 2000). These dated sites, in association with evidence from China, and the Jordan Valley (Figure 3.3), show that *Homo ergaster/habilis* spread northwards and eastwards out of Africa as early as 1.5 million years ago, and that this migration preceded the spread of hominids into Europe, possibly by as much as a million years.

3.2.4.2 $^{40}Ar/^{39}Ar$ dating of anatomically modern Homo sapiens *from Ethiopia.* There are two broad groups of theories about the origins of *Homo sapiens*. Some researchers continue to support the view that anatomically modern humans developed in geographically widespread areas throughout the course of the Pleistocene. This 'multi-regional' hypothesis has, in large measure, been replaced by the 'out of Africa' model which argues that, as with the earlier forms of hominid discussed in the previous section, modern humans originated in Africa and subsequently spread outwards from there. However, there are still considerable differences of opinion over the mode of modern human evolution; for example, whether it proceeded gradually or in 'fits and starts' (punctuated evolution), and from precisely which region modern humans originated

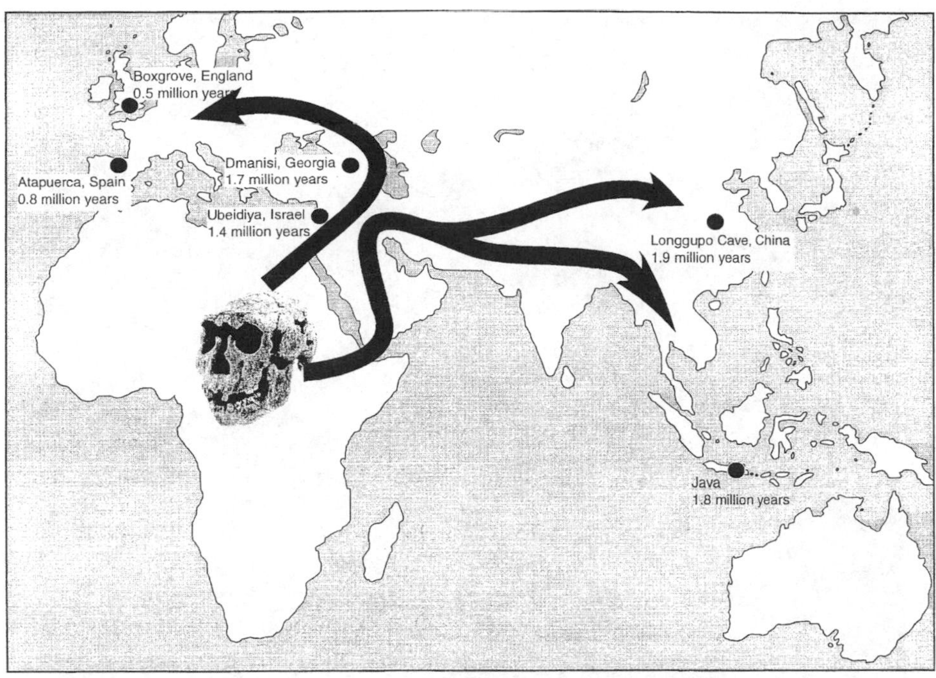

Figure 3.3 *Dmanisi, Georgia, Java and other key localities for hominid dispersal into Eurasia (after Mithen and Reed, 2002). Reproduced with permission of Elsevier*

(Stringer, 2003). A key question concerns the timing of the evolutional and dispersal events, for thus far there has been a lack of securely dated African hominid fossils between 100 000 and 300 000 years.

In the context of this ongoing debate, the recent discovery of anatomically modern human fossils, in association with artefacts and fossils, at Herto, Middle Awash, Ethiopia, is of considerable importance (Clark *et al.*, 2003; White *et al.*, 2003). The fossil crania (Figure 3.4) display characteristics suggesting that they are morphologically and chronologically intermediate between archaic African fossils and later anatomically modern Late Pleistocene humans. Significantly, $^{40}Ar/^{39}Ar$ dating provides a precise age for the fossils. The sandstone unit in which the artefacts and fossils were found also contains embedded pumice and obsidian (Chapter 6) clasts, and these yielded argon–argon ages of 163 000±3000, 162 000±3000 and 160 000±2000 years. Immediately to the south of the hominid site, a comparable suite of fossiliferous deposits is overlain by a tuff[3] which can be correlated with a tuff from the Konso region of southern Ethiopia where a laser fusion $^{40}Ar/^{39}Ar$ age of 154 000±7000 years has been obtained. The Herto hominids are therefore precisely dated to between ca. 160 000 and 154 000 years, which makes them the oldest definite record of what we currently consider to be modern *Homo sapiens*. Their anatomy and antiquity provide strong evidence for the hypothesis that modern humans emerged from Africa and, moreover, that East Africa in particular may have been the cradle for modern human evolution.

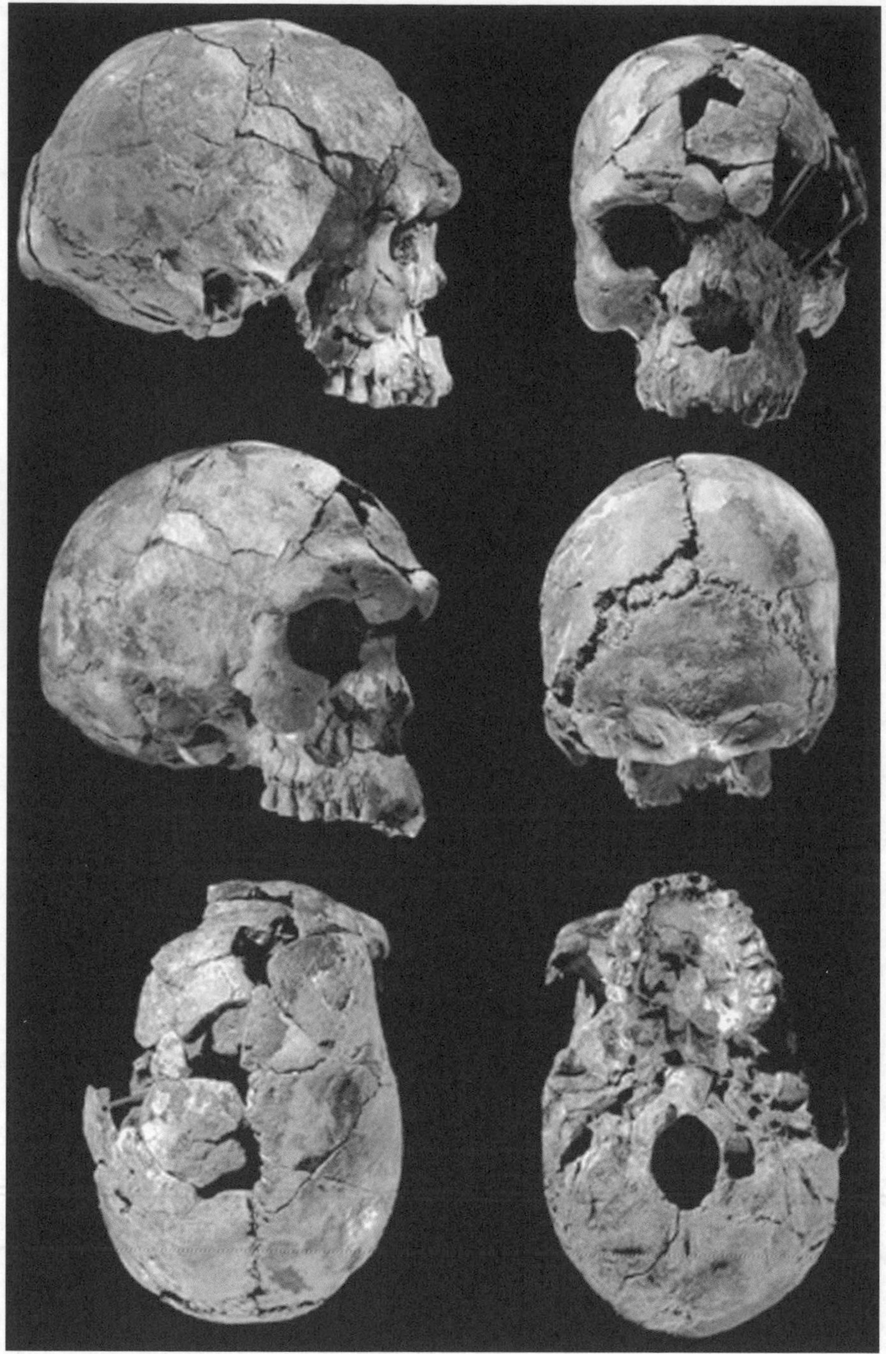

Figure 3.4 *Adult cranium (BOU-VP-16/1) from Herto, Ethiopia, dated by $^{40}Ar/^{39}Ar$ to between 160 000 and 154 000 years* BP *(photo: David Brill). Reprinted from Nature. Copyright 2003 Macmillan Magazines Limited*

3.2.4.3 $^{40}Ar/^{39}Ar$ *dating of historical materials: the eruption of Vesuvius in* AD *79.* One of the principal benefits to Quaternary science of the development of the $^{40}Ar/^{39}Ar$ technique is the potential it offers for the dating of relatively recent material. As noted earlier, single-crystal dating is the preferred procedure because it avoids problems arising from possible effects of xenocrystic contamination, but single crystals are often too small to yield reliably measurable quantities of radiogenic argon in samples of Late Holocene Age. In some circumstances, however, large crystals are available and meaningful ages can be obtained, even on samples from the historical period.

This approach has been applied to volcanic ejecta from the eruption of Vesuvius, which has been documented in the writings of Pliny the Younger. The volcano erupted on 24 August 1979 and buried Pompeii, Herculaneum and other neighbouring towns and cities under a layer of ash and pumice. Renne *et al.* (1997) obtained a large pumice clast from the recently excavated Villa of Poppea in Pompeii (Figure 3.5), which contained large phenocrysts of sanidine up to 8 mm in diameter. Twelve samples were irradiated, analysed by incremental heating, and an isochron age of 1925 ± 94 years was obtained. This is in excellent agreement with the Gregorian calendar-based age (1918 cal. years ago) of the eruption. Although excess argon (see above) was present in the sample, this could be detected by careful laboratory analysis of the sanidine and corrected for. This example shows that a sample less than 2000 years old can be dated with better than 5% precision, and validates the $^{40}Ar/^{39}Ar$ technique as a reliable geochronometer into the Late Holocene.

Figure 3.5 *The recently excavated Villa of Poppea, Pompeii. Remnants of the ash and pumice from the* AD *79 eruption of Vesuvius can be seen in the section to the right of the villa (photo: Barry Burnham)*

3.2.4.4 $^{40}Ar/^{39}Ar$ dating and geological provenancing of a stone axe from Stonehenge, England. An unusual application of argon–argon dating is described by Kelley *et al.* (1994) in relation to the discovery of a stone axe at Stonehenge in southern England. A fragment of the axe, which had been fashioned from rhyolitic tuff, was $^{40}Ar/^{39}Ar$-dated and 11 analyses produced an isochron age value of 341 ± 5 million years, in other words within the Lower Carboniferous period of the Upper Palaeozoic. On the basis of petrological analysis, the source of the rhyolitic material was thought to be Pembrokeshire, South Wales (Roe, 1990), which is also believed to be the origin of the doleritic 'bluestones' that form the inner circle of Stonehenge. However, argon–argon dating shows that the stone axe is younger than the volcanic rocks of South Wales, which are Precambrian and Lower Palaeozoic (Ordovician and Silurian) in age (i.e. >395 million years), and hence it cannot have come from that area. Trace element analysis of the Stonehenge axe indicates an affinity with rhyolitic tuffs in the north of Britain, possibly in the Midland Valley of Scotland. This, in turn, raises questions as to how this implement reached Salisbury Plain in the south of England. Glacial transport is a possibility, although the glaciation of Salisbury Plain is a controversial topic and the balance of evidence suggests it to have been unlikely (Scourse, 1997). Alternatively, human agency could have been involved, which raises interesting questions about trade routes and exchange. In either event, argon–argon dating has resulted in the provenancing of a potentially important archaeological artefact which would otherwise have been ascribed a different, and entirely erroneous, source area.

3.3 Uranium-Series Dating

After radiocarbon, uranium-series (U-series) dating has probably been the most widely employed radiometric dating technique in Quaternary science. The method was developed in the years following the Second World War and was initially applied to the dating of deep-ocean sediments, but since the 1960s it has been routinely used for the dating of both marine and terrestrial carbonates. As with radiocarbon, significant advances came in the 1980s through the application of mass spectrometry, and further refinements over the last ten years (see below) have led to further improvements in both precision and accuracy. The technique is based on two naturally occurring uranium isotopes, ^{238}U and ^{235}U, both of which decay to stable forms of lead (^{206}Pb and ^{207}Pb, respectively) through complex decay chains involving the creation of intermediate ('daughter') nuclides, each with very different half-lives (Figure 3.6). Of the two, ^{238}U is far more abundant, the activity ratio between the $^{238}U/^{235}U$ being 21.7:1 (Schwarcz, 1989), which means that the analysis of the ^{235}U decay chain is correspondingly more difficult. A third decay series is headed by an isotope of thorium, ^{232}Th, but the daughter isotopes have such short half-lives that this decay chain cannot be used as a basis for dating. However, ^{232}Th is valuable in U-series dating as it can be used to correct for the presence of detrital contamination in sample material (section 3.3.2). In the two uranium decay chains, the $^{230}Th/^{234}U$, $^{231}Pa/^{235}U$, $^{234}U/^{238}U$ daughter/parent combinations are used in Quaternary dating.[4] In addition, the $^{231}Pa/^{230}Th$ ratio can also be used to generate an estimate of age. General overviews of the method are provided by Schwarcz (1989), Smart (1991a) and Latham (2001), while more

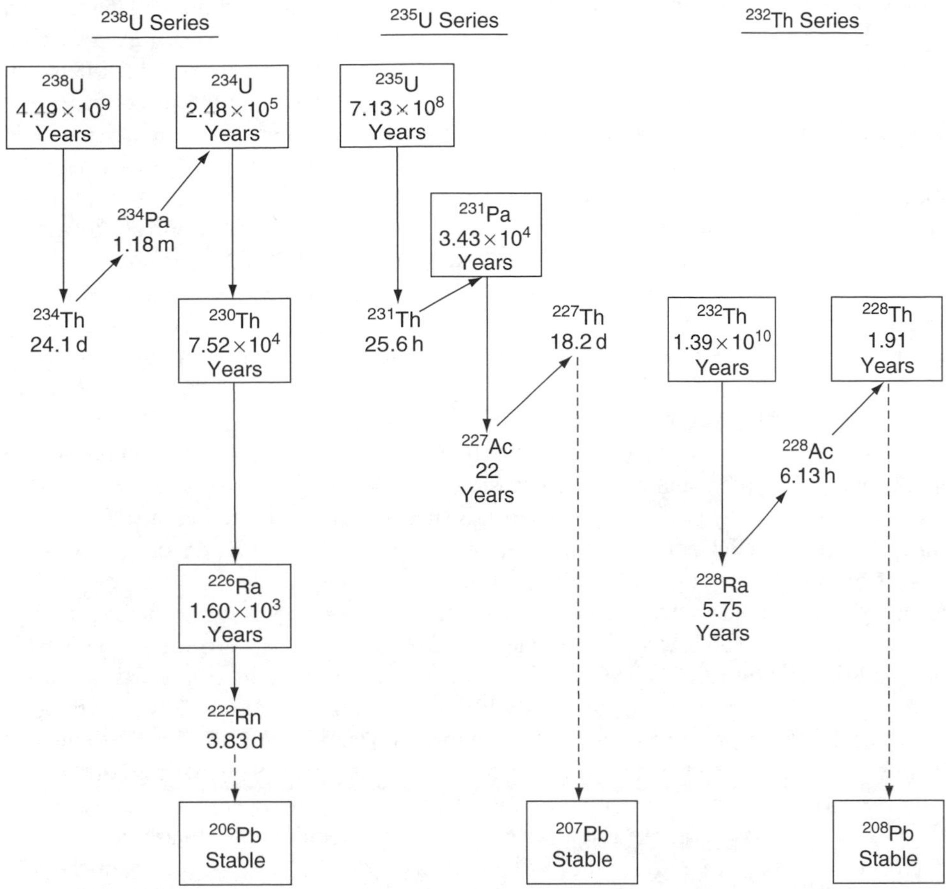

Figure 3.6 *The chain decay pathways and half-lives of ^{238}U, ^{235}U and ^{232}Th. The longer-lived isotopes are outlined by boxes, and the daughter/parent combinations used in Quaternary dating by double boxes. The vertical arrows indicate alpha decays, the diagonal arrows beta decays, and the dotted lines show incomplete series (d=days; m=months) (after Smart, 1991a). Reproduced by permission of the Quaternary Research Association*

in-depth accounts and reviews of many applications can be found in Ivanovich and Harmon (1995) and Bourdon *et al.* (2003a).

3.3.1 Principles of U-Series Dating

As can be seen in Figure 3.6, the half-lives of the parent isotopes at the head of the dating series, ^{238}U and ^{235}U, are significantly longer than any of their daughters, which means that the number of parent atoms remains essentially constant for several half-lives of the daughters. If a uranium mineral is left undisturbed for several million years, the **activity** of each of the daughter isotopes will come to be equal to that of the parent uranium isotope, as measured in numbers of disintegrations per unit time per unit weight of rock. In other words, if the system remains closed, a state of **secular equilibrium** will have been reached. However, if the system is not closed and daughter isotopes can escape, then there will be a break in the decay chain and a state of **disequilibrium** will exist between the nuclides above and below the break. In natural systems, disequilibrium between the longer-lived isotopes in the decay chain can arise in a number of different ways. For example, radon gas may be lost from the system by diffusion through rocks where the matrices are porous, while chemical differences between the various elements in the series can result in separation during weathering, transport and deposition in the hydrosphere. Once isotope migration ceases, however, and the system becomes closed again (e.g. where minerals are precipitated from solution, such as in cave speleothem), there is a slow return to equilibrium conditions, and it is this process that forms the basis for estimating age (Smart, 1991a). Insofar as determination of age rests on the interruption of the decay chain and the selective removal of some decay products, the technique is frequently referred to as **uranium-series disequilibrium dating**.

There are two ways in which disequilibrium in the U-series decay chain can be used to infer age, and both depend on the process of fractionation, in other words a selective separation of the decay products. These are known as the **daughter deficient (DD)** and the **daughter excess (DE)** methods. In the case of the former, the daughter nuclide is initially absent but increases with time until equilibrium is achieved, the most important DD technique being based on the measurement of $^{230}Th/^{234}U$ ratios. This works on the principle that as uranium is soluble, whereas the daughter products (such as ^{230}Th) are not, organisms that build a shell by secreting carbonate from ocean water, such as corals and molluscs, will take up dissolved uranium and this will be incorporated into the skeletal structure. There will, however, be little or no uptake of thorium. Hence, age of the fossil sample can be inferred by the extent to which the daughter, ^{230}Th, has grown back towards equilibrium with the parent ^{234}U. The same principle applies in the dating of other carbonate materials, such as speleothems and travertines, where the age of precipitation of carbonate can be determined by measuring the extent to which the decay product, ^{230}Th, has reappeared in the carbonate matrix. In the ^{235}U decay series, protactinium is also insoluble, and hence the $^{231}Pa/^{235}U$ ratio can similarly be used as an indicator of age.

In the DE methods, the daughter isotope is present in excess of the concentration at secular equilibrium before decaying over time. The initial disequilibrium state will have been caused either by preferential leaching of the parent isotope or by precipitation of the daughter. For example, the decay of uranium in a water column will result in the precipitation of the daughter nuclides ^{230}Th and ^{231}Pa which will accumulate on the floors of lakes or on the sea bed. The uranium, meanwhile, remains in solution. As a consequence, accumulating sediments will contain ^{230}Th and ^{231}Pa, but will be deficient in uranium. Hence the age of lake or ocean floor sediments can be determined by measuring the extent to which ^{230}Th or ^{231}Pa has decayed down the profile.

Because many of the key isotopes in the decay chain decay through the emission of alpha particles, the conventional approach to the estimation of U-series ages has been by alpha spectrometry following chemical extraction of uranium and thorium from the sample material. Each U–Th measurement counting usually takes a few days, and hence analysis is often carried out in batches (Latham, 2001). However, samples must contain a sufficient amount of uranium (typically more than 0.1 ppm). Moreover, because isotopic occurrence is being detected indirectly on the basis of the decay products (as in radiometric ^{14}C

dating), and because nuclear decay is a random process, an element of uncertainty will be associated with each age determination. U-series ages are therefore accompanied by a one standard deviation value which measures the precision of the age estimate. As this value reflects uncertainties in decay counting, the greater the number of counts, the lower will be the error value. Smart (1991a), for example, has noted that 10 000 counts are needed to give a 1σ error of ±1% for each isotope. More generally, the precision of U-series dates obtained using alpha spectrometry tends to be between ±5% and 10%.

The introduction of mass-spectrometric analysis in the 1980s, in which individual atoms as opposed to the less frequent alpha particles are counted, enabled much higher levels of analytical precision to be achieved, often on relatively small samples (milligrams rather than grams of sample) of material. As a consequence, meaningful ages could now be obtained on samples of 50–100 years, while the upper end of the dating range was extended to beyond 500 000 years. The improvements in counting statistics meant, for example, that a 2σ error of ±1000 could be routinely obtained on samples of last interglacial age (125 000 BP). One particularly valuable application of these developments was the paired U-series and ^{14}C dating of fossil corals which forms the basis for calibrating the radiocarbon timescale beyond 12 000 years BP (section 2.6). During the 1980s, **thermal ionisation mass spectrometry (TIMS)** became the standard measurement method (section 3.3.3.3), while more recent developments have seen the increasing application of **multi-collector inductively coupled-plasma mass spectrometers (ICP-MS)** (Goldstein and Stirling, 2003). The U-series technique is applicable over time ranges from a few hundred up to about 350 000 years using alpha-particle spectrometry, and up to about 500 000 years by mass spectrometry.

3.3.2 Some Problems Associated with U-Series Dating

Apart from possible sources of error arising from sampling and laboratory measurement, there are two principal potential problems associated with U-series dating. First, it must be assumed that there has been no loss or gain of nuclides since deposition: in other words, the parent to daughter ratio is solely a function of the radioactive decay process. Several materials that can be dated by U-series, however, show evidence of departure from such **closed-system behaviour** (Table 3.1). Dating of such materials necessitates the use of correction factors which, in turn, requires a detailed knowledge of the processes that give rise to the isotope disequilibrium. A particularly problematic material is bone, for it is well known that bone can continue to take up uranium after deposition or burial, and scientists can never be certain when a given sample of bone acquired its uranium. However, because bone (both human and animal) is such an important and relatively common fossil material, strenuous efforts have been made to resolve the problem of open-system behaviour in buried bone (e.g. Millard and Hedges, 1995; 1996; Pike *et al.*, 2002). These have involved the development of uptake models to try to describe how and when the uranium was absorbed into the bone. Such models, however, often require calibration by independent dating methods (such as electron spin resonance: section 4.3), and even then uranium loss cannot be modelled, while the possibility of a delay in uranium uptake after burial of the bones cannot be tested (Latham, 2001). Open-system behaviour is not confined to bone, however, for other materials, such as molluscs, may incorporate

Table 3.1 *Reliability/unreliability of uranium-series dates for terrestrial materials arising from deviations from closed-system behaviour and contamination by ^{230}Th and ^{234}U from detrital material (after Smart, 1991a)*

Reliability	Material	Closed system?	Contaminated?
Reliable	Unaltered coral	Closed	Clean
	Clean speleothem		Clean
	Volcanic rocks		–
	Dirty speleothem		Contaminated
Possibly reliable	Ferruginous concretions	Possibly closed	Contaminated
	Tufa		Contaminated
	Mollusc shells		Contaminated
	Phosphates		Contaminated
Generally unreliable	Diagenetically altered corals	Open	Clean
	Bone		?
	Evaporites		Contaminated
	Caliche		Contaminated
	Stromatolites		Contaminated
	Peat and wood		?

uranium into carbonate shells after their death (McLaren and Rowe, 1996), while some calcitic deposits (such as travertines[5]) may have been partly recrystallised after deposition. In the case of the latter, however, such open-system activity may be detectable by careful petrographic analysis of the calcite prior to dating (Schwarcz, 1989).

A second potential source of error relates to the assumption that the daughter isotopes in the measured sample are entirely radiogenic in origin. In other words, it is assumed that the ^{230}Th content of a sample of speleothem, for example, was 0 at the time of crystal formation, and hence the measured ^{234}U/^{230}Th activity ratio is a reflection purely of the reappearance of ^{230}Th through radioactive decay. However, it is not uncommon for carbonate materials to be contaminated by **detrital materials** (such as aeolian dust, or water-transported silts and clays) that already contain daughter nuclides. Such contamination can lead to U-series ages that are *older* than the true age of the sample. However, detritus may also carry ^{234}U and ^{238}U and, if uncorrected for, will result in ages that are *younger* than the true sample age. Fortunately, the effects of detrital contamination can be corrected for by measuring the activity of ^{232}Th that is present in the sample but which plays no part in the decay chain of uranium. This isotope is present in detritus, but not in pure calcite, and hence the ^{232}Th/^{230}Th ratio can be used to correct for detrital addition of ^{230}Th. This is normally done using the **isochron technique**, in which multiple-sample leaching analyses are undertaken and the activity ratios of the different isotopes in each of the samples are plotted against each other (Schwarcz and Latham, 1989). This will show the extent to which detrital contamination has influenced the ^{230}Th/^{234}U and ^{234}U/^{238}U ratios, the isotopic ratios from which ages are calculated (Figure 3.7). The correction for the former ratio can be found by plotting ^{230}Th/^{232}Th versus ^{234}U/^{232}U, while the latter ratio can be corrected by plotting ^{234}U/^{232}Th against ^{238}U/^{232}Th. These corrected ratios, which are reflected in the gradients of the lines on the isochron plots, are then inserted into the normal age equation to calculate the age of the sample.

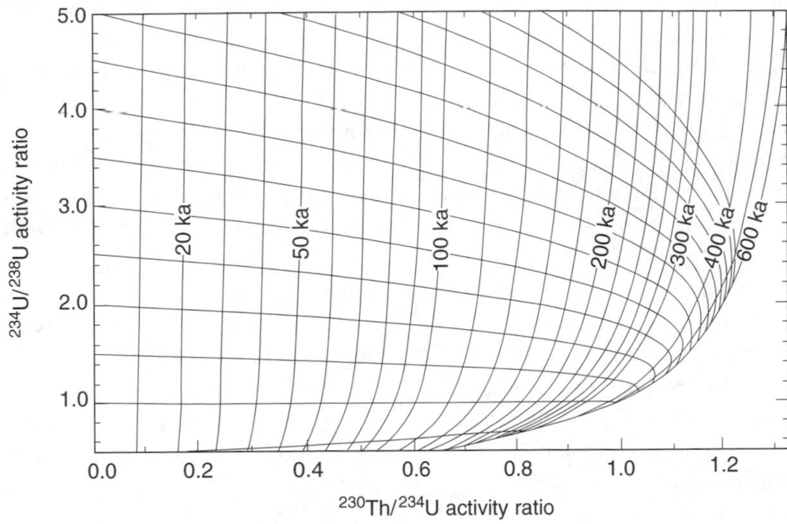

Figure 3.7 *Variation of $^{234}U/^{238}U$ and $^{230}Th/^{234}U$ activity ratios with time in a closed system where there is no initial ^{230}Th present. The near vertical lines are lines of contant age (isochrons), while the near horizontal lines show changes in nuclide activity with time for different initial $^{234}U/^{238}U$ activity ratios (after Heijnis, 1995)*

3.3.3 Some Applications of U-Series Dating

U-series dating has been applied to a wide range of carbonate materials in a variety of different contexts. It has been used to derive chronologies of sea-level change, particularly relating to interglacial high-stands of the sea, through the dating of corals (see following section) and molluscs (Jedoui *et al.*, 2003), and also sediments from submarine carbonate platforms (Henderson and Slowey, 2000). It has been extensively employed in the dating of speleothem (section 3.3.3.3), while U-series dates have also been obtained on a range of carbonate materials, including ferricretes[6] (Augustinus *et al.*, 1997) and travertines (Eikenberg *et al.*, 2001). Tufa has also been dated by U-series (e.g. Szabo *et al.*, 1996), although detrital Th contamination and open-system uranium loss make this a more problematic dating material (Garnett *et al.*, 2004). Other applications include the dating of organic sediments that contain carbonate residues, such as peats (Heijnis and van der Plicht, 1992) and lake deposits (Harle *et al.*, 2002), and the dating of teeth (Esposito *et al.*, 2002). More detailed examples are described in the following section.

3.3.3.1 Dating the Last Interglacial high sea-level stand in Hawaii. One of the most successful materials for the application of U-series dating is fossil coral. This is because after death coral skeletons act as effectively closed systems until the coral dissolves or changes to calcite. There is little or no inherited ^{230}Th, and hence problems of detrital contamination are minimised. Moreover, most corals contain sufficient uranium (typically 2–3 ppm) for the application of both the $^{230}Th/^{234}U$ and $^{231}Pa/^{235}U$ methods which enables an independent check to be made on calculated ages (Smart, 1991a). U-series dating of fossil coral reef complexes has therefore been widely employed in studies of Late Quaternary sea-level change, and has been used to gain an independent estimate of the duration of interglacial periods (Edwards *et al.*, 2003).

One area where there is extensive evidence of interglacial sea-level high-stands is the Hawaiian Islands in the Pacific Ocean (Szabo *et al.*, 1994). On Oahu, to the northwest of the main island of Hawaii (Figure 3.8), shorelines that formed during interglacial high-stands of sea level have been raised by tectonic activity and now occur up to 13 m above present-day sea level (Muhs *et al.*, 2002). TIMS U-series dating of coral samples produced ages of between ~134 000 and ~113 000 years, with the majority clustering in the range of ~125 000–115 000 years. These results showed that not only were the shorelines formed during the Last Interglacial, but that the Interglacial may have been of significantly longer duration (~136 000 to ~115 000 years) than hitherto considered. The conventional view of the duration of the Last Interglacial based, for example, on the MOI record is that peak interglacial conditions lasted for ca. 10 000 years. These data from Hawaii suggest that such conditions, when global ice volume was significantly less than present (and sea levels were commensurately higher), may have lasted very much longer.

3.3.3.2 Dating of early hominid remains from China. Many cave sites, especially in limestone areas, contain well-preserved fossil bones. This is because mineral salts in solution that percolate through the sediments are deposited in the vacant pore spaces within the bones, as a result of which the bone material becomes completely **permineralised**. However, because of the problem of post-mortem uptake of uranium by bone (section 3.3.2), the age of vertebrate bone assemblages is often determined through the dating of interbedded speleothem calcite which is more likely to act as a closed system for U–Th isotopes.

This latter approach was employed by Shen *et al.* (2001) to obtain a chronology for hominid remains discovered in the Zhoukoudian Cave, about 50 km southwest of Beijing, China. The assemblages include six fairly complete crania and other hominid fossils representing at least 40 individuals, 98 species of non-hominid mammalian fossils, and a substantial quantity of stone artefacts. The Zhoukoudian hominid specimens, commonly referred to as 'Peking Man', have been widely recognised as representatives of *Homo erectus*. Previous dating, using a variety of techniques, had placed the remains in the range between 230 000 and 500 000 years. More recent U-series dating of flowstones interbedded within the hominid-bearing horizons, however, indicates a much greater antiquity for the assemblages. The age of the number 5 skull, the best-preserved of the cranial finds, is closely constrained to between 400 000 and 500 000 years while hominid remains from the underlying strata in the cave are at least 600 000, and may be greater than 800 000 years old. These U-series dates show that human occupation of the Zhoukoudian cave site occurred much earlier and lasted for much longer (more than half a million years) than previously thought. The redating of the Zhoukoudian site also has a bearing on the ongoing debate on human evolution, for the new ages support the idea that Asian *Homo erectus* and African *H. heidelbergensis* may represent different stages of human evolution, the former predating the latter. This contrasts with the generally held belief that the two hominid species were broadly contemporaneous. In addition, a much older age assignment to the Zhoukoudian hominids means a greater interval of time between *H. erectus* and *H. sapiens* in eastern Asia, making the former more suitable as an

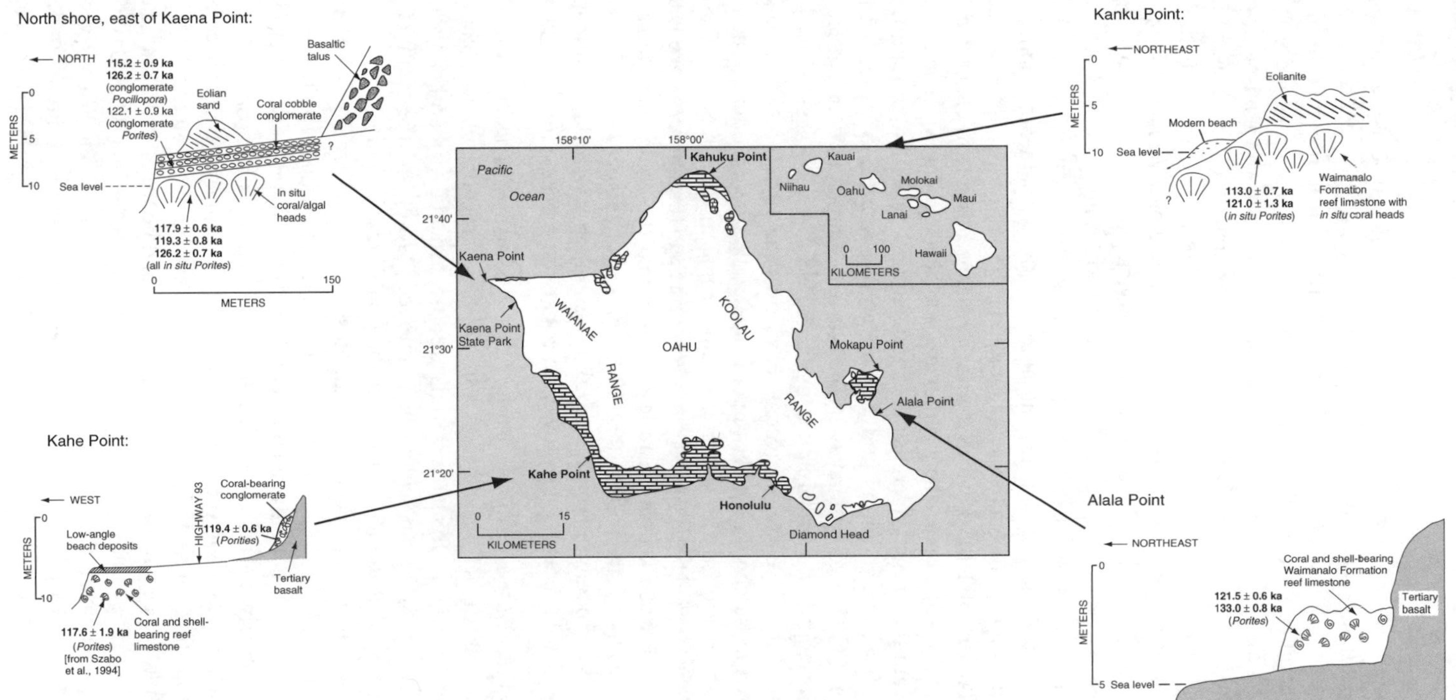

Figure 3.8 *The island of Oahu, Hawaii, and schematic cross-sections of some of the dated interglacial high sea-level locations (after Muhs* et al., *2002). Reproduced with permission of Elsevier*

ancestral form than *H. heidelbergensis*, which has, hitherto, been widely considered to be the direct ancestor of *H. sapiens*.

3.3.3.3 Dating of a speleothem from northern Norway. Cave speleothems are very important sources of palaeoclimatic data. Carbonate precipitation is strongly reduced or arrested during cold episodes and increases to a maximum during warm intervals. In mid- and high latitudes, therefore, speleothem development reflects the sequence of climatic changes during a glacial–interglacial cycle. In addition, because the oxygen isotope composition of cave seepage water is strongly influenced by temperature, the $\delta^{18}O$ signal in cave speleothem carbonate can be used as a palaeothermometer, in other words to provide a quantified record of past cave temperature changes (Lowe and Walker, 1997). Where organic residues have been incorporated into the carbonate, such records can sometimes be dated by AMS ^{14}C. More frequently, however, U-series dating is used to provide a timescale for climatic reconstructions from cave speleothems (Frumkin *et al.*, 1999; Williams *et al.*, 2004). A good overview of U-series chronology and environmental applications of speleothems can be found in by Richards and Dorale (2003).

With the development of TIMS U-series dating, it became possible to obtain precise age determinations on very young speleothem carbonate, and hence to date speleothem-derived climate records from the present interglacial. An example is provided by Lauritzen and Lundberg (1999) from a cave site in northern Norway. There the $\delta^{18}O$ record from a stalagmite sample (SG93) was calibrated against surface temperature and the resulting palaeotemperature record was underpinned by 12 TIMS U-series ages which ranged from $10\,409 \pm 21$ to 253 ± 2 years. The average error over the 10 000-year time span was 22 years. There is a close correlation between the speleothem temperature trend and the temperature record derived from the oxygen isotope signal in the Greenland GISP2 ice core, particularly in the period between 8000 and 4000 years ago (Figure 3.9A), while there is also a close match between known 'historical events' and the speleothem temperature reconstruction, especially for the period 2000–1000 years ago (Figure 3.9B). This suggests that not only is the temperature record from the speleothem reflecting hemispherical climate trends, but that U-series dating provides a coherent timescale for the sequence.

3.3.3.4 Dating of fluvial terraces in Wyoming, USA. The ages of coarsely clastic Quaternary deposits, such as fluvial terraces and alluvial fans, are difficult to obtain using conventional dating methods. Organic materials are rare, especially in arid and semi-arid areas, and hence radiocarbon can seldom be employed. Moreover, the relatively short half-life of ^{14}C means that even if organic residues are present, only relatively young deposits can be dated using this technique. Luminescence dating (sections 4.1 and 4.2) may be difficult to apply to such complex depositional materials, while cosmogenic nuclide dating (section 3.4) not only requires considerable analytical effort, but tends to generate ages of relatively low precision when used on old alluvial deposits.

One technique that can be employed, however, is U-series dating of pedogenic carbonate, the thin carbonate coatings or rinds that form on gravels and cobbles. Studies have shown that these can develop as geochemically closed systems, and hence $^{230}Th/U$ dating can be applied to such materials (Ludwig and Paces, 2002). Sharp *et al.* (2003) describe an example of this approach in the dating of a fluvial terrace sequence in the Wind River Basin of western Wyoming, USA (Figure 3.10A). Samples were obtained from the undersides of gravels (Figure 3.10B) in four terraces (WR-1–WR-4) and dated by TIMS. The youngest

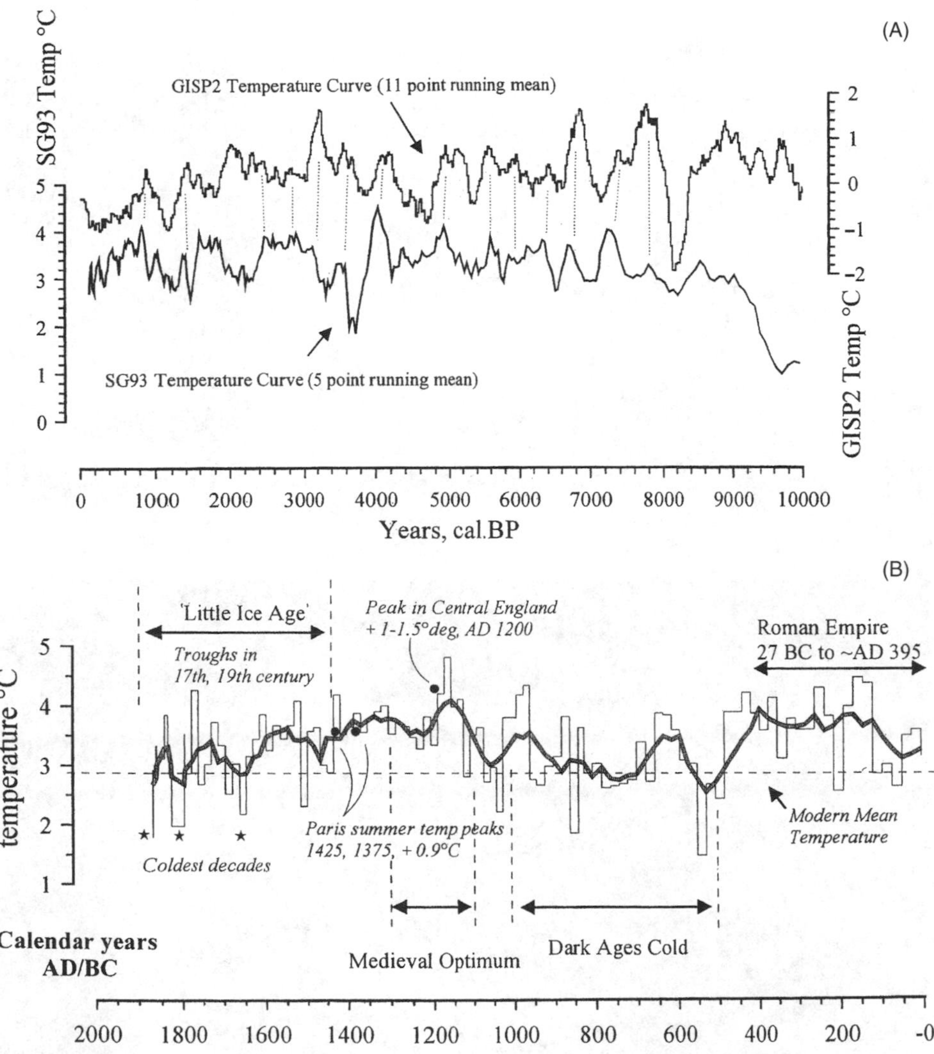

Figure 3.9 *(A) Temperature curves for the last 10 000 years from speleothem SG93 (lower curve) and GISP2 (upper curve). (B) Temperature history from speleothem SG93 inferred for the last 2000 years compared with known 'historical' events (after Lauritzen and Lundberg, 1999). Reproduced by permission of Arnold Publishers*

terrace, WR-1, was dated at 16 900 ± 900 years, which agrees well with cosmogenic ages of between 16 000 and 23 000 years for the landform. This, in turn, suggests that the time lag between deposition of the alluvial gravels and the formation of dateable clast rinds in the WR-1 terrace is relatively short, apparently of the order of 2000–5000 years. The ages for terraces WR-4 (167 000 ± 6400 years) and WR-2 (55 000 ± 8600 years) indicate a mean incision rate of 0.26 ± 0.05 m per thousand years for the Wind River over the last

(A)

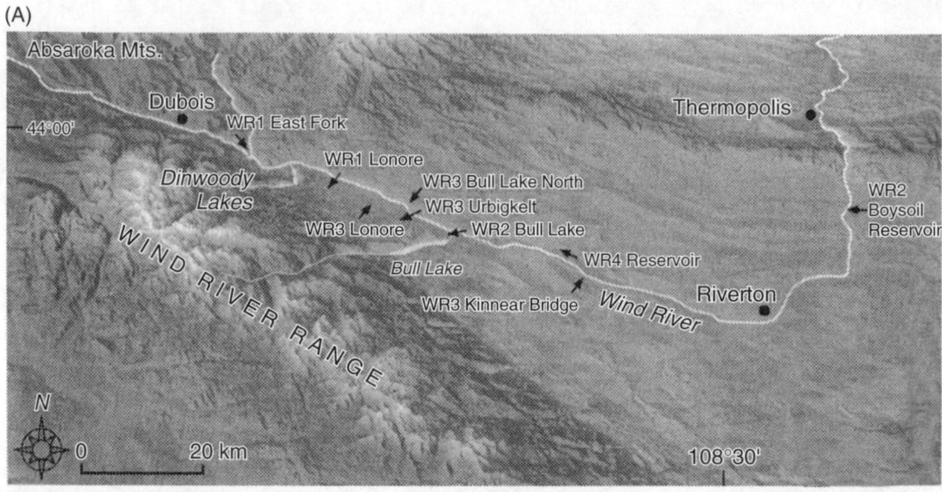

(B)

Figure 3.10 *(A) The Wind River basin, western Wyoming, showing the sampling localities for U-series dating. (B) Soil developed on terrace WR-1 showing pedogenic carbonate rinds (light-coloured coatings) on the bottom of the clasts (after Sharp et al., 2003). Reproduced with permission of Elsevier*

glacial cycle. Terrace WR-3, which is U-series-dated at 150 000±8300 years, is coeval with the final maximum glacial advance of the penultimate Rocky Mountain (Bull Lake) glaciation, and is broadly synchronous with the penultimate global ice volume maximum during MOI stage 6.

3.4 Cosmogenic Nuclide Dating

We have seen in Chapter 2 how cosmic rays reaching the earth from space interact with constituents of the earth's atmosphere to create the radioactive nuclide ^{14}C. Shortly after Libby's pioneering work in the use of radiocarbon as a basis for dating, a seminal paper was published in the *Annals of the New York Academy of Sciences* which suggested that radionuclides are also produced by the interaction of cosmic rays and rock minerals at the earth's surface, and that this process could have possible geological applications (Davis and Schaeffer, 1955). It was not until the 1980s, however, with the development of AMS, that the very low levels of cosmogenic nuclides formed in this way could be measured, and the technique of **cosmogenic nuclide (CN) dating**, sometimes termed **terrestrial *in situ* cosmogenic nuclide (TCN) dating**, began to be used as a routine chronometric method. In Quaternary science, its principal application has been in the determination of exposure ages of rock surfaces, but it has been employed more widely to determine rates of operation of geomorphic processes and long-term landscape evolution. Reviews of the principles and technical details of CN dating in the earth sciences are provided by Bierman (1994), Cerling and Craig (1994), Kurz and Brook (1994) and Gosse and Phillips (2001). Stuart (2001) considers the potential of the technique for archaeology, while geomorphological applications are reviewed by Cockburn and Summerfield (2004).

3.4.1 Principles of Cosmogenic Nuclide (CN) Dating

The basic principles underlying dating using cosmogenic nuclides[7] are relatively straightforward. High-energy cosmic rays entering the atmosphere collide with nuclei triggering a cascade of high-energy neutrons (and a smaller component of muons) that bathe the earth's surface. The collision between these neutrons and muons and target nuclei within certain minerals leads to the breaking apart of those nuclei into fragments (a process known as **spallation**) and the creation of new nuclides. The concentration of these accumulated nuclides within surficial rocks is therefore directly related to the time that the surface has been exposed to cosmic-ray activity. The greater the time that has elapsed since exposure, the greater will be the abundance of cosmogenic nuclides in rock-surface samples. By measuring the latter, an estimate can be made of the former. In essence, therefore, this aspect of CN dating is a type of **surface exposure dating** discussed in Chapter 6. However, the methods considered in that chapter only allow surfaces to be ranked in terms of *relative order of age*. Using CN dating, by contrast, a specific age can be assigned to the time of surface exposure.

In the above application, CN dating is being used to date specific exposure 'events', but the technique can also be employed in other contents. One example is the determination of 'incremental changes' that take place at the earth's surface, in other words changes that result from the progressive stripping away of small increments of material by the various processes of weathering and erosion (Cockburn and Summerfield, 2004). Quantifying denudation rates using surface concentrations of cosmogenic nuclides works on the principle that denudation involves bringing to the surface rock that was previously buried. In a steadily eroding rock outcrop, the cosmogenic nuclide concentration approaches saturation (or secular equilibrium) as a result of constant production on the one hand and losses by denudation as well as radioactive decay (in the case of radionuclides) on the other. Where enough time has elapsed following exposure for erosion to have removed a sufficient thickness of rock (typically 1–2 m), a measured surface concentration of cosmogenic nuclides can be accurately modelled in terms of a constant denudation rate representing an integrated rate for the minimum period of time required to reach secular equilibrium. This model is commonly referred to as the 'steady-state erosion model' and has been used to estimate site-specific denudation rates in a wide range of locations and geomorphic settings (Cockburn and Summerfield, 2004).

Where a rock surface has been selected for dating, rock samples are removed either from the bedrock itself or from boulders lying on the surface of the overlying deposits, by chiselling off a 1–2 cm thick layer of rock. The samples should ideally be taken from horizontal or near-horizontal surfaces and, most importantly, from surfaces that show no obvious signs of weathering or erosion (see below). In this way samples containing representative cosmogenic nuclide concentrations can be obtained.

A number of cosmogenic nuclides are now commonly used as a basis for dating (Table 3.2). Of these, the two 'noble gases' helium-3 (^{3}He) and neon-21 (^{21}Ne) are stable and can be measured using a mass spectrometer (Niedermann, 2002). The remainder, beryllium-10 (^{10}Be), carbon-14 (^{14}C), aluminium-26 (^{26}Al) and chlorine-36 (^{36}Cl), are radioactive and decay to daughter isotopes over time. They are also more difficult to detect and are measured using AMS. Each of these isotopes has a different half-life (Table 3.2) which means that they are applicable over very different time ranges. For the dating of young

Table 3.2 *Some cosmogenic nuclides used for exposure dating (after Kurz and Brook, 1994)*

Isotope	Half-Life (years)	Measurement method	Procedural comments	Approximate age range
^{3}He	Stable	Mass Spectrometry	Diffusive loss? high production rate; lowest detection limit; inherited He	1 ka to ca. 3 Ma
^{10}Be	1.5×10^{6}	AMS	Atmospheric contamination	3 ka to 4 Ma
^{26}Al	7.16×10^{5}	AMS	^{27}Al interference (must use Al-poor minerals)	5 ka to 2 Ma
^{36}Cl	3.08×10^{5}	AMS	No mineral separates; composition-dependent (produced by spallation and slow neutrons)	5 ka to 1 Ma
^{21}Ne	Stable	Mass spectrometry	Inherited neon; useful for old samples	7 ka to 10 Ma (?)
^{14}C	5730	AMS	Shortest half-life; atmospheric ^{14}C contamination	1 ka to 18 ka

surfaces (less than ~5000 years in age) current technology means that ^{3}He, ^{14}C and ^{36}Cl are the most appropriate nuclides to employ, whereas the last two would not be suitable for the dating of older Quaternary materials. The nature of the sample material will also vary. ^{14}C, ^{10}Be and ^{26}Cl, for example, are measured on pure quartz samples, whereas ^{36}Cl analyses are generally undertaken on whole rock samples. Cosmogenic ^{3}He and ^{21}Ne abundances are usually determined on a particular mineral type, such as olivine or garnet (Stuart, 2001).

It is also worth noting that cosmogenic nuclides are present in glacier ice. These are produced not at the earth's surface, but rather in the atmosphere by cosmic-ray spallation reactions (as in the case of ^{14}C) and they are subsequently incorporated into the ice through precipitation processes. Because of the long half-life of these cosmogenic nuclides (Table 3.2), they cannot be used to assign ages to ice less than 100 000 years old, but much older ice can be dated. For example, a ^{36}Cl age of ~760 000 years was obtained from near-bottom ice in the Guliya ice cap on the Qinghai–Tibetan Plateau (Thompson *et al.*, 1997). Cosmogenic nuclides can also be used to synchronise ice-core records. Changing atmospheric concentrations of cosmogenic nuclides will be reflected in downcore variations in those nuclides in the ice. As the atmospheric changes are global in terms of their effects, the cosmogenic nuclide 'signal' may be used as a basis for correlation between ice-core records from different regions of the world. A good example is the marked peak in ^{10}Be and ^{36}Cl concentrations at ~40 000 years, which has been detected in both Greenland and Antarctic ice cores, and which can be used to link the ice-core records from the northern and southern polar regions (Yiou *et al.*, 1997).

3.4.2 Sources of Error in CN Dating

As with radiocarbon dating, a number of assumptions underlie CN dating, and several of these constitute a potential source of error. First, it is assumed that the surface that is being dated has not inherited cosmogenic isotopes from previous exposure events. In other words, the 'exposure clock' has been effectively zeroed. Prior exposure, for example in glacial or fluvial sediments, can never be completely excluded and can only be corrected for by replicate measurements. Second, there has to be an assumption that the surface has not been eroded or weathered since the time of initial exposure. To some extent, this can be checked by careful field sampling, but it remains a source of uncertainty in certain contexts, for example in the dating of glacial moraines where erosion of the moraine surface and exhumation of fresh boulders can lead to serious underestimates of moraine age (Putkonen and Swanson, 2003). Underestimates of age may also be made if the dated surface has been shielded from cosmic radiation, either by snow cover or by soils and/or sediments (Benson *et al.*, 2004). Third, the surface must have acted as a closed system since exposure, so there has been no loss of nuclides or contamination by others. Fourth, it must be assumed that the surface selected for dating has been exposed for a sufficient length of time for a measurable quantity of the cosmogenic nuclides to have accumulated or, in the case of radioactive nuclides, for a short enough time for the concentration of radioactive nuclides not to have reached equilibrium, which is normally about five half-lives (Stuart, 2001).

It is also essential that the production rates of cosmogenic nuclides on rock surfaces can be established with a reasonable degree of accuracy. This is a vital component of

CN dating, but obtaining a reliable estimate of production rates may not be straightforward as these are known to vary with altitude, latitude, depth below the ground surface, and degree of shielding from cosmic rays (e.g. Stone, 2000). In addition, past changes in cosmic-ray flux have to be built into the calculations. In practice, therefore, determining an absolute surface exposure age requires scaling-calibration data for the latitude and altitude of the sample in question (Dunai, 2000), and such data may not be easy to obtain. Finally, problems may be caused by the natural presence of non-cosmogenic nuclides in rocks. For example, ^{36}Cl may be produced in terrestrial rocks by neutron activation of ^{35}Cl resulting from the decay of uranium and thorium isotopes (Zreda and Phillips, 1994). In addition, cosmogenic nuclides may be present that have not been generated by spallation reactions in the rock, but rather have been adsorbed from the atmosphere. In both cases, those nuclides that have not been generated *in situ* from cosmic-ray bombardment have to be identified and corrected for.

All of the above generate uncertainties in exposure ages which, at present, are in the range of 10–20%, although advances in our understanding of the factors contributing to cosmogenic nuclide production rates mean that the total uncertainty in exposure ages is continually improving (Gosse and Phillips, 2001). In addition, however, there are analytical errors arising from AMS measurements. For ^{10}Be, ^{14}C, ^{26}Cl and ^{36}Cl determinations, this is usually less than 7%, but for ^{3}He and ^{21}Ne this is typically better than ±3% (Stuart, 2001).

3.4.3 Some Applications of CN Dating

CN dating has developed rapidly over the past ten years with eight AMS laboratories routinely measuring ^{10}Be and ^{26}Al and four producing measurements of ^{36}Cl (Cockburn and Summerfield, 2004). The applications of the technique have been extremely varied, and it has been used to date such diverse phenomena as glacial advances, volcanic activity, meteorite impacts, seismic events, erosional histories and patterns of large-scale landform evolution (Gosse and Phillips, 2001). Depending on the surface preservation and exposure history, CN dating has an effective range from the Pliocene to the Late Holocene. Four applications relating specifically to the surface exposure dating of Quaternary events are considered in the following sections.

3.4.3.1 Cosmogenic dating of two Late Pleistocene glacial advances in Alaska. Some of the earliest applications of CN dating were in the field of glacial geology, where cosmogenic isotopes were used to determine surface exposure ages of moraines formed during successive phases of glacier advance and retreat (e.g. Phillips *et al.*, 1990). This approach has continued to be used as a basis for glacial chronologies (e.g. Phillips *et al.*, 2000; Owen *et al.*, 2003; Kelly *et al.*, 2004), as well as in the dating of periglacial features (Barrows *et al.*, 2004). An example is described by Briner *et al.* (2001) from the southwestern part of the Ahklun Mountains, Alaska, where two glacial advances, the Arolik Lake and the Klat Creek glaciations, are delimited by well-preserved moraines and associated features (Figure 3.11). Previous research in the area had tentatively assigned these to Early and Late Wisconsinan (Figure 1.4) glacial stages, respectively. Thirty-two cosmogenic ^{36}Cl exposure ages were obtained on boulders from one Arolik Lake moraine and six Klat Creek moraines. Four moraine boulders on the Arolik Lake moraine have a mean surface exposure age of 60 300 ± 3200 years, which confirms this

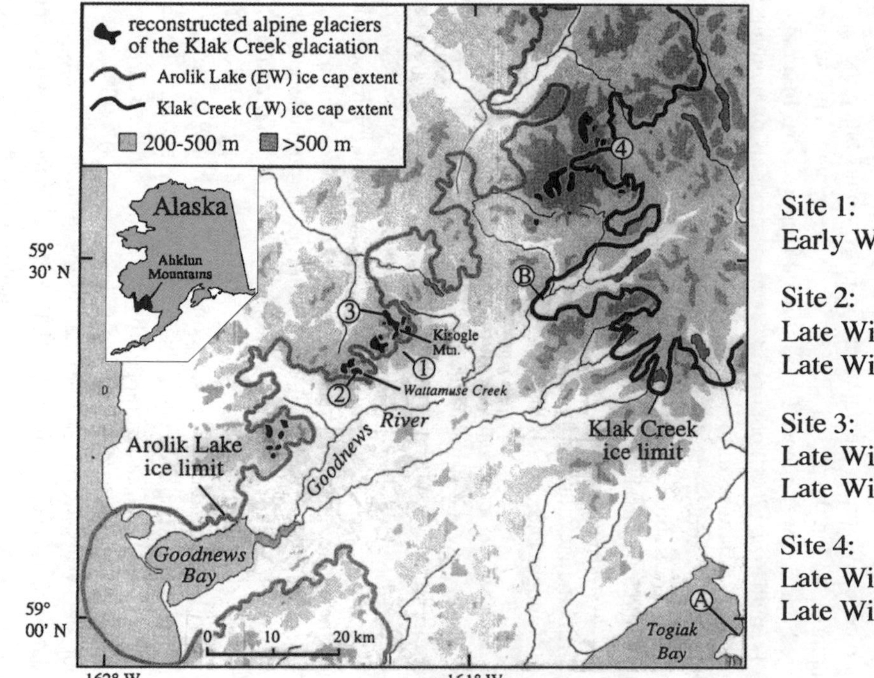

Mean surface exposure ages (yrs BP)

Site 1: Goodnews River valley	
Early Wisconsinan ice-cap outlet moraine:	60,300 ± 3200
Site 2: Wattamuse Creek valley	
Late Wisconsinan terminal moraine:	30,400 ± 5300
Late Wisconsinan recessional moraine:	26,000 ± 400
Site 3: Kisogle Mountain region	
Late Wisconsinan terminal moraine:	17,500 ± 900
Late Wisconsinan recessional moraine:	16,700 ± 1400
Site 4: Klak Creek region	
Late Wisconsinan terminal moraine:	17,800 ± 1900
Late Wisconsinan ice-cap outlet moraine:	19,600 ± 1400

Figure 3.11 *The Ahklun Mountains, southwestern Alaska, showing the extent of the Arolik Lake (Early Wisconsinan?) and Klak Creek (Late Wisconsinan?) glacial advances. Sample sites for cosmogenic nuclide dating are indicated by numbers. The weighted mean ^{36}Cl exposure ages for the moraines are shown on the right (after Briner et al., 2001). Reproduced with permission of Elsevier*

ice advance as being of Early Wisconsinan in age. A moraine deposited by an ice-cap glacier during the hypothesised Late Wisconsinan advance has a mean surface exposure age of 19 600 ± 1400 years (Site 4, Figure 3.11), while five moraines deposited by alpine glaciers have mean surface exposure ages ranging from 30 000 to 17 000 years, again confirming a Late Wisconsinan Age for this glacial phase. These results show that, in contrast with many other areas, the most extensive Late Pleistocene glaciation in this part of Alaska occurred during the earlier, rather than in the later, part of the Wisconsinan (Last) Cold Stage.

3.4.3.2 Cosmogenic dating of the Salpausselkä I formation in Finland. As the Fennoscandian ice sheet wasted at the end of the Last (Late Weichselian) Cold Stage, a large lake (the Baltic Ice Lake) was impounded between the northward and westward retreating ice margin and the land areas to the south. In many areas, readvances of the ice margin into this lake during overall deglaciation led to the formation of morainic ridges of glacial, glaciofluvial and glaciolacustrine materials. Some of the most spectacular are the Salpausselkä ice-marginal formations of southern Finland which comprise two sub-parallel ridges that can be traced over a distance of ca. 600 km. The timing of the readvance that created the Salpausselkä formations has been an important question in studies of the overall pattern of deglaciation in the Baltic region, and the currently accepted age based on varve chronology (section 5.3) ranges between 11 680 and 11 430 years BP, although a recent revision to the varve chronology would place the timeframe for deposition of Salpausselkä I in the period 12 700–11 700 years ago. This suggests that the readvance occurred during the Younger Dryas Stadial (Figure 1.5). In an attempt to test this hypothesis, Tschudi *et al.* (2000) obtained ^{10}Be measurements on boulders on the surface of the Salpausselkä I moraine to the west of Lahti, Finland (Figure 3.12). These gave four minimum surface exposure ages ranging from 11 050 ± 910 to 11 930 ± 950 years, producing an error-weighted mean age of 11 420 ± 470 years. Correcting for erosion on the surface boulders (estimated to be 5 mm/kyr), this generated a surface exposure age of 11 610 ± 470 years, an age estimate that accords very closely with the varve chronology and confirms the view that Salpausselkä I was formed by a readvance of the ice margin during the Younger Dryas cold stage.

3.4.3.3 Cosmogenic dating of Holocene landsliding, The Storr, Isle of Skye, Scotland. Major Holocene rock slope failures are a common feature of many parts of the Scottish Highlands, but precise dating of these events has often proved difficult and the causes of slope failure remain equally uncertain. In an attempt to determine the age of one of the most famous examples of slope failure in Scotland, The Storr on the Trotternish escarpment of the east coast of the Isle of Skye, Ballantyne *et al.* (1998a) obtained ^{36}Cl ages on rock samples from two separate landslide blocks near the upstanding pinnacle known as the 'Old Man of Storr' (Figure 3.13). Separate exposure ages of 6300 ± 700 and 6600 ± 800 years gave an overall age estimate of 6500 ± 500 years for rock slope failure at the site. This date is consistent with AMS radiocarbon dates on buried soil horizons on the summit plateau of The Storr. These contain particles that have been weathered from the headwall of the landslide and blown upwards to be trapped by vegetation at the scarp crest. The radiocarbon dates place the onset of aeolian accumulation after 7200–6900 cal. years BP but before 5600–5300 cal. years BP, thereby bracketing the surface exposure age of ca. 6500 years BP. The dates are also consistent with relative dating evidence which indicates that most rock slope failures in the Scottish Highlands occurred during the Early

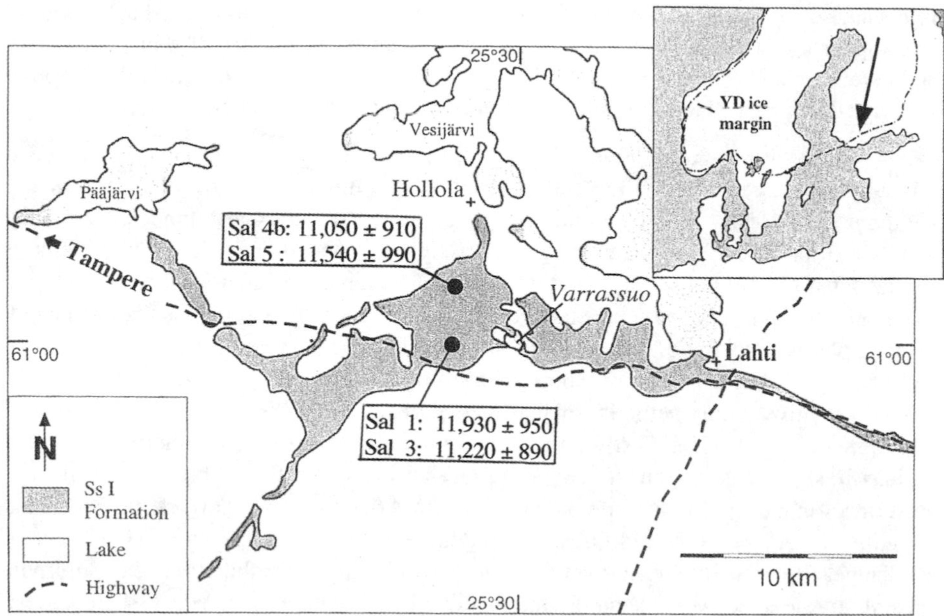

Figure 3.12 *The Salpausselkä I moraine formation in the Lahti area, southern Finland, showing the locations of ^{10}Be measurements (after Tschudi* et al., *2000). Reproduced by permission of Taylor & Francis AS, www.tandf.no/boreas*

Figure 3.13 *The upper part of the Storr landslide, Isle of Skye, Scotland, showing the 'Old Man of Storr' (centre) and other pinnacles and landslide blocks. The sampling sites for ^{36}Cl exposure dating were located on rock ridges near the centre of the slipped mass, immediately to the left and right of the Old Man of Storr (photo: Colin Ballantyne)*

Holocene, several thousand years after the disappearance of the last ice sheet. This, in turn, suggests that a glacial or periglacial cause is unlikely, but rather that progressive joint extension and shearing of rock bridges may have resulted in slope failure, although slope instability related to regional seismic activity cannot be completely excluded.

3.4.3.4 Cosmogenic dating of alluvial deposits, Ajo Mountains, southern Arizona, USA. A characteristic feature of the landscape of the southwestern United States is the piedmont slope. This typically comprises an upper erosional zone connected with the mountain front, a middle zone of transition, and a lower zone of aggradation that includes alluvial fans. Landforms of the piedmont slope and their associated surfaces developed during an interval of relatively constant conditions, but environmental changes associated, for example, with tectonics, climate and base-level changes create a new generation of landforms and surfaces. Establishing chronologies for these surfaces has usually relied on relative dating methods, such as degree of soil development and associated soil development indices (Chapter 6), but these have not always proved satisfactory.

Liu *et al.* (1996) describe an alternative approach using ^{36}Cl dating of boulders on alluvial fan surfaces in the Ajo Mountain region of southern Arizona. Previous work had identified two alluvial fan units (Qf1 and Qf2), two fluvial terrace units (Qt1 and Qt2), and active fan channels (Qfc) within the study area (Figure 3.14A), and a combination of geomorphic and soil-stratigraphic evidence suggested that Qt2 is the youngest surface, Qt1 the oldest, and the terraces Qt1 and Qt2 are of intermediate age. In order to test this hypothesis, samples for ^{36}Cl dating were taken from boulders and cobbles on the surfaces of the alluvial units, and organic material was also recovered from the terrace sediments for ^{14}C dating to provide a check on the ^{36}Cl dates. The results of ^{36}Cl dating confirmed the relative chronology outlined above, but the marked difference in age between ^{14}C and ^{36}Cl dates (the latter being significantly older), and the considerable spread in ^{36}Cl exposure ages on each of the surfaces, suggested that repeated cycles of erosion and deposition have resulted in an inherited ^{36}Cl signal in many of the dated samples. The effects of erosion may also have influenced the apparent ^{36}Cl ages of some samples. These problems were corrected for by using the apparent ^{36}Cl age of a modern debris flow and channel deposits, and an alluvial history was reconstructed for the area (Figure 3.14B). The oldest preserved alluvial fan surface apparently stabilised at least 440 000 years ago, but it may have been reactivated between 330 000 and 230 000 years BP. Incision of Qf1 and deposition of Qf2 sediment apparently dates to between 180 000 and 100 000 years BP. Both alluvial fan surfaces seem to have experienced at least minor modifications between 60 000 and 30 000 years followed by incision and deposition of the older terrace sediments (Qt1) around 18 000 years. Finally, the channels were incised again and the younger terrace sediment (Qt2) was deposited as a shallow fill in the channels around ca. 2500 years (based on ^{14}C dating). The ^{36}Cl data clearly show that these surfaces experienced extremely complex histories during the Mid- and Late Quaternary, more so than had been evident in previous investigations.

3.5 Dating Using Short-Lived Isotopes

Two isotopes with very short half-lives, ^{210}Pb (22.26 years) and ^{137}Cs (30 years), have been used to date very recent sediment sequences spanning, at most, the last two or three

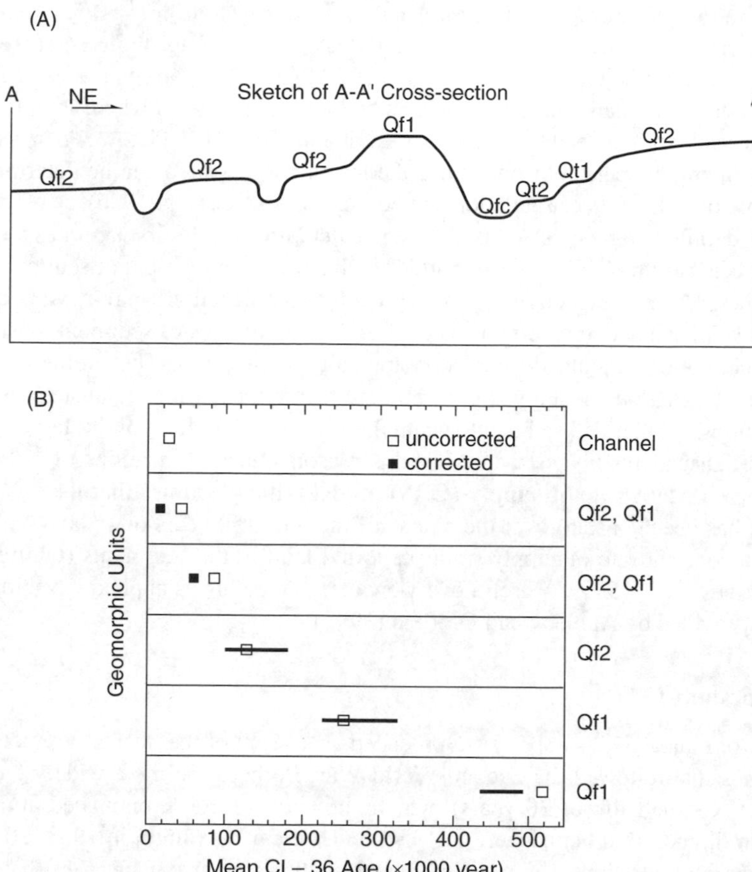

Figure 3.14 *(A) Cross-section through the major surfaces in the Ajo Mountain region, southern Arizona. Qf1 and Qf2 represent alluvial fan surfaces and Qt1 and Qt2 represent terraces. (B) Mean cosmogenic ^{36}Cl ages for the different geomorphic units in A. The open squares represent uncorrected ages; the filled squares, the ages of the terraces corrected for inherited ^{36}Cl alone; and the black bars, the age ranges given by correcting for inherited ^{36}Cl and erosion (after Liu* et al.*, 1996). Reproduced with permission of Elsevier*

centuries. These techniques have, in the main, been applied to the dating of lake sediments, but other depositional contexts have also been dated (Appleby and Oldfield, 1992). Attempts have also been made to use a third short-lived isotope, ^{32}Si, and if technical difficulties relating to counting can be overcome (see below), this could also prove to be a valuable dating tool for the last millennium.

3.5.1 Lead-210 (^{210}Pb)

The method of ^{210}Pb dating was developed in the 1960s and has since been used primarily in the dating of lake sediments within the time range of 1–150 years. It is based on the escape of radon gas from the earth into the atmosphere. Radon (^{222}Rn) is part of the uranium-series decay chain (section 3.3.1) and decays via a series of daughter isotopes to ^{210}Pb. This unstable isotope, with a half-life of 22.26 ± 0.22 years, is removed from the atmosphere by precipitation and becomes embedded in sediments where it subsequently decays to a stable form of lead (^{206}Pb). By measuring the ratio of ^{210}Pb to ^{206}Pb in a sequence of lake sediment, the period of time since the lead was deposited can be determined and thus the rate of sediment accumulation can be established (Olsson, 1986).

In ^{210}Pb dating, two models are used for age calculations. The first assumes a **constant initial concentration (CIC)** of unsupported ^{210}Pb per unit dry weight of sediment at each depth in the profile, irrespective of whether or not any variations may have occurred in the rate of sediment accumulation. Hence, in undisturbed cores of sediment, unsupported ^{210}Pb concentrations should decline monotonically with depth. The problem with this model is that variations in sediment accumulation rate will influence unsupported ^{210}Pb concentrations; accelerated sediment accumulation, for example, will dilute ^{210}Pb values and lead to an underestimate of age. For this reason, most applications of ^{210}Pb dating employ the **constant rate of supply (CRS) model** which assumes that the unsupported ^{210}Pb flux has been constant over the course of the past 150 years or so, and hence there has been a constant rate of supply of unsupported lead to the sediments (Oldfield *et al.*, 1978; Appleby *et al.*, 1979). A useful overview of ^{210}Pb dating as applied to sedimentation studies is provided by Appleby and Oldfield (1992).

3.5.2 Caesium-137 (^{137}Cs)

A number of radioactive isotopes were produced as a result of the nuclear weapons testing programmes that followed the Second World War. Perhaps the most widely recognised has been ^{137}Cs (half-life of 30 years), which showed the first pronounced atmospheric increase in the northern hemisphere in 1954, and a clear maximum in 1963, after which atmospheric concentrations declined significantly with successive nuclear test-ban treaties. The 1963 maximum in atmospheric ^{137}Cs is reflected in lake sediment sequences, and forms a distinctive time-stratigraphic marker horizon (Olsson, 1986). Other artificially generated radioactive isotopes, such as plutonium ($^{239+240}Pu$: half-lives of 2.4×10^4 years and 6.5×10^3 years, respectively) and americium (^{241}Am: half-life of 432.7 years), show similar 1963 maxima, and have also been used to date recent sediment sequences (see below). In Europe, the ^{137}Cs peak due to the Chernobyl fallout (1986) is larger than the weapons peak, but the latter is global whereas the Chernobyl signal is regional.

3.5.3 Silicon-32 (^{32}Si)

^{32}Si is produced in the atmosphere through cosmic-ray interactions, and is transported rapidly to the earth's surface by precipitation. It accumulates in lake and marine sediments, and in snow, and therefore constitutes a dating method not only for sediments but also for glacier ice. The half-life of ^{32}Si is around 140 years (Morgenstern *et al.*, 1996), although studies based on varved sediment sequences suggest a slightly longer half-life of 178 ± 10 years (Nijampurkar *et al.*, 1998). ^{32}Si dating is applicable over a timescale of 30–1000 years, which means that it bridges the time gap between the present and the younger end of the ^{14}C range. In addition to providing a chronology for natural processes, therefore, the technique is of particular interest to archaeologists in dating events over the course of the last millennium, notably the Medieval Climatic Optimum, the Little Ice Age between AD 1650 and 1850, and the impact of human colonisation and industrialisation during the last 150 years (Morgenstern *et al.*, 2001).

3.5.4 Some Problems in Using Short-Lived Isotopes

One of the main problems with ^{210}Pb dating is that most sediments contain minerals that incorporate small amounts of uranium or uranium daughter isotopes, and hence there is a continuous supply of ^{210}Pb throughout the sediment column. It is therefore necessary to determine the amount of this 'supported' ^{210}Pb, and then subtract it from the total ^{210}Pb, in order to find the atmospherically generated ^{210}Pb (the 'unsupported' lead) that forms the basis for dating. The amount of 'supported' lead in the dated sediment column can be determined by measuring the lead content at levels older than 150 years (working on the assumption that all the ^{210}Pb in those sediments will be 'supported', as all the 'unsupported' lead will have decayed) and then applying that value as a correction factor to the levels younger than 150 years. This, however, may not be straightforward, and may be complicated by a number of factors, including disequilibrium between radium and lead in the sediments, which means that the amount of supported ^{210}Pb may vary with depth. Further problems can arise as a result of agricultural activity around a lake catchment, leading to the inwashing of clays and other mineral particles and hence lower ^{210}Pb concentrations per gram of sediment, while organic matter in lakes may retain lead and prevent release into the sediments, also resulting in lower ^{210}Pb, values (Olsson, 1986). Bioturbation and reworking of lake sediments are further processes that can lead to problems in the use not only of ^{210}Pb, but of other short-lived isotopes. Particular difficulties are encountered in the use of these isotopes in the dating of peats (Oldfield *et al.*, 1995). These arise, *inter alia*, from erosion and sediment loss in the upper parts of peat profiles, and the downwash of ^{210}Pb in those parts of the profile above the water table. As a result, ^{210}Pb of peats tends to require corroboration by independent methods (Appleby *et al.*, 1997).

The principal problems with ^{32}Si dating relate to the detection of ^{32}Si due to its extremely low natural specific activity and the vast excess of stable silicon (i.e. low ^{32}Si/Si ratio). This means that in the dating of sediments, very selective radiochemical purification procedures are required. Measurement of natural ^{32}Si is carried out using either beta radiation (radiometry) or AMS, although the latter approach has so far been applicable only to samples of rain, snow and ice, for in limnic and marine sediments where biogenic silica (diatoms and radiolaria[8]) constitute the dating media, the ^{32}Si/Si ratio is below the detection limit for AMS (Morgerstern *et al.*, 2000; 2001).

3.5.5 Some Dating Applications Using Short-Lived Isotopes

As noted above, ^{210}Pb, ^{137}Cs and the other short-lived radioactive isotopes have been most widely employed in the dating of recent lake sediments, particularly where the history of human impact on lake catchments and ecosystems is under investigation (Varvas and Punning, 1993; Oldfield *et al.*, 2003). Notwithstanding the problems noted above, ^{210}Pb has also been used, along with other techniques, in the dating of recent peat sequences (Roos-Barraclough *et al.*, 2004; Hendon and Charman, 2004), ice cores (Stauffer, 1989) and marine sediments (Jensen *et al.*, 2004; van der Kaars and van den Bergh, 2004), while ^{137}Cs has been employed, *inter alia*, in investigations of ground retreat on mining spoil heaps (Higgitt *et al.*, 1994), and in the assessment of rates of soil erosion on archaeological sites (Davidson *et al.*, 1998). Although less widely applied, ^{32}Si has been used in the dating of glacier ice (Clausen, 1973; Nijampurkar and Rao, 1992), and in the dating of recent lacustrine and marine sequences (Nijampurkar *et al.*, 1998; Suckow *et al.*, 2001). Three examples are considered below.

3.5.5.1 Dating a record of human impact in a lake sequence in northern England. The various physical, chemical and biological properties of lake sediment records provide evidence of changes that have taken place not only within the lake ecosystem, but also on the slopes around the catchment (Lowe and Walker, 1997). One aspect of lake sediment analysis is that it can provide a valuable record of human impact on the landscape in the vicinity of the lake. Oldfield *et al.* (2003) describe such a study in which the history of human activity around Gormire Lake in northern England was obtained from a range of data including pollen analysis (to reconstruct vegetation), mineral magnetic measurements (to infer erosional episodes) and organic biogeochemical analyses (to detect lake ecosystem changes). AMS radiocarbon dating proved problematical because of the influence of older carbon residues in the bulk sediment samples (section 2.5.1), but the upper part of the core could be dated by ^{210}Pb and ^{137}Cs. These two approaches produced a coherent chronology for the top 20–25 cm of the sequence (Figure 3.15) and formed the basis for correlation between two cores (A and B) from the lake. The rate of sedimentation that could be estimated from these ^{210}Pb records enabled deeper horizons to be dated by extrapolation, where additional 'pinning points' for the chronology were obtained from known variations in atmospheric stable Pb concentrations resulting from the smelting of lead. The results showed two distinct phases of deforestation and catchment erosion: the first dating from the Iron Age/Romano British period (ca. 200 BC to ca. AD 600), and the second from the time of Medieval monastic activity (ca. AD 1200) and which continued to the nineteenth century.

3.5.5.2 Dating a 500-year lake sediment/temperature record from Baffin Island, Canada. Laminated sediments have been widely used as a basis for dating (section 5.3). They consist of distinctive layers or laminae that are often annually deposited, in which case they are referred to as varves. The thickness of the varves is related to sediment input into the lake basin. At Soper Lake, Baffin Island, Hughen *et al.* (2000) used this relationship to reconstruct past temperatures, working on the basis of previous studies which have shown that in Arctic areas, run-off, suspended sediment concentration in lake water, and resulting varve thickness, varies as a function of temperature during the snow or glacier melt season. The sediments in the lake comprised alternating bands of light and dark sediments, and the latter were believed to represent deposition during the snowmelt season. In order to test the hypothesis that each light/dark couplet did indeed represent an annual cycle of deposition, ^{210}Pb dating was applied to a laminated lake sediment sequence. This showed a close agreement between ^{210}Pb and varve age (Figure 3.16A), thereby confirming that the varves are indeed annually deposited sediments. As a further check, $^{239+240}$Pu measurements were made and these showed a prominent peak corresponding to varve dates of 1962–1964, in excellent agreement with the known atmospheric $^{239+240}$Pu concentration

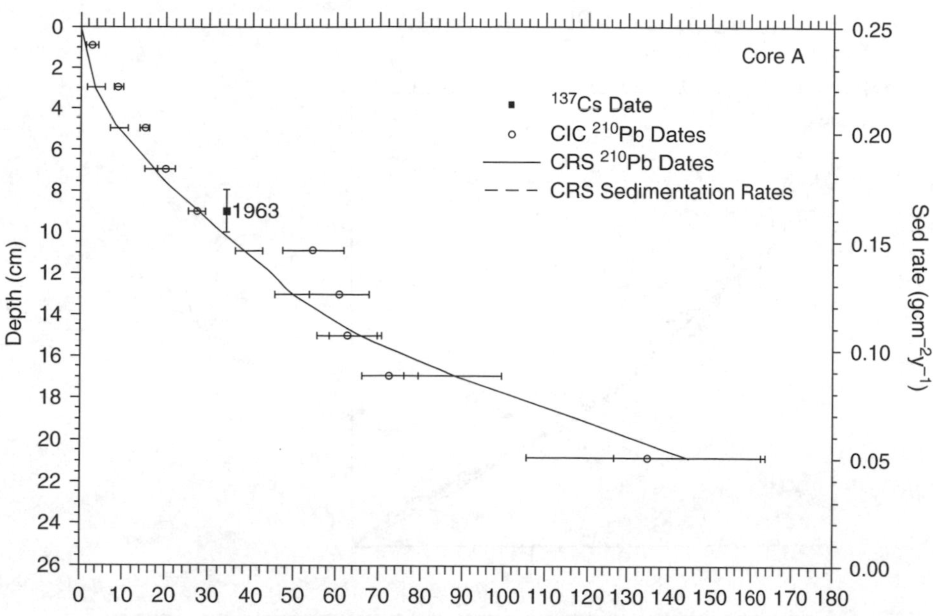

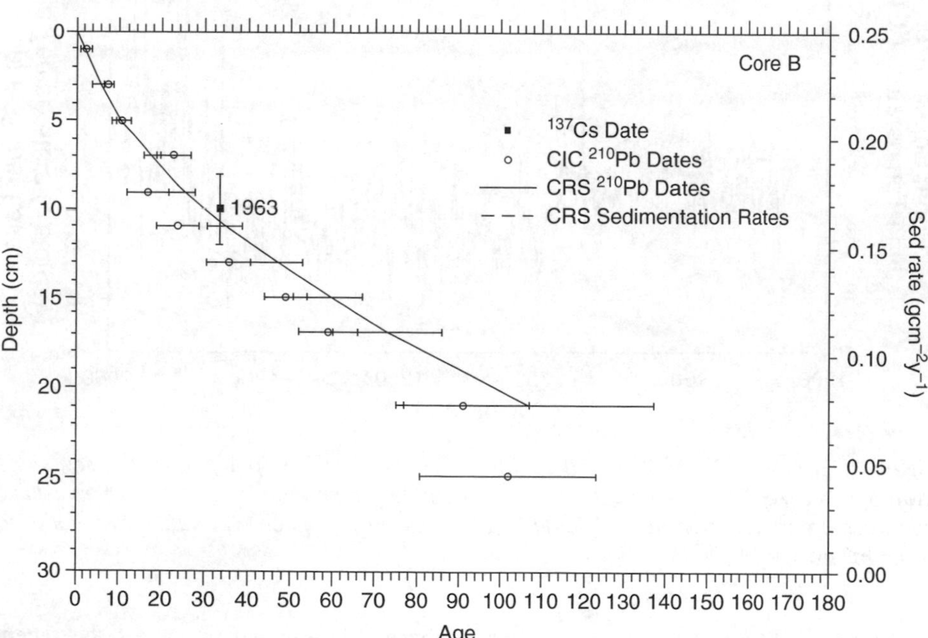

Figure 3.15 *Age–depth profiles for two cores from Gormire Lake, northern England, based on ^{210}Pb chronologies. Note the ^{137}Cs date of 1963 which acts as a check on the ^{210}Pb timescale (after Oldfield* et al.*, 2003). Reproduced by permission of Arnold Publishers*

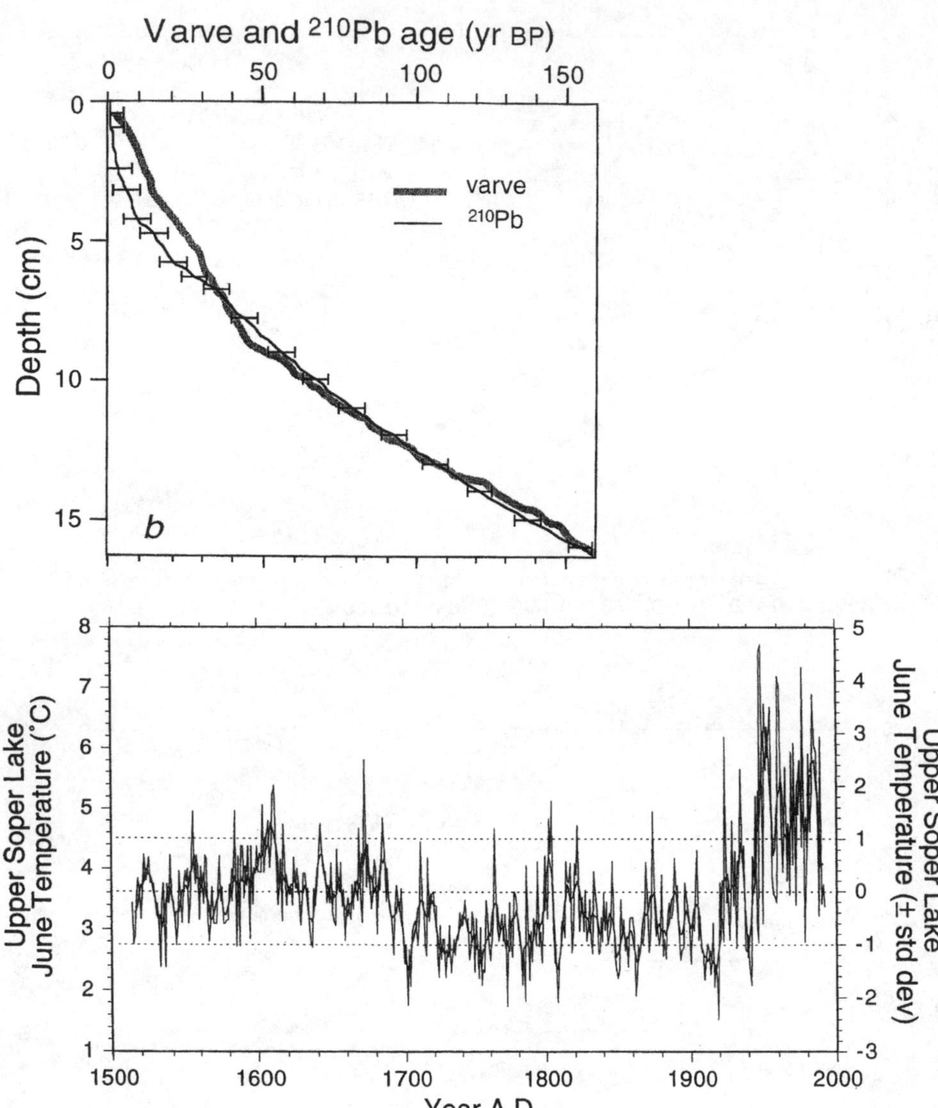

Figure 3.16 *(A) ^{210}Pb ages plotted against varve age in the upper part of the sequence from Soper Lake, Baffin Island. (B) The average June temperature record from Soper Lake constructed from variations in dark laminae thickness (after Hughen* et al.*, 2000). Reproduced by permission of Arnold Publishers*

maximum in 1963 (section 3.5.2). Having established the annual nature of the recent varved sequence using these two short-lived isotopes, Hughen *et al.* (2000) were able to calibrate varve thickness for each year against twentieth-century temperature data from a nearby recording station, and then use these data to extrapolate back through the sequence to reconstruct a temperature record for the past 500 years (Figure 3.16B).

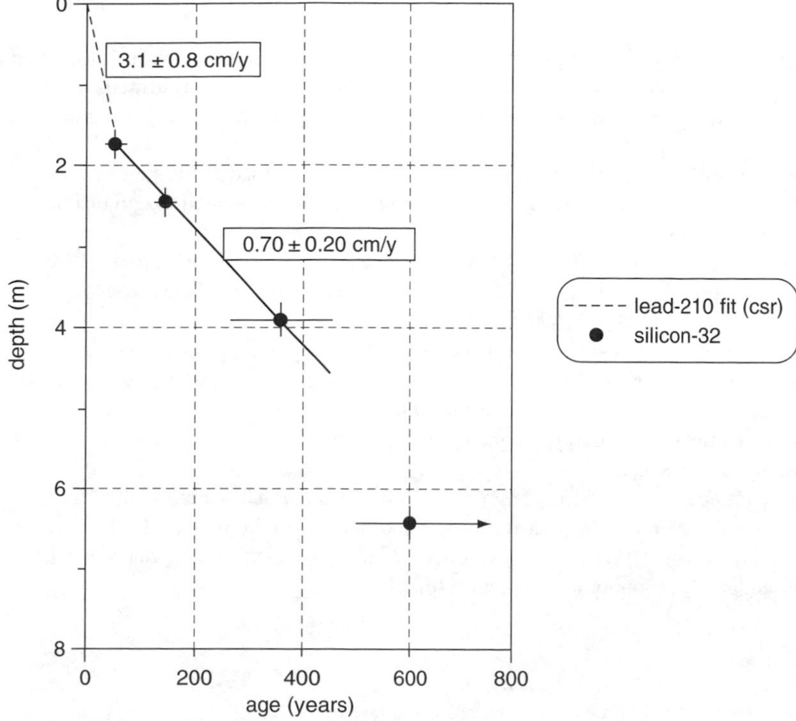

Figure 3.17 *^{32}Si depth profile for a sediment core from the Bangladesh continental shelf. The sedimentation rates for the part of the core older than 50 years (solid line) are derived from the ^{32}Si data, while the dotted line between 0 and 50 years is based on ^{210}Pb dating (after Morgenstern* et al., *2001). Reproduced by permission of Radiocarbon*

3.5.5.3 ^{32}Si dating of marine sediments from Bangladesh. As noted above, ^{32}Si has been used to date marine sediment records and Morgenstern *et al.* (2001) describe the results of such a study from the Bangladesh continental shelf. There, an investigation of the sediment budget near the mouth of the Ganges–Brahmaputra River system employed ^{32}Si to provide a chronology of the deeper part of the submarine delta foreset beds that lay beyond the dating range of other short-lived isotopes such as ^{210}Pb and ^{137}Cs. Although the studied core represents extremely difficult analytical conditions for ^{32}Si dating, there was good agreement between ^{32}Si and ^{210}Pb dates for the past 50 years, and a coherent age depth profile was obtained for the past 400 years (Figure 3.17). This shows a mean sedimentation rate of 0.7 ± 0.2 cm year^{-1} for the period from 50 to several hundred years BP, and 3.1 ± 0.8 cm year^{-1} for the past 50 years. The fourfold increase in rate of sedimentation over the last 50 years may reflect increased sediment loads in the rivers due to increasing human colonisation of the two drainage basins.

Notes

1. **Feldspars** are a rock-forming group of crystalline minerals, white or pink in colour, and consisting of aluminosilicates of potassium, sodium, calcium and barium. **Sanidine** is a member of the alkali feldspar group and is the name given to high-temperature potassium feldspar or potassium–sodium feldspar.
2. **Xenocrysts** are crystals in an igneous rock that have not formed in the place in which they are now found. In other words, they have been introduced into the melted magma from some extraneous source.
3. **Tuff** is a volcanic material that has been consolidated (welded) to form a rock.
4. The recommended half-lives for the three isotopes used in U-series dating are ^{234}U – 245 250±490 years; ^{230}Th – 75 690±230 years; ^{231}Pa – 32 760±220 years (Bourdon *et al.*, 2003b).
5. **Travertine** is a form of calcareous tufa deposited by certain hot springs in volcanic regions.
6. **Ferricrete** is an iron-pan or near-surface zone of iron oxide concentration that develops in tropical or arid regions through weathering and soil formation.
7. The term **cosmogenic** is used throughout this discussion to refer to any nuclide that is produced by cosmic-ray particles.
8. **Diatoms** are microscopic (5 μm–2 mm) unicellular algae, which secrete a silicious shell or frustule, and which are found in both freshwater and marine environments. **Radiolaria** are marine amoebic protozoans (100 μm and 2000 μm in diameter) which secrete elaborate skeletons also composed largely of amorphous (opaline) silica.

4

Radiometric Dating 3: Radiation Exposure Dating

Footprints on the sands of time

Henry Wadworth Longfellow

4.1 Introduction

In this group of dating methods, a chronology is obtained by measuring the cumulative effect of nuclear radiation on the crystal structure of minerals or fossils, hence the term **Radiation Exposure Dating**. Four techniques are considered, three of which derive age estimates by freeing and counting electrons that have become lodged at structural defects or impurities (electron traps) within crystal lattices following exposure to radiation, while the fourth measures the amount of damage caused to irradiated crystal structures and glass, and uses this as a basis for determining age. All work on the same principle, namely that the larger the number of trapped electrons or amount of crystal damage, the longer has been the time of exposure to radiation, and hence the greater the age of the host material. Electrons can be freed from traps by heating (**Thermoluminescence**) or by exposure to a light source (**Optically Stimulated Luminescence** or **Optical Dating**). Together these two techniques are known as **Luminescence Dating**. A third approach to trapped electron measurement involves the use of a magnetic field to 'excite' and detect the electrons, and this is termed **Electron Spin Resonance Dating**. Because all three methods use the number of free electrons (or charged particles) trapped within a crystal lattice as a basis

Quaternary Dating Methods M. Walker

for inferring age, these are collectively referred to as techniques of **Trapped Charge Dating** (Grün, 2001). In the fourth method, the degree of radiation-induced damage will be reflected in the number of 'damage tracks' or 'trails' within a crystal structure, and this forms the basis for the method of **Fission Track Dating**. These techniques can provide ages on a wide variety of materials ranging from rocks, grains of sediment or speleothem calcite, to tooth enamel, flint artefacts and pottery.

4.2 Luminescence Dating

Luminescence dating works on the principle that materials containing naturally occurring radioactive isotopes, such as uranium, thorium or potassium-40, or which lie in close proximity to other material containing these radioactive elements (in buried contexts, for example), are subject to low levels of radiation. In mineral crystals, this leads to ionisation of the atoms in the host material and freed electrons may become trapped in structural defects or 'holes' in the mineral crystal lattice. These electrons can be released in the laboratory by heating under controlled conditions, and where the trapped charges recombine at luminescence centres in the crystal (a special type of structural defect or impurity), an emission of light occurs that is proportional in intensity to the number of trapped electrons. This emission of light is known as **thermoluminescence (TL)** and is the basis of **thermoluminescence dating**. Alternatively, the electrons can be released from traps by shining a beam of light onto the sample, and again the luminescent signal is a reflection of the number of electrons trapped within the crystal lattice. This is **optically stimulated luminescence (OSL)**, and hence the technique is generally referred to as **optically stimulated luminescence dating**, although the abbreviated term **optical dating**, first introduced in the seminal paper by Huntley *et al.* (1985), is also used.

There are many kinds of electron trap. All can be emptied by exposure to sufficient heat, but only some are sensitive to light. TL dating was initially developed for quartz[1] and feldspar grains but, in principle, other minerals may also be used. In the environment, the 'luminescence clock' can be reset by heating, for example the firing of pottery effectively zeroes the clock within the minerals of interest. In the case of sediments, the clock is zeroed by exposure to heat from a campfire or forest fire, or from exposure to sunlight, a process often referred to as 'bleaching'. In both instances, most of the electron traps are effectively emptied, but then progressively filled when the sediment is buried (Figure 4.1). The 'natural luminescence signal' therefore provides an indication of the time that has elapsed since firing (in pottery, for example) or, in the case of a body of sediment, since burial and removal from sunlight. In volcanic rocks, of course, zeroing of the 'electron clock' will be an inevitable consequence of both heat and mineral formation (recrystallisation). Further details on luminescence dating can be found in Aitken (1985; 1998), Grün (2001), Lian and Huntley (2001) and Duller (2004), and in recent conference proceedings (Grün and Wintle, 2001; 2003).

4.2.1 Thermoluminescence (TL)

TL was the earlier of the luminescence techniques to be developed and was originally applied to the dating of fired pottery or other forms of baked sediment, notably brick and tile.

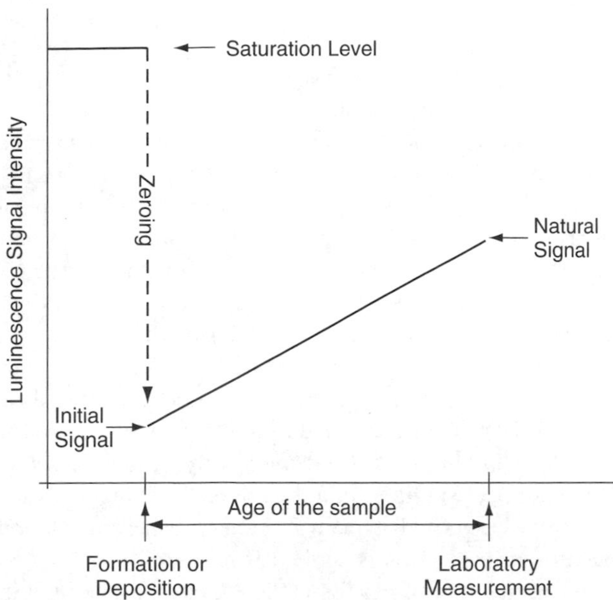

Figure 4.1 *The principle of luminescence dating as applied to a body of sediment. Exposure to daylight, during erosion, transport or deposition, 'bleaches' the sample and effectively removes any latent luminescent signal acquired at some time in the past. Over time the latent signal builds up again as a result of weak natural radiation. The laboratory measurement therefore provides an indication of the time that has elapsed since the last bleaching event (after Aitken, 1998). By permission of Oxford University Press*

Since then, however, it has been used to date a wide range of media, including other fired materials, such as burnt flint artefacts, burnt stones and volcanic products, as well as unburnt samples of, for example, cave speleothem carbonate, and sediments such as loess,[2] glacial lake deposits and sediments from the deep ocean floor (Aitken, 1985; Berger, 1988; 1995).

The intensity of TL released by heating a piece of pottery or sample of sediment is proportional to the quantity of natural radiation that has been absorbed by particular minerals since the time of firing, or the last exposure to sunlight. TL measurements are carried out on a sample of mineral material, usually a separated quartz or feldspar fraction. This is heated to temperatures in excess of 500 °C, and as light (photons) is emitted from the luminescence centres, the photons are converted to electric pulses using a photomultiplier tube, an instrument that is a very sensitive detector of light. The light emission (TL intensity) is then plotted against the heating temperature to produce a **glow curve**, in which the peaks are reflective of the thermal lifetimes of the various electron trap populations in the sample. Of particular interest to the dating specialist are the traps with long thermal lifetimes (so-called 'deep traps') because electrons within them will remain there over long periods of geological time. The 'natural' TL signal is compared with the 'artificial' signals obtained from portions of the sample to which known doses of radiation have been administered from a calibrated laboratory radioisotope source. This allows an evaluation of the **equivalent dose** (D_E), which is a measure of the amount of radiation that would be needed to generate a TL signal equal to that which the sample has acquired subsequent to the most recent firing (zeroing) event or exposure to sunlight. The D_E is sometimes referred to as the **palaeodose**, although this usage is not strictly correct as it is a 'palaeodose equivalent' that is actually being determined.

Variations of two approaches are employed to determine the D_E: the **additive-dose method** and the **regeneration method**. In the former, a number of nearly equal portions of the sample (*aliquots*) are divided into groups; one is reserved for measurement of the natural TL signal only (N in Figure 4.2A), while the others are given various doses of laboratory radiation, generating more trapped electrons and increasing the TL signal. In the simplest case, all the sample aliquots are heated ('preheated') together to empty electrons from thermally unstable traps that have been filled during laboratory irradiation (leaving only electrons in thermally stable traps), and then all the aliquots are measured together and the luminescence intensity is plotted against laboratory radiation dose; this forms the sample's dose response. The D_E is determined from the intercept of the fitted line with the dose axis (Figure 4.2A). This ***total bleach method*** is only valid for cases where the sample has been heated or exposed to sunlight for a long duration (Singhvi *et al.*, 1982). In most cases, however, some electrons will remain in hard-to-bleach traps, and here a second dose–response curve is needed. This is formed from the TL measured in aliquots that have also been given a brief, and spectrally restricted, exposure to light in the laboratory. The second curve is used to correct for the presence of electrons in the hard-to-bleach traps (the ***partial bleach method*** of Wintle and Huntley, 1980), the D_E being determined at the point where the two dose–response curves intersect above the dose axis (Berger, 1988; Lian and Huntley, 2001).

In the **regeneration method**, in the simplest case, all except the natural aliquots are zeroed by heating and then given laboratory doses of radiation. The D_E by regeneration is obtained by interpolation, in other words a direct comparison is made of the natural TL with the TL resulting from laboratory irradiation (Figure 4.2B). In most cases, however, the regeneration curve is compared with an additive-dose curve, and the additive-dose curve is shifted along the dose axis until it is in alignment with the regeneration curve, the magnitude of the shift being proportional to the D_E. This technique is often referred to as the ***Australian slide method*** (Huntley *et al.*, 1993a; Prescott *et al.*, 1993; Lian and Huntley, 2001). The advantage of the regeneration method over the additive-dose method is that no extrapolation is involved, and so uncertainties that might arise from non-linearity in the dose–response curves, for example, are reduced. The principal disadvantage is that if there is a change in sensitivity (TL per unit dose) between measurement of the natural and regenerated TL signals, the palaeodose estimate will be in error (Huntley *et al.*, 1993a; Lian and Huntley, 2001).

4.2.2 Optically Stimulated Luminescence (OSL)

Although TL continues to be used in luminescence dating, recent developments in this field have been mainly associated with optical dating. Many of the principles of TL dating underlie OSL dating, but the major difference is that during measurement the trapped electrons are released by light rather than by heat. The idea that visible light could be used to stimulate electrons, and that this approach could be used to establish the time of deposition of sediments, was first demonstrated 20 years ago (Huntley *et al.*, 1985). In that

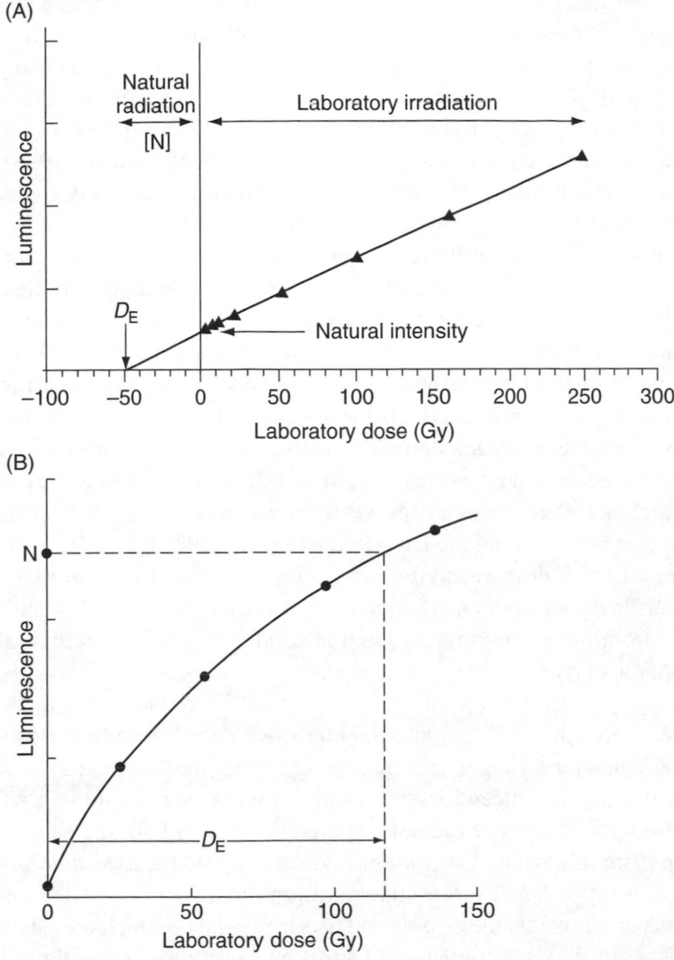

Figure 4.2 *(A) Additive method of palaeodose evaluation. After measurement of natural TL intensity using different aliquots, samples are irradiated in the laboratory, the triangles representing values for increasing levels of irradiation. Fitting the data points to an exponential function enables a value for equivalent dose* (D_E) *to be obtained by extrapolation. The irradiation dose rate is in grays (Gy) (after Grün, 2001). Reproduced by permission of John Wiley & Sons Ltd. (B) Regenerative method of palaeodose evaluation. The natural TL of the sample (N) is compared to the TL signal from aliquots which have been bleached to low levels and then progressively irradiated. The equivalent dose* (D_E) *can then be read off from the plotted curve. Note that for OSL dating, the dose response can be constructed using multiple aliquots or by using a single aliquot (after Aitken, 1998). By permission of Oxford University Press*

study, a beam of green light from an argon ion laser was employed, but light sources other than lasers have subsequently been used, including filtered halogen lamps and high-powered light-emitting diodes (LEDs). A further development has been the employment of **infrared stimulated luminescence (IRSL)**, although this approach can be applied only to feldspar grains, as electron traps in quartz are insensitive to infrared (IR) stimulation (Hütt *et al.*, 1988). For feldspars, IRSL has a number of advantages over the use of green light, including the fact that it tends to generate a stronger luminescence signal, electron traps sampled by IRSL are more effectively bleached at deposition than luminescence from traps stimulated by green light, while adequate stimulation power can be provided by an array of LEDs, which are both cheap and convenient (Aitken, 1998). For quartz, expensive lasers and unreliable halogen lamps have now been largely replaced by high-powered blue-green LEDs.

Determining D_E using optical dating is similar to that for TL dating, but there is one notable difference, for in OSL dating there is no way of selecting only thermally stable traps during laboratory measurement. To circumvent this problem, sample aliquots are heated after laboratory irradiation, but before the final measurement. This 'preheat' is designed to empty thermally unstable traps that have been filled during irradiation. Unfortunately, the preheat also leads to the transfer of some electrons from thermally unstable traps to the thermally stable light-sensitive traps that are to be used for dating. However, when the additive-dose method is used, this thermal transfer can be corrected for by constructing a second dose–response curve from aliquots that have been given various laboratory doses and then a long bleach (Huntley *et al.*, 1993b). The D_E is proportional to the point where the two dose–response curves intersect over the dose axis. It should be noted, however, that this thermal-transfer correction method used for optical dating is very different from the partial bleach method used in TL dating, which also employs two dose–response curves.

In order to arrive at an estimate of age using either TL or OSL, one further parameter has to be determined; this is the environmental **dose rate** (or **annual dose**), and it is a measure of the radiation dose per unit of time absorbed by the mineral of interest since the zeroing of the luminescence clock by firing or by exposure to sunlight. The dose rate is calculated from an analysis of the radioactive elements in both the sample and its surroundings: these are referred to as the *internal dose rate* and the *external dose rate*. These are determined using the measured concentrations of radioactive elements (uranium, thorium, potassium-40) within the sample and its surroundings, which are, in turn, converted into dose rates using standard conversion factors and formulae. The determination of the total radiation dose that is absorbed by a sample is more difficult, as the sample, if buried, will not only have been affected by alpha, beta and gamma radiation from the surrounding sediments, but will also have been influenced by cosmic rays. The intensity of cosmic rays depends on the sample depth below the ground surface and can be estimated using the formula of Prescott and Hutton (1994). A further complication is that water and organic matter, if present in the sediment matrix, absorb radiation differently from mineral sediment, and these have to be accounted for in the dose-rate calculations (Lian *et al.*, 1995). In practice, both field measurements using gamma-ray spectrometer or a TL dosimeter, and laboratory elemental analysis of sediment samples to determine the concentration of radioactive elements, are necessary in order to obtain a measure of external dose rate. Overall, this will usually be measured with a precision of around 5% (Grün, 2001).

Once the dose rate has been established, then TL/OSL age can be calculated from the following:

$$\text{TL/OSL age} = \frac{\text{equivalent dose}}{\text{dose rate}}$$

The increasing use of OSL in the dating of sediments has stimulated a range of methodological developments in recent years. These include the **single aliquot regeneration (SAR) method**, which employs repeated measurements on a single sample, rather than using multiple aliquots from a sample as in TL, to determine the D_E (Wintle and Murray, 2000). Not only does this avoid the problems of differing intrinsic luminescence sensitivity (arising, for example, from grain-to-grain variations between aliquots from a single sample), but it also means that smaller samples of material can be dated. Another significant advance has been the ability to obtain OSL measurements from single grains of quartz or feldspar (e.g. Roberts *et al.*, 1998; 1999). Progress towards dating single aliquots of feldspar has also been made, although devising an adequate correction for sensitivity of change remains a problem (e.g. Wallinga *et al.*, 2000), as also does the effect of anomalous fading (section 4.2.3). The ability to date single grains is important because it makes it possible to determine whether all of the grains in a sample of sediment have the same apparent luminescence age (Duller, 2004). Grains within a sample may have different apparent ages, either because some of them were not exposed to daylight for a sufficient period of time at deposition to reset their luminescence signal (see below), or because the sediments have been disturbed and older material incorporated into younger horizons or vice versa (e.g. Bateman *et al.*, 2003; Jacobs *et al.*, 2003a).

Despite the fact that OSL is a relatively recent development in luminescence dating, it has effectively replaced TL in the dating of sediments. There are a number of reasons for this. In OSL dating only the most light-sensitive electron traps are sampled, and this allows samples to be dated that have been exposed to only a few seconds of direct sunlight. Residual signals are usually much smaller for OSL dating than for TL dating, and this permits younger samples to be dated. Moreover, because OSL samples do not have to be heated to high temperatures, the apparatus can be made much simpler (Lian and Huntley, 2001).

4.2.3 Sources of Error in Luminescence Dating

Some of the problems associated with luminescence dating have already been touched upon. A number of potential difficulties centre on the zeroing of the luminescence clock. We have already noted that in most TL measurements there may be a residual TL component which, if undetected and/or unaccounted for, will influence the calculation of D_E and generate an aberrant age. Although this is not a problem in the majority of cases where OSL is employed, the younger the sample, the more stringent the zeroing requirement and, as Aitken (1998) has noted, 'there are depositional contexts in which "absolute zero" is an unattainable ideal'. The depositional environment may also create difficulties, particularly in water-lain deposits where there may be considerable grain-to-grain variability in the extent of bleaching. Reworking of sediments and intermittent exposure to sunlight are further possible sources of error in luminescence dating of sediments. Variations in luminescence signal may also be found between individual mineral grains; in feldspar samples, for example, bleaching is less rapid than in quartz, sometimes by a factor of about 10. Some samples may contain large amounts of residual OSL, and while a few minutes of exposure to bright sunlight may be sufficient to reduce this to an acceptably low level, under lower light intensities (cloudy conditions) the zeroing process would be markedly less effective. Further problems may arise through the leakage of electrons from thermally stable electron traps, a problem that affects both TL and OSL of feldspars (but not quartz), and is usually referred to as **anomalous fading** (Aitken, 1998; Huntley and Lamothe, 2001), through disequilibrium in the uranium decay chain (Chapter 3) and variations in past water content of sediments, the latter being often very difficult to assess (Wintle, 1991). Additional sources of error include systematic errors in the calibration of laboratory radiation sources, possible light contamination during field sampling, and the accurate determination of the environmental dose rate (section 4.2.2).

4.2.4 Some Applications of Luminescence Dating

The age range of luminescence dating varies, at its lower end, with the sensitivity of the sample material and the efficiency of the zeroing process. At the upper end of the dating range, a point is reached at which an additional radiation dose results in no further increase in the luminescence being measured (**saturation**), and this along with the stability (thermal lifetime) of the TL/OSL signal and the magnitude of the environmental dose rate appear to be the principal constraints on age. Recent work on very young sediments has shown that in depositional contexts where quartz grains have been very well zeroed prior to deposition and burial, it is possible to obtain OSL dates within the last 200–300 years that are entirely consistent with well-attested dates from historical sources (e.g. Bannerjee *et al.*, 2001; section 4.2.4.5). TL ages of around 100 years have also been obtained from fired pottery (Aitken, 1985). At the upper end of the range, luminescence dating of feldspars has tended to be limited to around 100 000–150 000 years, due largely to the problems of anomalous fading (see above), although recent studies using far-red IRSL may enable this age range to be extended (Lai *et al.*, 2003). Quartz, on the other hand, does not suffer from anomalous fading, but the luminescence emitted usually saturates at much lower doses than does the luminescence from feldspar. Nevertheless, quartz may also have a current upper age range similar to that of feldspar. Luminescence ages older than 150 000 years have been obtained on other materials, however. For example, TL dating of burnt flint has yielded ages >300 000 years (Mercier *et al.*, 1995). In the following section, we consider eight different applications of luminescence dating across these various age ranges, four involving TL dating of different types of material, and four describing OSL dating of sediments, again involving different materials and contexts. Other illustrative examples are described by Roberts (1997) and Aitken (1998).

4.2.4.1 TL dating of Early Iron Age iron smelting in Ghana. The Iron Age, which is believed to have been initiated in Armenia in the fourth millennium BP, spread to Cyprus and Greece by 3200 BP, and by various routes to West Africa by the third millenium BP. There the majority of the early iron-working sites stretch from Senegal to Chad and from the Sahel to the rainforests where the earliest phase of iron working dates from between 600 BC and AD 600.

In Ghana, evidence for early iron working comes from only a few sites, and at most of these dating is problematical, although the vast majority of Iron Age occurrences in Ghana appear to post-date 1000 BP. Godfrey-Smith and Casey (2003), however, provide TL evidence that confirms an Early Iron Age date for iron working in northern Ghana. At the Birimi site, in the far north of the country, TL dating was carried out on quartz grains extracted from the walls of three iron smelters (Figure 4.3), two of which yielded statistically indistinguishable ages of 1080 ± 70 and 1090 ± 60 years, while the third provided an

Figure 4.3 *Smelter 1 at Birimi, northern Ghana. Dashed white curves follow the outside of the visible wall remnants. The sample for TL dating was taken from above the left end of the lower curve. A hammer and African hoe are shown for scale (after Godfrey-Smith and Casey, 2003). Reproduced with permission of Elsevier*

earlier age estimate of 1600±100 years. All are significantly older than the sole radiocarbon date of 460±90 years (550–330 cal. years BP) obtained on a furnace in the northern region of Ghana. The TL ages point to the existence of a full-blown smelting technology in northern Ghana during the Early Iron Age, between 1600 and 1080 years ago. This time period predates any previous ages for iron smelting in this part of Ghana, and shows that the technology flourished in the northern region of the country several centuries earlier than previously suspected.

4.2.4.2 TL and AMS radiocarbon dating of pottery from the Russian Far East. Some of the oldest pottery in the world is found in eastern Asia. In Chapter 2, AMS radiocarbon dating of carbonaceous residues in pottery was considered and an example was provided of the

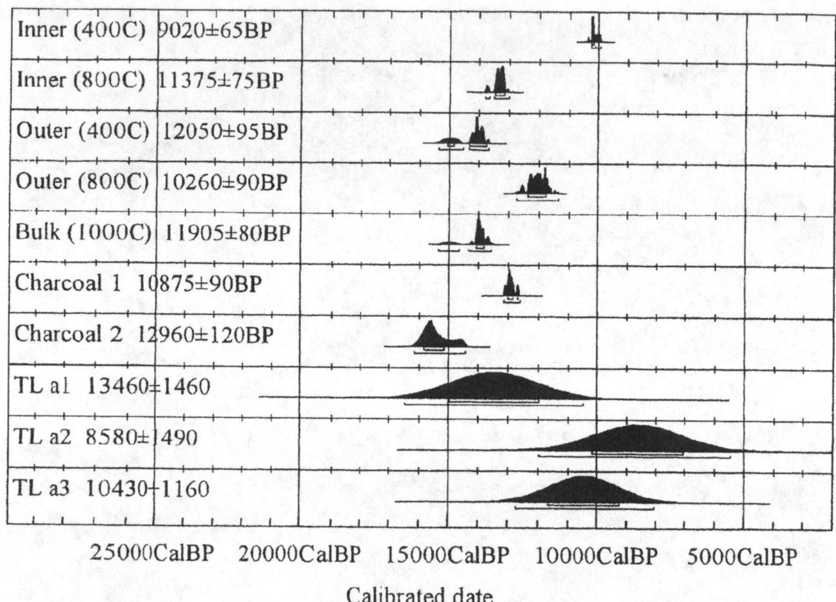

Figure 4.4 *AMS radiocarbon and TL dates from the Gasya site in the lower Amur Valley, Russian Far East. Note the broad measure of agreement between the AMS dates on organic residues from pottery and TL dates from pottery sherds. The timescale is in calibrated years* BP *(after Kuzmin* et al., *2001). Reproduced with permission of Elsevier*

use of this approach to date the earliest pottery in Japan (section 2.7.2.6). Kuzmin *et al.* (2001) describe a similar study of a very Early Neolithic site at Gasya in the lower Amur Valley in the far east of Russia, where samples of pottery containing organic materials and associated charcoal fragments were AMS-radiocarbon-dated. In this case, however, TL dates were also obtained on the pottery fragments to provide an independent check on the radiocarbon age determinations. Five radiocarbon dates span the time interval from ca. 10 000 to 13 000 cal. years BP, while the TL age range (three dates) is from 8590 to 13 460 years (Figure 4.4). Because of uncertainties relating to the external dose rate (section 4.2.2), the errors on the TL dates are of the order of 1500 years compared with ~100 years for the radiocarbon dates. Nevertheless, the TL and radiocarbon dates do correspond well, the TL results confirming the reliability of the AMS radiocarbon dates. With ages in excess of 8500–10 000 years, the data indicate that this Early Neolithic pottery is among the oldest in the world, and it is possible that the Russian Far East may represent an independent centre of ancient pottery production within eastern Asia. Further work is now needed in order to determine how this early Russian pottery compares in terms of age with other examples from the Far East, such as the Jomon pottery from Japan (10 085–12 700 years BP) or the earliest Chinese pottery from south of the Yangtse River (13 700–14 600 years BP).

4.2.4.3 TL dating of burnt flint from a cave site in France. Burnt flint and burnt stones have been a popular medium for TL dating as both of these are often found in archaeological

Table 4.1 *TL age estimates for burnt flint specimens from Combe-Capelle Bas, southwestern France*

Sample	TL age (years BP)
CC10	57 400 ± 4200
CC9	55 600 ± 4400
CC6	48 200 ± 3300
CC7	48 900 ± 3800
CC8	56 900 ± 7800
CC3	52 900 ± 4600
CC2	36 600 ± 2700
Weighted mean of samples 1–6	51 800 ± 3000

contexts, especially in cave sites. The TL method is particularly useful for dating Middle Palaeolithic sites whose ages lie beyond the range of radiocarbon dating, and the technique has provided age estimates for a number of important Mousterian[3] sites in Europe and the Near East. One example is Combe-Capelle in the Couzer Valley, a tributary of the Dordogne in southwest France (Valladas *et al.*, 2003a). Successive excavations have yielded a rich stone tool industry, but dating had proved to be problematic, the most recent estimates based on geological evidence and correlation with the marine isotope record suggesting that the formation of the principal Mousterian deposits date prior to MOI stage 6 and were most likely emplaced during MOI stage 8 or even stage 10 (Chapter 1). However, TL dating of burnt flint suggests that these dates are significant overestimates of the age of the deposits. Seven flint samples were dated and the ages range from ca. 37 000 to ca. 57 000 years, but six of these fall within the 50 000–60 000 age bracket (Table 4.1). The mean age of 51 800 ± 3000 indicates that the fires in which the flints were burnt belong to the early part of MOI stage 3. This age estimate is in good agreement with evidence from other sites in southwest France, many of which show evidence of Mousterian industries that post-date the last (Eemian) interglacial.

4.2.4.4 TL dating of the first humans in South America. In Chapter 2, the question of early human colonisation of the Americas was raised in the context of radiocarbon dating of skulls from a site in Mexico (section 2.7.1.7). We noted there that the most widely accepted model for human colonisation of the Americas is one that envisages migration across the Bering Land Bridge and then southwards through the present-day United States after the wastage of the last ice sheets some time after ca. 12 000 years ago.

Increasingly, however, sites are being discovered with dates older than 12 000 years BP and which pose a challenge to this model (Marshall, 2001; Fiedel, 2002; Dillehay, 2003). The evidence from South America has proved to be especially enigmatic, with the Monte Verde site in Chile in particular possibly pointing to human occupation prior to 30 000 years BP (Dillehay, 1997). However, this interpretation has not been widely accepted by American archaeologists. In the context of this debate, the TL dates reported by Watanabe *et al.* (2003) are of some interest. In the Toca da Bastiana rockshelter site in the Serra da Capivara National Park, Piaui, northeastern Brazil, layers of calcite have formed over ancient rockwall cave paintings, and samples of this material were subjected to both TL and ESR (Electron Spin Resonance) dating (section 4.3). The results were consistent, the

TL analysis producing an age estimate of 35 900 ± ~1550 years and ESR generating a date of 35 350 ± ~5000 years. If these dates are correct, they indicate that the people who painted the rockwalls did so prior to ca. 36 000 BP and an early human presence in South America would be confirmed. Radiocarbon dates of up to 48 500 years BP have been obtained on charcoal from other rock shelters in the area, while TL age estimates on burnt quartz pebbles range from 30 000 to more than 100 000 years ago (Valladas *et al.*, 2003b). While neither case has independent evidence for the burning being associated with human activity, when taken in conjunction with the dates on the rock paintings, these data do pose further questions about the validity of the currently accepted model for the initial peopling of the Americas.

4.2.4.5 OSL dating of young coastal dunes in the northern Netherlands. Although OSL is now widely used as a means of dating sediments, relatively few studies have been carried out to test the accuracy of the technique for deposits younger than a few hundred years. One which has is the work of Ballarini *et al.* (2003) on coastal dune from the southwest coast of the island of Texel in the Netherlands (Figure 4.5), where the ages of the dune ridges are known from historical sources. Samples of quartz from the dunes were OSL-dated using the SAR procedure (section 4.2.2). The youngest of these yielded ages of less than 10 years, showing that the OSL signal of the quartz grains is very well zeroed prior to deposition and burial. OSL ages on five samples taken from a 250-year-old dune ridge (samples TX02-28 to 32: Figure 4.5) are statistically indistinguishable and accord perfectly with the date from historical sources. Overall, 17 out of the 20 OSL ages are in excellent agreement with the well-established independent age controls. This study highlights the potential of using quartz OSL dating as a chronometric tool for reconstructing recent coastal evolution, and for providing geomorphological data that are important for coastal management.

4.2.4.6 OSL dating of dune sands from Blombos Cave, South Africa: single and multiple grain data. Blombos Cave is one of a number of cave sites along the southern coast of Africa (Figure 4.6A) that have been found to contain evidence of Middle Stone Age occupation in the form of finely worked bifacial points, a range of bone tools and, more recently, shell beads (Henshilwood *et al.*, 2001; 2004). It is located 300 km to the east of Cape Town and is currently ca. 100 m from the shore and ca. 35 m above sea level. A large Quaternary sand dune, which is banked against the coastal cliffs, also extends into the cave where it forms a continuous, archaeologically sterile layer that separates the Later Stone Age deposits from the artefact-rich Middle Stone Age deposits beneath (Figure 4.6B). The aeolian nature of the sand makes it a good candidate for optical dating, and this is important because the stratigraphical position of the sand means that such a date would provide a minimum age for the Middle Stone Age deposits. Jacobs *et al.* (2003a,b) carried out both single- and multiple-grain dating on three samples from the dune sand. The ages obtained on the basis of single-grain measurements were 67 300 ± 3800, 65 600 ± 2800 and 68 300 ± 3000 years. These values are ca. 4% less than the age determinations obtained using the conventional multiple-grain methods (69 200 ± 3900, 69 600 ± 3500 and 70 900 ± 2800 years) but, given the relatively large errors, the dates are in good agreement. Collectively, the data confirm the antiquity of the artefacts found in the Middle Stone Age deposits, and suggest a minimum age of 65 000–70 000 years ago. These dates have subsequently been broadly confirmed by OSL dating of quartz grains

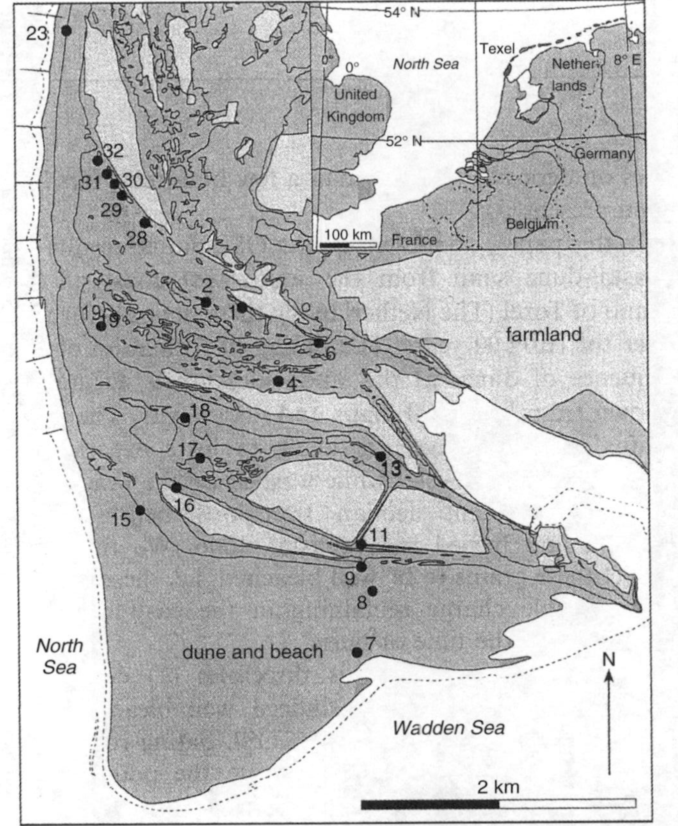

Sample Nos	OSL age[d] (years)	OSL age[d] (AD)	Independent age range (AD)
1	248 ± 11	1754 ± 11	1738–1749
2	220 ± 12	1782 ± 12	1774–1795
4	145 ± 9	1857 ± 9	1855–1863
6	163 ± 10	1839 ± 10	1795–1838
7	6 ± 2	1996 ± 2	1996–2002
8	7.4 ± 1.3	1995 ± 1	1996–2002
9	17 ± 3	1985 ± 3	1976–1989
11	25 ± 2	1977 ± 2	1963–1996
13	62 ± 4	1940 ± 4	1896–1930
15	13 ± 2	1989 ± 2	1976–1989
16	19 ± 1	1983 ± 1	1962–1980
17	36 ± 3	1966 ± 3	1942–1960
18	26.2 ± 1.4	1976 ± 1	1925–1939
19	160 ± 9	1842 ± 9	1855–1863
23	73 ± 23	1929 ± 23	2001–2002
28	264 ± 18	1738 ± 18	1738–1749
29	240 ± 14	1762 ± 14	1738–1749
30	266 ± 12	1736 ± 12	1738–1749
31	253 ± 14	1749 ± 14	1738–1749
32	267 ± 15	1735 ± 15	1738–1749

Figure 4.5 *The southwestern coast of the island of Texel, the Netherlands, showing the locations of samples for OSL dating. The results of OSL dating compared with ages of the dunes based on historical sources are listed on the right (after Ballarini* et al.*, 2003). Reproduced with permission of Elsevier*

(A)

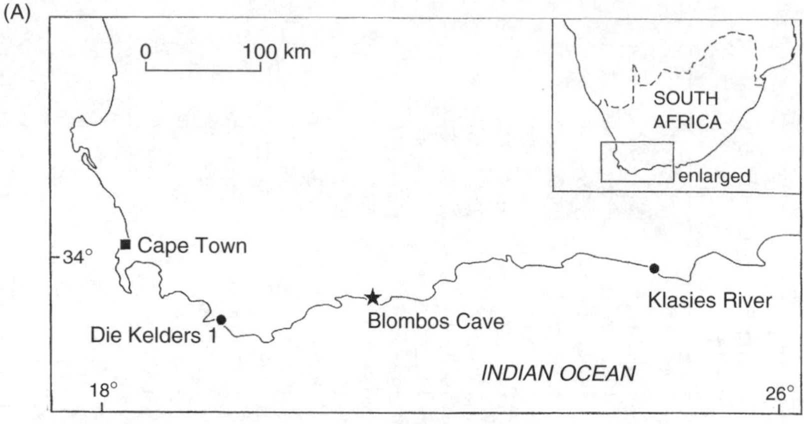

(B)

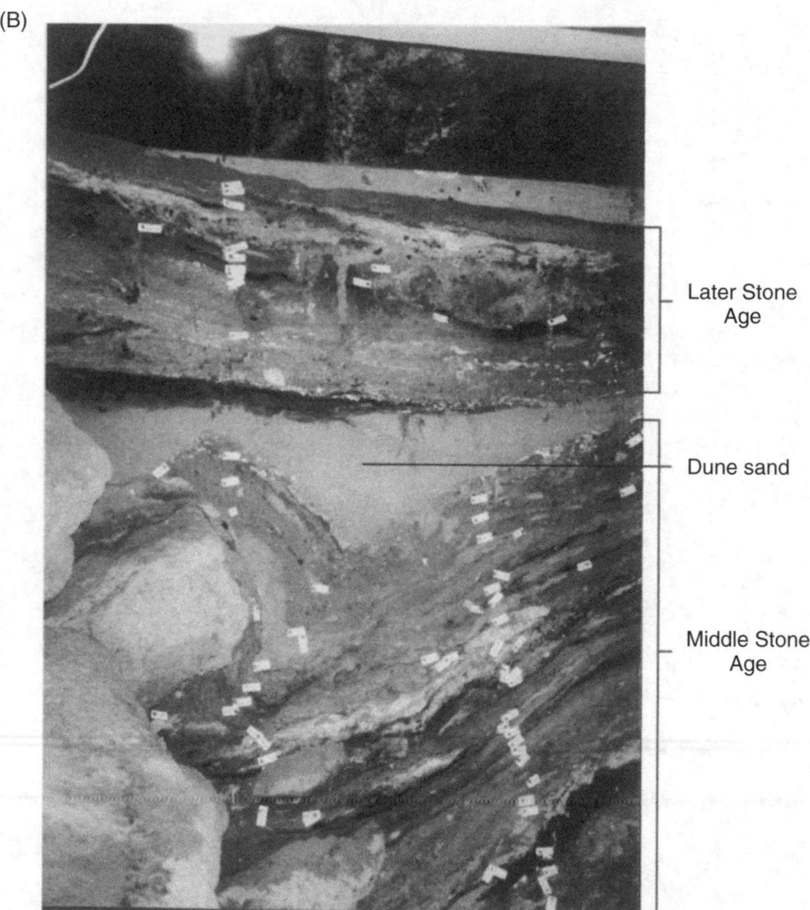

Figure 4.6 *(A) The location of Blombos Cave and other important Middle Stone Age cave sites in South Africa. (B) The stratigraphic sequence in Blombos cave showing dune sand overlying the Middle Stone Age deposits (after Jacobs et al., 2003b). Reproduced with permission of Elsevier*

(75 600±3400 years) and TL dating of burnt flint (77 600±6000 years) from the Middle Stone Age levels that yielded the shell bead ornaments (Henshilwood *et al.*, 2004).

4.2.4.7 OSL dating of fluvial deposits in the lower Mississippi Valley, USA. Although OSL has been most widely employed in the dating of aeolian sediments, it has also been successfully employed to date other sedimentary materials, such as coarse-grained fluvial deposits (Fuchs and Lang, 2001), river alluvium (Lian and Brooks, 2004), colluvial sediments (Fuchs *et al.*, 2004) and glacial deposits (Spencer and Owen, 2004). These may be difficult to date by radiocarbon because they are often devoid of organic material, because organic materials that are present may have been reworked, or because the sedimentary sequences under investigation lie beyond the range of the radiocarbon method.

One area where OSL dating has been used to develop a chronology of river channel changes is the northern lower Mississippi Valley immediately south of the confluence with the Ohio River (Rittenour *et al.*, 2003). Geomorphological mapping had identified three cross-cutting channel belts, from highest (oldest) to youngest (lowest) designated Pve, Pv 2 and PV 1 (Figure 4.7). Samples of quartz from the fluvial sands on each of these

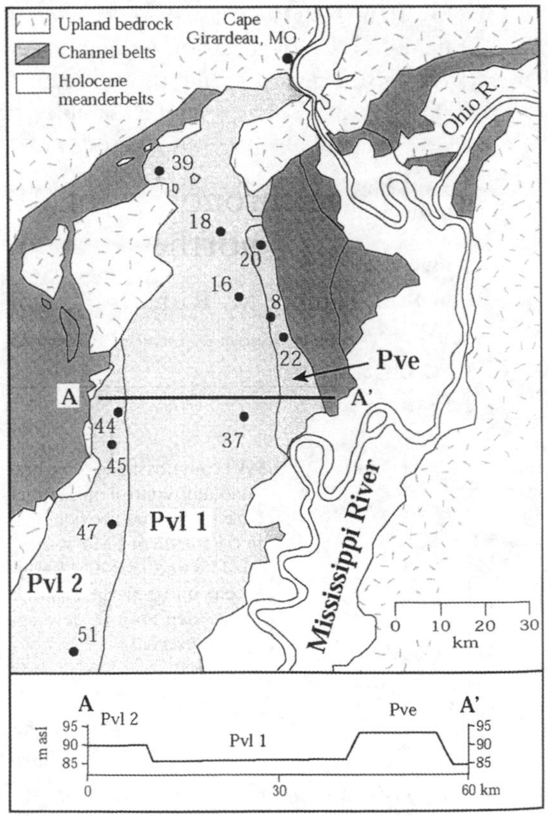

Figure 4.7 *The northern part of the lower Mississippi Valley showing the confluence of the Ohio and Mississippi Rivers, the three channel belts (Pve, PV 2 and PV 1) in light shading, and a schematic cross-section through the channel belts. Also shown are the locations of the sample sites for optical dating (after Rittenour* et al.*, 2003). Reproduced with permission of Elsevier*

three surfaces were analysed using the SAR method which produced optical ages of 19700–17800 (Pve: three samples), 16100–15000 (Pv 2: four samples) and 12500–12100 (Pv 1: four samples) years, respectively. These results are entirely consistent with the relative ages of the channel belts suggested by geomorphic relationships, and confirm that the sequences formed during the Late Pleistocene under conditions of high meltwater discharge from the wasting Laurentide Ice Sheet to the north. These optical ages provide the first detailed chronology of fluvial changes in this part of central USA and are the first step towards developing a chronology of river channel changes for the entire lower Mississippi Valley.

4.2.4.8 OSL dating of marine deposits in Denmark. Optical dating can also be applied to marine sand, and Murray and Funder (2003) describe an experimental study to show that this can produce reliable results. At Gammelmarke, on the coast of Jutland, Denmark, deposits dating from the last (Eemian) interglacial are exposed. These consist of marine sands and clays, rich in the shells of marine molluscs (*Cyprina*, *Mytilus* and *Tapes* or *Paphia*), and which accumulated during the marine transgression and subsequent high sea-level stand of the interglacial. The marine sequence overlies till of the penultimate glaciation and is, in turn, overlain by glaciofluvial sands and till deposited by the Weichselian ice sheet (Figure 4.8A). On the basis of correlations with sites elsewhere, the body of sediment between OSL samples 4 and 18 on Figure 4.8A was deposited between 125 000 and 132 000 years ago. IRSL dating of 25 samples, using SAR protocols and based on large aliquots of ~5000 grains, generated a series of age estimates, 22 of

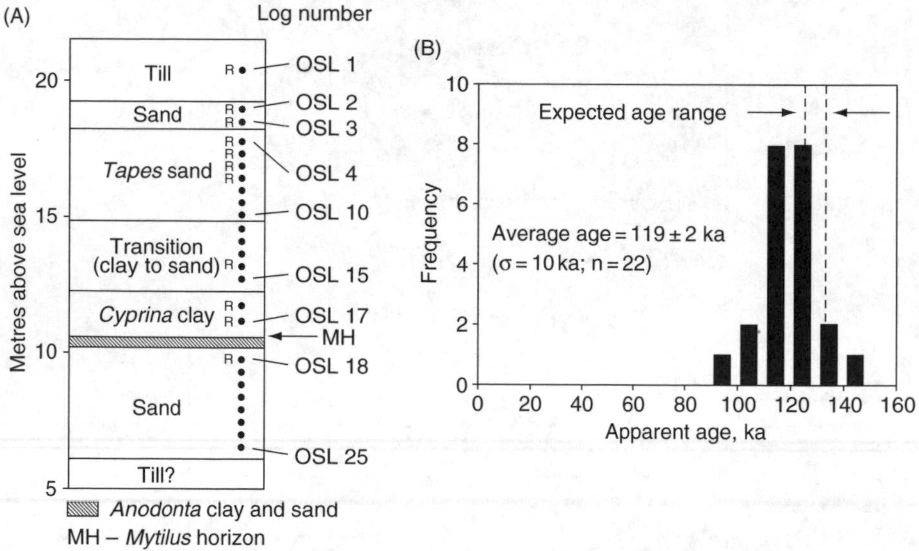

Figure 4.8 *(A) The cliff section at Gammelmarke, southern Jutland, Denmark, showing the sequence of deposits that accumulated during the Eemian marine transgression, and the location of the OSL samples. (B) The distribution of those OSL ages considered to be reliable in relation to the expected age range of the Gammelmarke marine deposits (after Murray and Funder, 2003). Reproduced with permission of Elsevier*

which were considered to be reliable (Figure 4.8B). These suggest an average age for the deposit of 119 000 ± 2000 years but, if systematic uncertainties arising from the calculation of environmental dose rate and other factors are taken into account, the overall uncertainty increases to ~6000 years. This means that the OSL age is very close to the expected age for the deposit. Moreover, the OSL evidence suggests that the marine transgression took place at 121 000 ± 7000 years ago, again an estimate that is close to that of 132 000–128 000 years based on independent evidence. These results confirm the view that reliable age estimates can be obtained from quartz-rich marine sands using OSL dating.

4.3 Electron Spin Resonance Dating

Electron spin resonance (ESR) spectroscopy was initially outlined in the 1930s (Gorter, 1936; Gorter and Kronig, 1936) and was used to investigate the magnetic defects in crystalline and amorphous materials. It was first successfully employed as a Quaternary dating technique by Ikeya (1975) who showed that it could be used to determine the age of speleothem calcite, but since then its applications have expanded to include the dating particularly of tooth enamel, as well as a range of other media including coral, molluscs, quartz-bearing rocks and sediments, and burnt flint (Rink, 1997). **ESR Dating**, sometimes referred to as **Electron Paramagnetic Dating (EPR)**, has a greater time range than most of the other Quaternary dating methods described in this book, extending from a few thousand years to about 2 million years in the case of tooth enamel. In general, the most important applications lie in the time range between 40 000 and 200 000 years, although a number of valuable age estimates of up to 500 000 years have also been obtained. Useful reviews of ESR dating are provided by Rink (1997), Grün (1997; 2001) and Blackwell (2001a), while various examples of the applications of the method are described in the papers in recent conference proceedings (Grün and Wintle, 2001; 2003).

4.3.1 Principles of ESR Dating

The basic principles of ESR dating are very similar to those for luminescence in that it is based on measurements of the trapped electrons in crystal lattices of rocks, sediments or other materials. In this case, however, the electrons are not released by heat or by light. Rather their abundance is estimated on the basis of their paramagnetic properties. The sample is placed in a strong, steady magnetic field and exposed to high-frequency electromagnetic radiation. The magnetic field is slowly changed and at a certain frequency the electrons (or **spins**) become 'excited' and resonate. The strength of this resonance signal can be determined using an ESR spectrometer. When the electrons are in resonance, electromagnetic power is absorbed in direct proportion to the number of electrons present, and hence the greater the number of electrons, the greater the absorption (Aitken, 1990). The latter is a reflection of the time that has elapsed since the onset of electron trapping, and hence is a measure of age.

As in luminescence dating, it is necessary to establish the sensitivity of the sample to radiation, and again the additive-dose method (Figure 4.2) is widely used to determine the **palaeodose** or **equivalent** dose. Recall that this is a measure of the amount of radiation that would be needed to generate an ESR signal equal to that which the sample has acquired subsequent to the most recent zeroing event. In addition, an estimate has to be made of the **dose rate** (or **annual dose**), and which is a measure of radiation dose per unit of time since the zeroing of the electron clock. Again, this involves an estimate of both the internal and the external dose rates (section 4.2.2). Once these parameters have been determined, the age of a sample can be obtained, as in luminescence dating, by dividing the value for equivalent dose (D_E) by the dose rate. One important difference between ESR and luminescence dating, however, is that in ESR dating the electron traps are not emptied as is the case in TL and OSL dating. This means that in ESR dating, replicate measurements can be made on a single sample of material.

4.3.2 Sources of Error in ESR Dating

Many of the error sources that were considered in relation to luminescence dating (section 4.2.3) apply equally to ESR dating. These include non-zeroing of the electron clock so that a residual ESR signal is retained in the sample material, leakage from electron traps, and problems associated with depositional contexts (reworking or materials; effects of groundwater movements, etc.). The last-named in particular poses difficulties in the determination of the annual dose rate. Ideally, both the surrounding deposits and the dated sample should form an unchanging and geochemically closed system with regard to the relevant radioisotopes. In practice, however, this is seldom the case. Erosion or deposition of relatively radioisotope-rich sediments surrounding the sample may result in marked changes in the environmental radiation flux over time (Smart, 1991b). Equally problematic is the fact that a number of materials conventionally dated by ESR, including teeth, bones and mollusc shells, show post-depositional uranium uptake which further complicates dose-rate determinations. The process of uranium uptake cannot normally be established, and models have to be employed to try to simulate this process (Grün, 2001). Hence, while there are good grounds for believing that many ESR dates are *accurate* (in that they give a reasonable approximation of the 'true' age of the sample), the *precision* of ESR age estimates is often low, principally because of the uncertainties surrounding calculation of palaeodose and annual dose rate. The numbers of measurements needed will also vary with the nature of the sample, tooth enamel for example, requiring many more parameters to be determined than speleothem calcite. In addition, precision will be affected by the strength of the ESR signal; a weak signal, for instance, could result in a precision of no better than ±100%. Overall, ESR dating is a method that seldom provides ages with quoted uncertainties of <10% (Rink, 1997).

4.3.3 Some Applications of ESR Dating

As noted above, ESR is considerably more versatile than luminescence dating in that it can date a much wider range of materials. It can also date materials of much greater antiquity. For example, ESR dates of ~463 000 and ~710 000 from glacial sediments on the margin of the Tibetan Plateau provide evidence for a glacial advance in western China during MOI stages 12, and 16 or 18, respectively (Zhou *et al.*, 2002). In this section we consider applications of ESR dating applied to four different media: tooth enamel, coral, mollusc shell and quartz.

4.3.3.1 ESR dating of teeth from the Hoxnian Interglacial type locality, England. The palaeolithic site at Hoxne, Suffolk, England, is one of the most important in the British Quaternary sequence. The interglacial lake sediments at the site have provided a wealth of palynological, mammalian and artefact data and it constitutes the type site for the Hoxnian Interglacial (Singer *et al.*, 1993; Figure 1.4). Dating of the site has proved problematical, however, for while a correlation with MOI stage 11 has been suggested on the basis of biostratigraphic (Schreve, 2001) and U-series dating evidence (Rowe *et al.*, 1999), amino acid data (section 6.6) suggest an equivalence with MOI stage 9 (Bowen *et al.*, 1989).

Two teeth from the deposits believed to have been deposited towards the end of the Hoxnian Interglacial were dated by ESR (Grün and Schwarcz, 2000). Four age determinations were made on one of the samples and two on the other. Calculations of ESR age involved the use of two different models: an open-system U-series model that assumes a continuous uranium uptake from the surrounding depositional environment since the teeth were buried (US-ESR model), and a closed-system model that assumes a short-term uranium uptake event at the apparent U-series age (CSUS-ESR model). The latter model provides the maximum open system age. The results are shown in Figure 4.9. The best age estimate for the teeth can be derived from the weighted means of samples 145 and 293: 404 000 + 33 000/–42 000 years (US-ESR) and 437 000 ± 38 000 years (CSUS-ESR). When these results are plotted against the MOI curve (Figure 4.9), it is clear that the older age is unlikely to be correct as the CSUS-ESR estimate is centred on MOI stage 12 (a glacial stage) and overlaps with the previous interglacial, MOI stage 13. Hence,

Sample	Dose (Gy)	US-ESR (ka)	CSUS-ESR (ka)
145A1	545 ± 99	$356\pm^{115}_{209}$	397 ± 184
145A2	500 ± 21	$336\pm^{65}_{127}$	371 ± 114
145B1	496 ± 49	$373\pm^{55}_{75}$	397 ± 66
145B2	457 ± 47	$331\pm^{46}_{51}$	350 ± 52
293A	541 ± 26	$425\pm^{52}_{59}$	451 ± 46
293B	524 ± 9	$417\pm^{41}_{43}$	440 ± 35

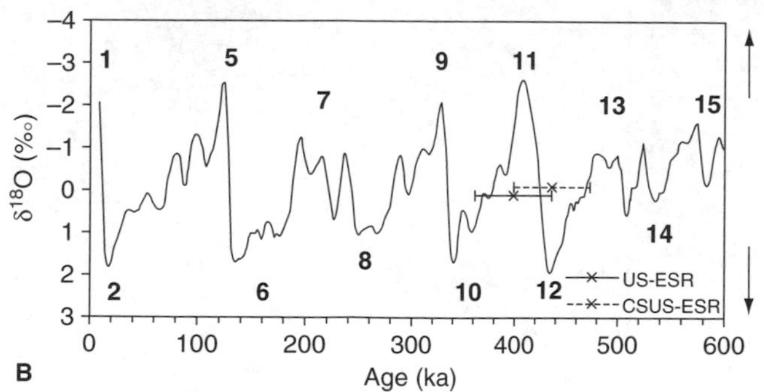

Figure 4.9 *ESR ages for the teeth from the Hoxnian Interglacial site, eastern England, and plots of the weighted mean values against the MOI curve (after Grün and Schwarcz, 2000). Reproduced with permission of Elsevier*

the younger of the two dates (US-ESR) is considered the more accurate and confirms an MOI stage 11 equivalence for the Hoxnian Interglacial.

4.3.3.2 ESR dating of mollusc shells from the Northern Caucasus and the earliest humans in eastern Europe. Although there are problems in the ESR dating of mollusc shells because of open-system behaviour in relation to uranium uptake, results have been obtained that are comparable with those based on other dating techniques (Rink, 1997). In the majority of cases, marine molluscs have been the preferred dating medium, but Molodkov (2001) describes the results of an ESR dating programme using terrestrial molluscs. The samples were obtained from stratified deposits in the Treugolnaya (Triangular) Cave in the northern Caucasus that contains a long sequence of palaeontologically and archaeologically rich sediments. Eight samples of aragonite[4] shell of the terrestrial mollusc *Monacha caucasicala* (Lindh.) were taken from two horizons that contained Late Acheulian[5] stone artefacts (Figure 4.10). The mean ESR age (two dates) of the upper layer was 393 000±27 000

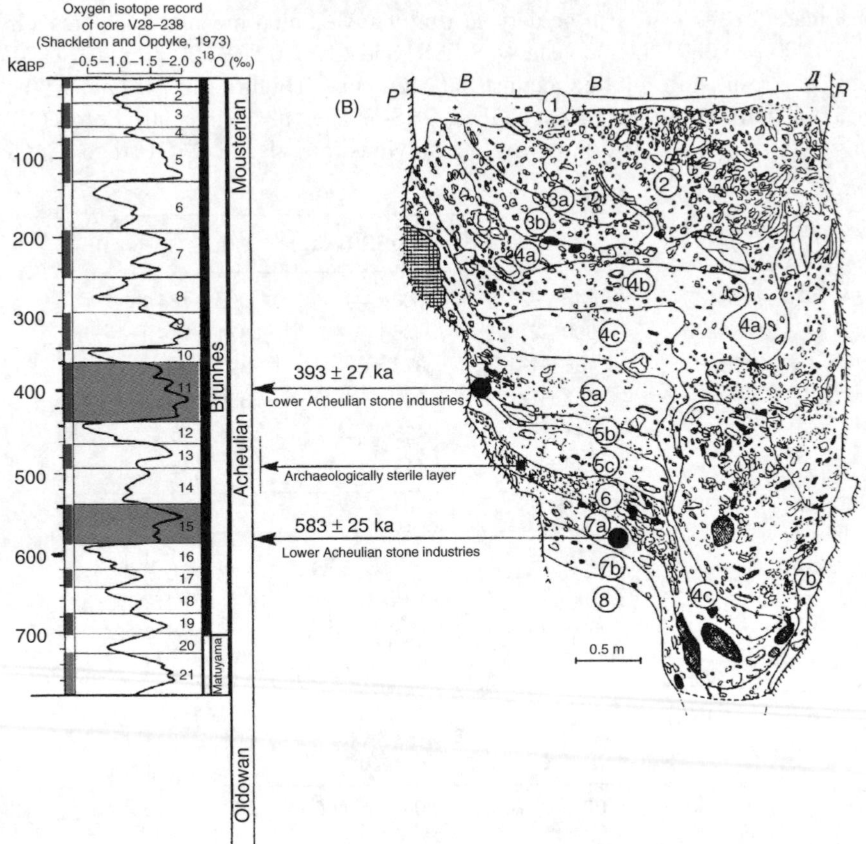

Figure 4.10 ESR ages of Treugolnaya Cave, northern Caucasus, culture-bearing horizons and suggested correlation with the MOI record (after Molodkov, 2001). Reproduced with permission of Elsevier

years and the mean for the lower horizon (six dates) was 583 000 ± 25 000 years. Independent biostratigraphic evidence from the cave deposits indicates that both horizons are associated with warm interglacial conditions, and comparison with the MOI record suggests correlations with MOI stages 11 and 15, respectively (Figure 4.10). These age estimates are of considerable significance because, if correct, they imply that humans had reached the Caucasus by the beginning of MOI stage 15, which is much earlier than had hitherto been believed.

4.3.3.3 ESR dating of Holocene coral: an experimental approach. Coral has been widely used as a medium for ESR dating, and well-preserved raised coral reef terraces up to 500 000–600 000 years in age have been dated by the ESR method (e.g. Pirazzoli *et al.*, 1991). However, problems relating to the determination of both the equivalent dose (D_E) and the past and present radiation doses continue to impose limitations on the dating of fossil coral, and especially of younger samples. However, new protocols for D_E determination (e.g. Schellmann and Radtke, 2001) have enabled much more precise and accurate ages to be obtained from aragonitic coral, even on samples of Holocene Age.

As a test of the reliability of these new approaches, Radtke *et al.* (2003) ESR-dated samples of coral from the Netherlands Antilles that had been transported onshore by tsunami or hurricane events from the living fringing reefs around the islands. Radiocarbon dates were also obtained from the coral samples to provide an independent check on the ESR age determinations. The ESR ages vary between 76 and 3601 years BP, and the calibrated and marine-reservoir corrected radiocarbon dates range from 0 to 3644 years BP (Figure 4.11). The mean age difference between ESR age and calibrated radiocarbon age (excluding one outlier shown on Figure 4.11) is 73 years, the ESR ages being on average slightly older. The standard deviation of the difference is 250 years, indicating that the two dating methods are compatible within the limits of the respective error margins of the two dating series. The results of this experimental programme show that ESR not only has great potential for the dating of Pleistocene corals, but can also be applied to much younger material that was formed only a few hundred years or, in some cases, only decades ago.

4.3.3.4 ESR dating of quartz: the Toba super-eruption. Many volcanic rocks contain quartz crystals, and because the ESR clocks in the quartz are zeroed by the volcanic activity, ESR dating potentially offers an alternative to K–argon or argon–argon as a means of dating volcanic events. However, with very few exceptions, previous studies have shown a high degree of internal inconsistency, and no widely accepted ESR dating protocol has emerged (Rink, 1997). Nevertheless, Wild *et al.* (1999) have shown that in the case of Mt Toba, ESR ages on the most recent eruption are consistent with independent evidence for the timing of that volcanic event.

Covering an area of 30 000 km^2, the Toba volcanic caldera in northern Sumatra, Indonesia, is the largest Quaternary caldera in the world. A number of eruptive events have occurred over the past 1.2 million years, the most recent of which produced the Youngest Toba Tuff. This ash flow, with an estimated volume of almost 3000 km^3, reflects one of the largest eruptions in geological history, and may have been the greatest single volcanic cataclysm in the Quaternary (Oppenheimer, 2002). The tuff had previously been dated by K–argon (73 500 ± 3000 and 74 900 ± 12 000 years) and $^{40}Ar/^{39}Ar$ (weighted mean age of 74 000 ± 4000 years) (Chesner *et al.*, 1991). ESR dating of aluminium (six samples) and

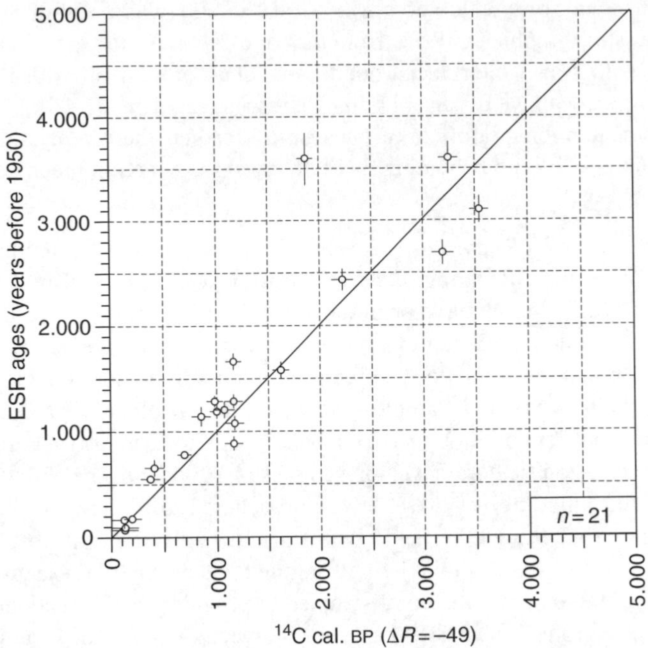

Figure 4.11 Comparison between radiocarbon and ESR ages obtained from coral samples from the Netherlands Antilles. The close correspondence between the results from the two dating methods suggests that ESR dating can provide a reliable basis for the dating of young coral samples (after Radtke et al.*, 2003). Reproduced with permission of Elsevier*

titanium (also six samples) centres within quartz phenocrysts[6] yielded ages ranging from ca. 53 000 to ca. 104 000 years, with an overall average age of 81 000 ± 17 000 years. While there is a degree of scatter in the ESR dates, the mean age is in good agreement with the previously obtained age for the Youngest Toba Tuff based on argon-isotope dating. These results suggest, therefore, that accurate dating of volcanic events may indeed be possible using ESR.

4.4 Fission Track Dating

The technique of **fission track dating** is based on the infrequent, but predictable, break-up of the most abundant isotope of uranium, ^{238}U, in a fission reaction. The process leads to high-energy collisions between the fission fragments and neighbouring atoms and creates **damage tracks** or **trails** in the enclosing crystal lattice. Precisely how this occurs is not completely understood, but it seems that as two positively charged fission fragments are driven apart, they strip electrons from atoms in the host lattice and, after passage of the fission fragment, a zone of positively charged ions remain which mutually repel each other. These are forced into the interstices of the crystal lattice thereby creating a damage zone or latent fission track, which is typically between 10 and 20 μm in length and a few angstrom units[7] wide (Hurford, 1991). The number of tracks is a function of the original uranium content of the sample material and time. The rate of spontaneous radioactive fission decay of ^{238}U is known, and hence the length of time during which fission tracks have been forming in the host material can be determined by counting the number of tracks that have been created. Fission track dating is particularly applicable to volcanic extrusive rocks including basalts, tephras and tuffs where, as in TL, the fission track clock is zeroed by the heating event. It has also been applied to volcanic glasses, such as obsidian and pumice. Detailed accounts of fission track dating are provided by Hurford (1991), Wagner and van den Haute (1992), van den Haute and Corte (1998) and Wagner (1998).

4.4.1 Principles of Fission Track Dating

The procedure of fission track dating involves the counting of spontaneous ^{238}U fission tracks under an optical microscope after the surface of the sample has been polished and etched with a chemical etchant to enlarge the fission tracks. In order to use these data to infer age, it is necessary to know the original uranium content of the sample. Clearly, this cannot be determined by direct measurement, but it can be obtained indirectly by measuring the amount of the less abundant uranium isotope, ^{235}U, that is contained within the sample material, exploiting the fact that thermal neutrons induce fission of ^{235}U. After the ^{238}U fission tracks have been counted, therefore, the sample is irradiated with thermal neutrons and fission in the atoms of ^{235}U generates a new set of fission tracks. These 'induced' fission tracks are etched and counted, their number reflecting ^{235}U abundance and neutron fluence. As the latter is known, the former can be calculated. This, in turn, enables the original ^{238}U content of the sample to be determined based on the known $^{238}U/^{235}U$ ratios in volcanic rocks and other materials.

Two techniques are employed to determine areal track density. In the **population method**, the sample is divided into separate aliquots, the spontaneous fission tracks are etched and counted in the first aliquot, while the second aliquot is irradiated and both the spontaneous (ρ_s) and induced tracks (ρ_i) are counted. The value of ρ_i is then determined by $(\rho_s + \rho_i) - \rho_s$. This assumes, of course, that the two aliquots have the same uranium concentration, and hence the method is employed using apatites and glasses in which the uranium concentration tends to be uniform. A variant of this approach is to heat the aliquot used to measure ρ_s in order to remove the spontaneous tracks, and then irradiate that aliquot to determine ρ_i. However, this is less satisfactory as the thermal treatment of the sample will affect the etching characteristics of the material and introduce an error into the value derived for ρ_i. In the **external detector method**, spontaneous tracks are counted in the etched mineral, while the induced tracks are counted in an external detector of low uranium muscovite (mica) held against the mineral during irradiation and subsequently etched for counting. This means that both spontaneous and induced tracks are measured in exactly matching areas from the same planar surface of an individual crystal and hence uranium inhomogeneity, both within and between crystals (as is found in zircon and sphene, for example), is of negligible consequence (Hurford, 1991).

In theory, any uranium-bearing material can be dated and the age range of the technique extends from a few hundred to many millions of years. In practice, however, factors such as uranium content and abundance, grain size and degree of crystallinity, and track retention place constraints on the applicability of the method. Initial uranium content is of paramount importance. For example, if an areal density of 10 tracks per square centimetre is presupposed as a minimum, then the uranium content must be at least 1 $\mu g\,g^{-1}$ in order to determine an age of 100 000 years with any degree of precision (Wagner, 1998). For ages younger than this significantly higher initial uranium concentrations are needed. Zircon is by far the most widely used mineral, particularly in the younger age range, because of its naturally high uranium content and also its widespread occurrence in volcanic rocks.

4.4.2 Some Problems Associated with Fission Track Dating

One of the major difficulties in fission track dating relates to the natural process of healing or erasure of the fission tracks. This is known as **fading** (or **annealing**) and may arise as a result of heating of the host material or through the spontaneous diffusion of ions. The result is that the tracks become narrower and shorter over time until they eventually disappear completely. Partial fading within a sample is reflected in a reduced number of fission tracks and will lead to an underestimate of age. Not all materials show the same tendency to anneal. The crystalline structure of zircon and titanite, for example, means that in these minerals the fission tracks are relatively stable and fading is less of a problem. Apatite is much less stable, however, while in glasses the stability tends to decrease with decreasing silica content. In some cases, however, it may be possible to identify the partially faded fission tracks on the basis of their reduced size, and several experimentally established correction procedures can be applied to counter this problem. The most common is the ***plateau correction procedure***. Here, the paired samples containing natural and induced tracks are heated at progressively higher temperatures until the ratio of the spontaneous track density in the natural sample to the induced track density in the irradiated sample reaches a plateau level. This plateau value provides a corrected age for the sample. A variant on this method is the ***isothermal plateau fission track (ITPFT) technique*** in which the paired samples are subjected to a single heat treatment until a pre-determined temperature is achieved (e.g. 150 °C) and which is then sustained for a specified period (Westgate, 1989).

4.4.3 Some Applications of Fission Track Dating

The fission track method is one of the most versatile of Quaternary dating methods in terms of its potential age range. At one extreme it can provide dates on ancient tuffs with an age of 1–2 million years, while at the other it can generate realistic ages on uranium-rich glass manufactured in the nineteenth century (Wagner, 1998). However, it must be acknowledged that for most materials measurable within hundreds of years, the low levels of analytical precision mean that the age estimates are of limited value. In Quaternary studies, the fission track method has been most widely (and successfully) employed in the dating of tephras and other volcanic products, such as obsidian. Three examples of the applications of fission track dating are considered in the following sections.

4.4.3.1 Fission track dating of glacial events in Argentina. In the western United States of America, and also in South America, fission track has been used to date ashes interbedded with glacial deposits, thereby providing limiting ages on individual glacial events (Espizua and Bigazzi, 1998; Colgan, 1999). A good example is Espizua *et al.*'s (2002) study of the glacial history of the Rio Grande Valley in the lee of the Andes Mountains in western Argentina. In that region, several glacial episodes have been distinguished on the basis of glacial landforms (moraines) and drift sequences, at least two of which, the Seguro and the Poti-Malal, are believed to predate the last glaciation in the area. Espizua and co-workers describe a site in the Rio Grande Valley where the till is overlain by a tephra layer which is, in turn, overlain by outwash from the Segura Glaciation. Fission track dating of samples of glass from the tephra horizon yielded an age of 226 000±25 000 years. This indicates that the overlying deposits from the Segura Glaciation equate with MOI stage 6 (ca. 127 000–186 000 years), while the Poti-Malal till must be at least as old as MOI stage 8 (pre-242 000 years), and could be even older – perhaps dating from the Early-Middle Pleistocene.

4.4.3.2 Fission track dating of a Middle Pleistocene fossiliferous sequence from central Italy. Marcolini *et al.* (2003) describe a site in Tuscany, central Italy (the Campani Quarry), where several palaeosols (fossil soils) believed to be of Middle Pleistocene age are exposed in sequence (Figure 4.12). The uppermost of the palaeosols contains a rich molluscan fauna and also an assemblage of small mammals, both of which are indicative of a woodland environment and a relatively mild climate. This suggests that the palaeosol formed during an interglacial period, but there are no indications in the faunal records as to which interglacial this might be. Immediately above the palaeosol, however, is a tephra horizon. Fission track dating on apatite grains from the Campani Tephra gave an age of 460 000±50 000 years. The date is within one standard error of $^{40}Ar/^{39}Ar$ ages of 403 000±30 000 and 419 000±20 000 years on a tuff from the Vico Volcano to the south, which has a glass composition identical to the Campani Tephra. This suggests that the Campani Tephra is also associated with the Vico Volcano. On the basis of the fission track and $^{40}Ar/^{39}Ar$ ages, therefore, the underlying palaeosol with its interglacial faunal and molluscan assemblages can be correlated with MOI stage 11, i.e. between ca. 364 000 and 427 000 years ago (Bassinot *et al.*, 1994).

4.4.3.3 Dating of obsidian in the Andes, South America, and the sourcing of artefacts. Obsidian is a raw material of the lithic industry found in many archaeological sites of the Andean countries. Determining the provenance of this material for sites located far from volcanic centres would make an important contribution towards the understanding of goods procurement and ancient trade routes and traditions in pre-Hispanic times.

One way in which source can be established is on the basis of geochemical composition of the glass. A second approach, as described by Dorighel *et al.* (1998), is to use fission track dating of obsidian artefacts found in archaeological contexts. These authors analysed 25 samples of obsidian collected from archaeological sites in Ecuador and Colombia, the majority of which range in age from ca. 500 BC to AD 1500. The dated obsidian fragments are largely residues discarded after flaking, although some are small flakes and blades. On the basis of fission track dating and geochemical analysis, it proved possible to divide the artefacts into eight distinct groups, with ages ranging from ca. 190 000 years for the youngest obsidian samples to more than 4.8 million years for the oldest. Two sources could be determined from these data: the Sierra de Guamani and the Rio Hondo obsidians (Figure 4.13). Interestingly, two of the coastal sites contained obsidian

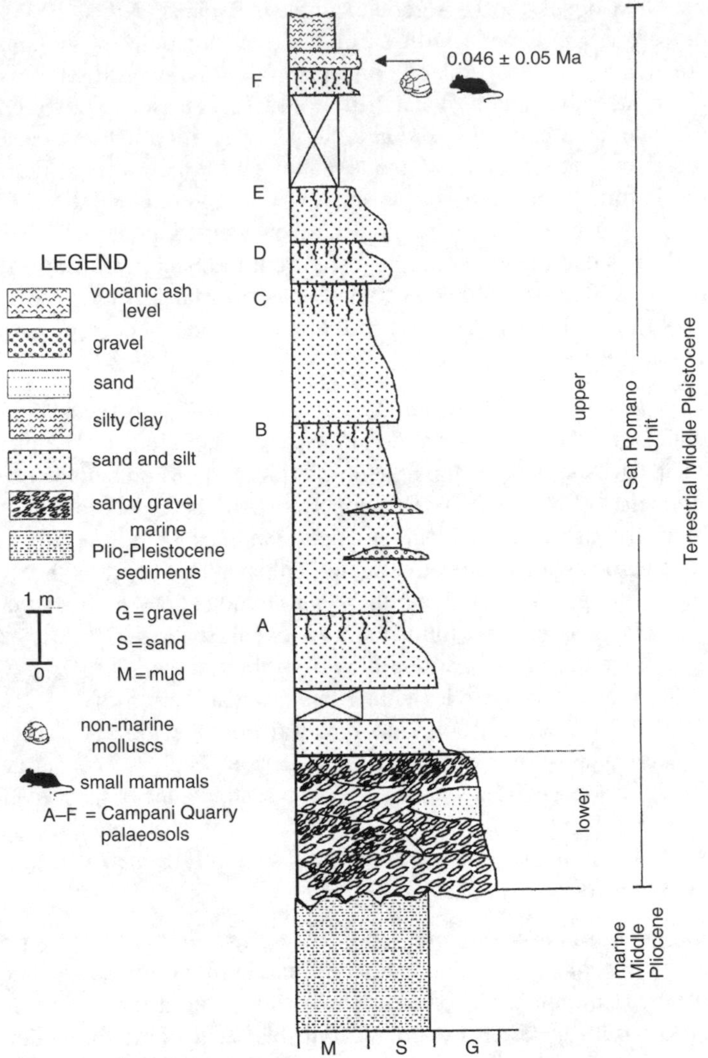

Figure 4.12 *The stratigraphic sequence in the Campani Quarry, Tuscany, central Italy, showing the sequence of palaeosols and the dated tephra layer overlying palaeosol F which contains the interglacial faunal and molluscan assemblages (after Marcolini* et al.*, 2003). Reproduced by permission of John Wiley & Sons Ltd*

from the Sierra de Guanami, indicating transport of artefact source material from this inland site to the coast. Although the other obsidian groups have yet to be related to known source, the results of this work are considered to be very encouraging as, hitherto, little has been known about obsidian sources for artefacts from this region of South America. There may, therefore, be considerable potential for fission track dating in this particular field of archaeology.

Figure 4.13 *Location of archaeological sites and obsidian sources dated by fission tracks in Colombia and Ecuador. The triangles show groups of sites in particular areas (after Dorighel et al., 1998, Fig 1). With kind permission of Kluwer Academic Publishers*

Notes

1. **Quartz** is a member of the silica group of minerals and is common in all kinds of rocks and mineral veins.
2. **Loess** is fine-grained, wind-blown (aeolian) sediment with a grain size in the range of 2–64 μm.
3. **Mousterian** industries are usually associated with Neanderthals and typically date from the Last Cold Stage. The industry is named after the type site at Le Moustier, Dordogne, southwest France (Wymer, 1982).
4. **Aragonite** is a crystalline form of calcium carbonate which comprises the shell of most molluscs, although after death some molluscan shells may invert to calcite. This poses problems for ESR dating as calcite is more likely to display open-system behaviour in relation to uranium uptake than is the case in the tighter crystal structure of aragonite.

5. **Acheulian** industries are characteristic of the Lower Palaeolithic. The typical products are core tools, such as handaxes and bifaces, which are made by flaking down a nodule of flint or other fine-grained rock into the desired form. The term comes from the site at St Acheul near Amiens in northern France where many discoveries of handaxes were made during the nineteenth century (Wymer, 1982).
6. **Phenocrysts** are large crystals that are found set in a finer-grained groundmass in volcanic rocks.
7. An **angstrom unit** is a unit of length equal to 10^{-8} cm.

5

Dating Using Annually Banded Records

A few more years shall roll,
A few more seasons come . . . . .

Horatius Bonar

5.1 Introduction

This group of techniques is based on the annual addition of material to organic tissue or to sequences of sediment. Perhaps the best known of these methods is **dendrochronology** (tree-ring dating), but **varve chronology** (based on annual accumulations of sediment in lakes or in the sea) has also been widely employed. Less common, perhaps, is **lichenometry** which uses differences in lichen size to establish the age of exposure of rock surfaces. In high-latitude and high-altitude regions, seasonal variations in snowfall produce **annual layers in glacier ice**, and these too form a basis for dating. Other contexts where regular additions of material can be used for dating purposes are cave sites, where **speleothems** (stalagmites and stalactites, for example) may show evidence of annual banding, and marine environments, where short-term chronologies (known as **sclerochronologies**) have been developed from annual growth bands in **corals** and **molluscs**. This family of dating methods forms the subject matter for this chapter.

Quaternary Dating Methods M. Walker

5.2 Dendrochronology

Dendrochronology is the technique which employs annual growth increments in the trunks of trees as a basis for a chronology. In most softwood (coniferous) trees, new water- and food-conducting cells are added to the outer perimeter of the trunk during the spring and summer months, following an inactive growth period during the winter. The demand for water and food is greater during the spring growth season, and hence these cells tend to be larger, whereas smaller cells with thicker walls develop later in the summer and autumn. The result is a series of clearly defined 'lines' or **tree rings**, and counting of these enables the age of the tree to be established. In hardwood (deciduous) trees, however, growth trends are more variable. Some display a marked contrast between cells formed in spring and summer, with markedly larger vessels forming earlier in the year. These are known as **ring-porous types** and include ash, elm and oak. In other trees, however, such as alder, beech, birch and lime, there is little difference in pore size between spring and summer growth increments. These are termed **diffuse-porous** trees. The result is a considerable difference in the nature of tree rings between species, and as some do not display clearly defined annual increments, not all trees are suitable for tree-ring dating. The most widely employed in dendrochronology are oak (*Quercus*) and certain coniferous species, mainly pine (*Pinus*), but also *Sequoia* and Douglas fir .

Dendrochronology as a systematic dating technique was developed in the United States, principally through the work of Andrew Douglass who established the world's first tree-ring laboratory at the University of Tucson, Arizona, in 1937. Douglass and his successors, notably Edmund Schulman and Hans Fritts, were responsible for the development of the first long tree-ring records, and they also recognised the potential of trees as archives of climatic information, a field of study known as **dendroclimatology** (Douglass, 1919; Schulman, 1956; Fritts, 1976). In the 1960s and 1970s, dendrochronologists from the Tree-Ring Laboratory in Tucson collaborated with radiocarbon scientists in pioneering work on the calibration of the radiocarbon timescale (Suess, 1970; Ferguson and Graybill, 1983; section 2.6). Their research primarily involved coniferous tree species, and particularly the remarkably long-lived bristlecone pine (section 5.2.3). In Europe, dendrochronology has been a more recent development, but a number of tree-ring laboratories have now been established, most notably at Belfast in northern Ireland and in several universities (e.g. Göttingen, Heidelberg, Hohenheim) in Germany. In Ireland, the majority of dendrochronological work has been carried out on oak, while in Germany chronologies based on both oak and pine have been developed. Good overviews of both the principles and the applications of dendrochronology can be found in Fritts (1976), Baillie (1982; 1995), Schweingruber (1996) and Briffa and Matthews (2002).

5.2.1 Principles of Dendrochronology

Dendrochronology can be applied to wood from a variety of contexts. The tree may be standing or felled, it may form part of a building, or it may be buried on an archaeological site or in a natural deposit such as peat. In some cases the wood may come from a secondary context, such as a pile of timbers from a demolished building or from some drainage operation (Baillie, 1995). Standing trees are sampled using an increment corer, a hollow metal tube which extracts small-diameter cylinders of wood from the tree trunk. Dead or subfossil[1] wood can be cut so that the ring sequences are exposed in section. In the laboratory, the wood samples are cleaned and mounted, and counting is carried out visually on a moving stage under a microscope. Other approaches involve the use of electronic measuring equipment (Cook and Kariukistis, 1990), or **X-ray densitrometry** which determines annual variations in wood density (Schweingruber, 1988).

Because tree growth is closely dependent on climate, the width (or wood density) of each annual ring will vary depending on whether climatic conditions in any one year have been favourable for, or inimical to, tree growth. This means that within a given area, ring widths will vary in response to local or regional climatic changes. The result is a characteristic ring-width pattern within which distinctive rings (representing a particularly good or bad growth year), or groups of rings, form **markers** and these can be used as a basis for cross-matching or **cross-dating** between wood of overlapping age range (Figure 5.1). As Baillie (1995) has observed, cross-dating is the art of dendrochronology, for it enables living trees to be linked to dead wood and those dead wood samples to be matched with even older materials. In this way, the tree-ring record can be extended back in time and a **master chronology** developed. Where dead wood is recovered from ancient buildings, for example, or from peat bogs, counting of the tree rings enables a **floating chronology** to be established, and this can be matched to the master chronologies using the distinctive ring-width patterns. Hence, samples of wood of hitherto unknown date can be assigned a precise calendar age.

5.2.2 Problems Associated with Dendrochronology

Because tree growth is more rapid in young trees, there tends to be a reduction in ring width with age. This can pose problems in cross-dating because it is possible for ring-width variations that reflect climatic or environmental factors to be masked by ring-width variations due to age. A further problem is that tree-ring widths vary with the height of the trunk, and there is no way of knowing in a piece of old or subfossil wood precisely where on the trunk the sample originated. One way in which these difficulties can be surmounted is by **standardising** the ring-width series to generate **ring-width indices** (Baillie, 1982). A range of statistical techniques, including regression analysis (Figure 5.2), is now available to generate such indices.

Other problems are more difficult to resolve. In some localities for example, where trees are growing in situations where there is little or no variation in climate, or where climatic fluctuations may have been of such low amplitude that they impose minimal levels of stress on the trees, there may be almost no variation in ring width through time. Tree-ring sequences of this type are termed **complacent series** and are of little value in dendrochronology, because marker rings or marker series of rings that form the basis for cross-dating cannot be identified. It is for this reason that dendrochronology is most readily applied to trees that grow or have previously grown in areas where some degree of climatic stress (low temperatures, shortage of moisture, etc.) generates **sensitive series** of tree rings. In such situations, however, difficulties may also arise for

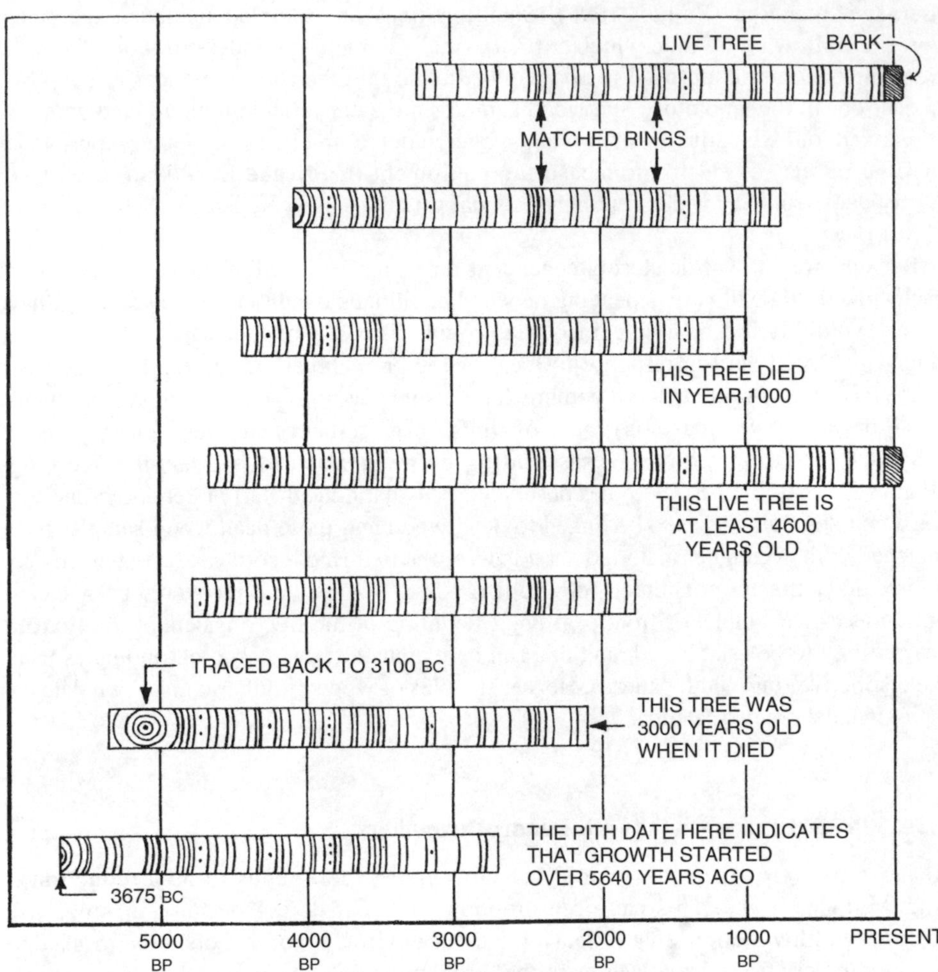

Figure 5.1 *Cross-dating of tree rings. Distinctive ring-width patterns enable living trees to be connected to dead wood, and dead wood to sub-fossil wood, thereby developing a continuous chronology which may extend back over several thousand years after Johnson [1999]*

if the stress is too extreme the tree may simply not produce a ring in a given year, or may only produce new cellular material on a part of the trunk. In these cases, therefore, there may be **missing** or **partial rings** in the tree-ring series. On the other hand, where the spring growth period is interrupted by, for example, severe late frosts, it is possible for more than one set of cellular structures to develop in one year. These are referred to as **false rings** or **intra-annual growth bands**, and may not be easy to recognise in individual tree-ring series. The only way in which missing and false rings may be identified and corrected for is by replication of such records through careful and systematic cross-dating.

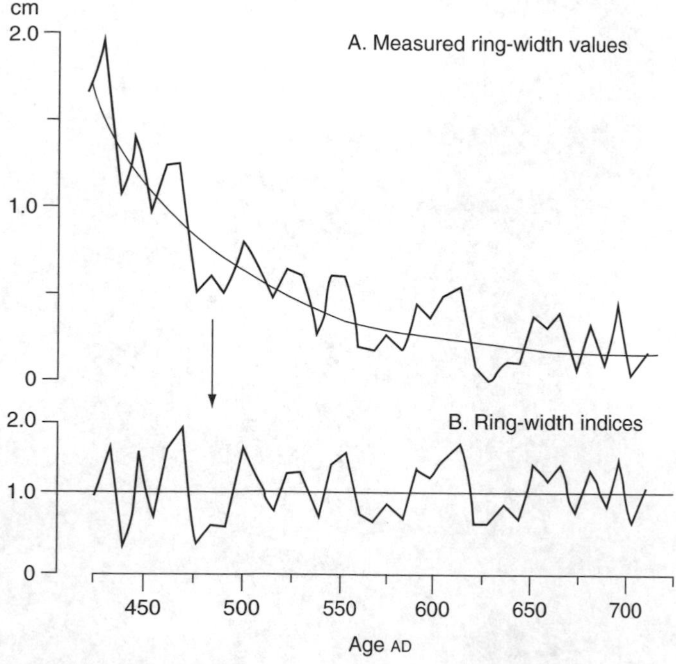

Figure 5.2 *Standardisation of ring-width measurements to produce ring-width indices. In (A) a regression line has been fitted to the measured ring widths and this provides an indication of the general decline in ring width with age. The value for each year is then divided by the value for that year obtained from the regression curve. (B) This generates a series of* ***ring-width indices*** *which have been corrected for any ageing effect, and hence the fluctuations in the tree-ring curves reflect the influence of environmental factors only (after Baillie, 1982). Reproduced by permission of Croon Helm*

5.2.3 Dendrochronological Series

The early dendrochronological work of Douglass and Shulman referred to in section 5.2 was based very largely on an extraordinary tree, the bristlecone pine. These trees grow throughout the southwestern United States, and thrive on dry and rocky sites at altitudes of up to 4000 m. They are particularly adapted to arid environments, and in the White Mountains of California, in the rain shadow of the Sierra Nevada, large numbers of twisted and stunted bristlecone pines are found (Figure 5.3), the oldest of which is more than 4700 years old. This makes them the oldest living tree in the world. These long-lived trees grow under particularly stressed conditions and form thin, highly sensitive tree-rings (Figure 5.4). This combination of longevity and sensitivity produces a perfect record of past climatic variations (Johnson, 1999). By cross-dating between living trees and dead wood, a continuous master chronology has been developed that extends back to 8681 years BP (Ferguson and Graybill, 1983). Other long tree-ring chronologies from North America include those on *Sequoia*, Douglas fir and pine. The longest of these, on foxtail pine (*Pinus balfouriana*), goes back to 3031 years, but may eventually be extended to 6000 years (Scuderi, 1987; 1990).

Figure 5.3 *Bristlecone pines growing in the White Mountains of California (photo: Mike Walker)*

There are no trees in western Europe that have a lifespan that is equivalent to the bristlecone pine, and hence long dendrochronological series have only been reconstructed through the systematic cross-dating of subfossil wood samples with ring patterns of less than 200 years. There are continuous records for England back to 6939 years (Baillie and Brown, 1988) and for Ireland back to 7429 years (Brown and Baillie, 1992). Both of these

Figure 5.4 *A section through a fallen bristlecone pine showing the very narrow growth rings (photo: Mike Walker)*

are based on oak (*Quercus robur* and *Q. petraea*). In central Europe, subfossil remains of oak and pine (*Pinus sylvestris*) found in the deposits of large rivers in southern and eastern Germany have been used to build a continuous Holocene oak chronology back to 10 430 years, while a pine chronology, linked to this oak chronology, extends the dendrochronological record back to 11 919 years (Spurk *et al.*, 1998; Friedrich *et al.*, 1999). In addition, a 1051-year-long 'floating pine chronology' has been developed for the Lateglacial period, which covers the interval between ca. 11 350–12 300 years BP of the radiocarbon timescale (Friedrich *et al.*, 2001). It is anticipated that, in due course, this chronology will be linked to the Holocene oak and pine chronology, and this would then provide a continuous record extending back over 14 300 years. Long continuous pine-based chronologies have also been obtained from northern Sweden (Grudd *et al.*, 2002), and northern Finland (Eronen *et al.*, 2002), the latter extending back 7519 years.

The most important repository for tree-ring data is the International Tree-Ring Data Bank (ITRDB) which is housed at the World Data Center-A for Paleoclimatology at the National Geophysical Data Center in Boulder, Colorado, USA. Founded in 1974, the ITRDB contains more than 3275 tree-ring chronologies from over 1500 sites worldwide, and makes its holdings freely available to all scientists working on dendrochronology or related fields (Grissino-Mayer and Fritts, 1997).

5.2.4 Applications of Dendrochronology

Dendrochronology has been applied in a wide range of contexts. Because long and continuous records can be generated, dendrochronology enables precise ages to be determined for a range of Holocene events. These include, *inter alia*, volcanic eruptions, major shifts in climate, and the rise and fall of civilisations. The technique has proved especially valuable in archaeology, allowing ages to be assigned to such diverse objects as wooden buildings, prehistoric trackways buried beneath peats, and ships' timbers and other maritime artefacts (Baillie, 1995). Tree rings contain a range of potential measures of past climate (McCarroll *et al.*, 2003), and dendroclimatology is providing new insights into Holocene climate change which are not only important from the point of view of historical climate reconstruction (Briffa, 2000), but the data constitute important baselines for atmospheric scientists who are modelling past and future climatic behaviour (Briffa and Matthews, 2002). Perhaps most important of all, however, has been the calibration of the radiocarbon timescale by dendrochronology, which was discussed in section 2.6.1. Some examples of these various applications of dendrochronology are considered in the following sections.

5.2.4.1 Dating a 2000-year temperature record for the northern hemisphere. Precisely dated dendroclimatological records are now available for many parts of the world, and Figure 5.5 shows a composite of several such curves from high-latitude regions in the northern hemisphere (Briffa, 2000). The individual records upon which this composite is based are from Mongolia, Siberia, the northern Urals, northern Sweden, northern North America and the Canadian Rockies. The normalised composite curve provides an indication of relative temperature changes (a mixture of predominantly summer and some annual signal) at high latitudes, mostly between 60°N and 70°N, throughout the last 2000 years. This exclusively tree-ring-based 'Northern Chronology Average' shows a number of clearly defined climatic phases and shifts, many of which have been identified in other proxy climate records. These include the pronounced fall in temperature during the sixth century AD and the subsequent period of cooler climate during the Dark Ages, significantly warmer conditions during the Medieval Warm Period from the tenth to the twelfth centuries, the Little Ice Age cooling of the post-Medieval, and the marked temperature rise that began in the nineteenth century and became even more pronounced during the course of the twentieth century.

5.2.4.2 Dating historical precipitation records. In addition to providing data on temperature change, tree-ring records can be used to infer past changes in precipitation and,

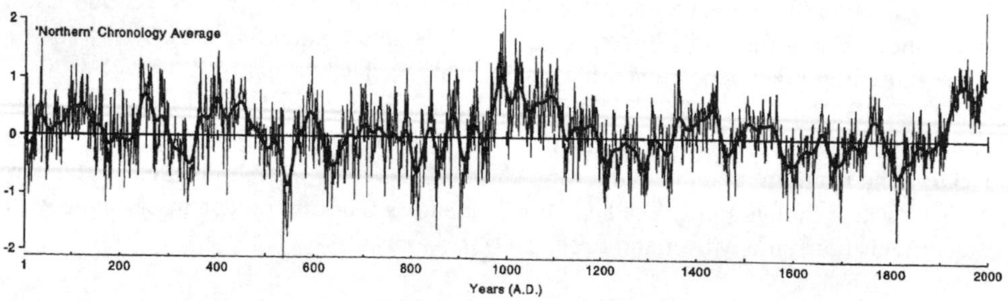

Figure 5.5 *Northern 'high-latitude' average temperature changes over the last 2000 years based on dendrochronological data from a number of sites in northern Asia, northern Europe and North America (after Briffa, 2000). Reproduced with permission of Elsevier*

once again, these palaeoprecipitation reconstructions can be precisely dated using dendrochronology. Watson and Luckman (2001) describe such a record from the southern Canadian Rockies. Seven Douglas fir (*Pseudotsuga menziesii*) tree-ring chronologies were developed from sites in Banff and Jasper National Parks, and from near Cranbrook in British Columbia. By comparing measured ring-width variations with temperature and precipitation records from nearby meteorological stations, a close statistical relationship was established between tree growth (as reflected in ring-width changes) and precipitation. It also proved possible to calibrate ring-width variation with precipitation levels so that a quantitative estimate of past changes in precipitation could be determined. Long-term trends in the three records are very similar to each other and also to other reconstructions from adjacent areas. They show that regionally extensive drier intervals extended over the southern Rockies through the AD 1700s (especially in the 1760s and 1790s), the 1850s–1860s, the 1890s, and during the historically documented drought of the 1920s–1940s. In the longest of the records, that from Banff (Figure 5.6), significantly drier intervals also occurred in ca. AD 1470–1510, in the 1560s–1570s and the 1630s–1650s. These reconstructions indicate regionally coherent precipitation patterns that fluctuate on decadal timescales and may be linked to changes in atmospheric circulation patterns.

5.2.4.3 Dating volcanic events. It has often been speculated that volcanic activity can exert an influence on climate (Lamb, 1995). Initially it was thought that a reduction in global temperature would result from the screening out of incoming solar radiation by particles of volcanic dust, but it is now believed that the ejection into the atmosphere during a volcanic eruption of large quantities of sulphur is likely to result in the creation of a more effective and long-lasting radiation screen (Bluth *et al.*, 1993). Indeed, recent research suggests that sulphur volatiles from volcanic eruptions can lead to a cooling of global climate by 0.2–0.3 °C for several years after the eruptive event (Zielinski, 2000).

Such cooling may well be reflected in tree-ring data. In the 1970s, work on the bristlecone pine revealed the presence of tree rings that exhibited signs of frost damage, reflective of unusually severe winters, and the fact that these tended to coincide with the

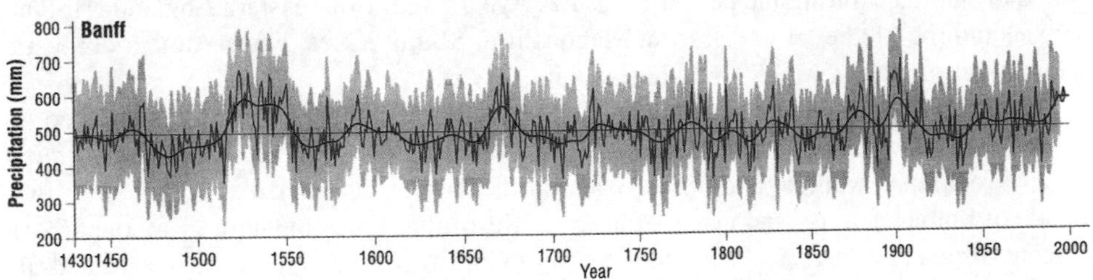

Figure 5.6 *Precipitation reconstruction for Banff, Alberta, Canada, for the period 1430–1994 based on dendroclimatological data. The mean precipitation for the period is indicated by the horizontal line, while a 25-year filter (smooth continuous line) reflects the long-term precipitation trend. A 2 standard error (2σ) confidence interval (shaded area) on the annual precipitation reconstructions is also shown (after Watson and Luckman, 2001). Reproduced by permission of Arnold Publishers*

known dates of volcanic events led to the suggestion that the trees were recording the climatic effects of large-scale volcanic eruptions (LaMarche, 1974). More recent tree-ring data from a range of northern hemisphere sites show a very close correlation with the GISP2 ice-core record (section 5.5) for the last 600 years, the lowest tree-ring density values (reflecting colder conditions) corresponding with episodes of explosive volcanism that are clearly recorded in the ice core, and which are especially marked during several decades of the 1600s and 1800s (Briffa *et al.*, 1998). Similar relationships between tree-ring width and volcanic episodes have been noted for earlier periods (Baillie and Munroe, 1988).

A good example of the detection and dating of a particular volcanic eruption in tree-ring series is the AD 536–545 event (Baillie, 1995). Tree-ring records from both Europe and North America show a significant decrease in ring widths during this time period, which appears to coincide with a pronounced decline in temperature (Figure 5.7). There is also a range of other proxy climatic and archaeological evidence for a climatic anomaly at this time, and in the Dye-3 ice core a marked acidity layer (indicating sulphuric acid fallout from the atmosphere) is dated at 540 ± 10 years BP. Although there have been suggestions of other causal factors for this period of short-lived climatic deterioration (the effects of cometary impact, for example), the evidence points strongly towards a massive volcanic eruption. The most likely candidates appear to be either Rabaul in New Britain or White River in Alaska, both of which have been radiocarbon-dated to ca. AD 525–540. In either case, the climatic consequences of this eruptive event are clearly registered in northern hemisphere dendrochronological records.

5.2.4.4 Dating archaeological evidence. Dendrochronology can be very useful in archaeology because it enables events to be dated to a precision that may be unattainable using conventional methods such as typology or even radiocarbon (Brunning, 2000). For example, dendrochronological dating of the Sweet Track, a Neolithic trackway found beneath buried peats on the Somerset Levels in southwest England, showed that the wood used to construct the trackway was cut in winter 5757 years ago or spring 5756 years ago, and that the Post Track, a structure that ran on a course parallel to the Sweet Track, was 31–32 years older (Hillam *et al.*, 1990). Dendrochronological dating has also been carried out on boat timbers. British examples include the Iron Age Hasholme logboat (timbers felled or that died during the period 2272–2227 years ago) from eastern England (Hillam, 1992), and the Medieval wreck from Magor Pil in South Wales, whose timbers indicate a construction date of AD 1233–1278 (Nayling, 1995).

A good example of an archaeological application of dendrochronology is provided by work on the Severn Estuary Levels in South Wales (Hillam, 2000). The Levels are an area of reclaimed wetland and on these and in the immediate intertidal area there is evidence of human activity and occupation from Mesolithic times onwards (Rippon, 1996). Recent excavations near Goldcliff to the east of Cardiff (Figure 7.15A) have revealed the presence of wooden trackways and a group of eight timber buildings (Figure 5.8) believed to be of Iron Age date (Bell *et al.*, 2000). Dendrochronological records were obtained from the timbers of some of the buildings and trackways, and these were matched against dated records from elsewhere in Britain. Cross-dating enabled ages to be assigned to the timbers of three of the buildings and one of the trackways (Table 5.1), which not only confirmed an Iron Age date for the structures, but showed that human activity may have continued into the Romano-British period.

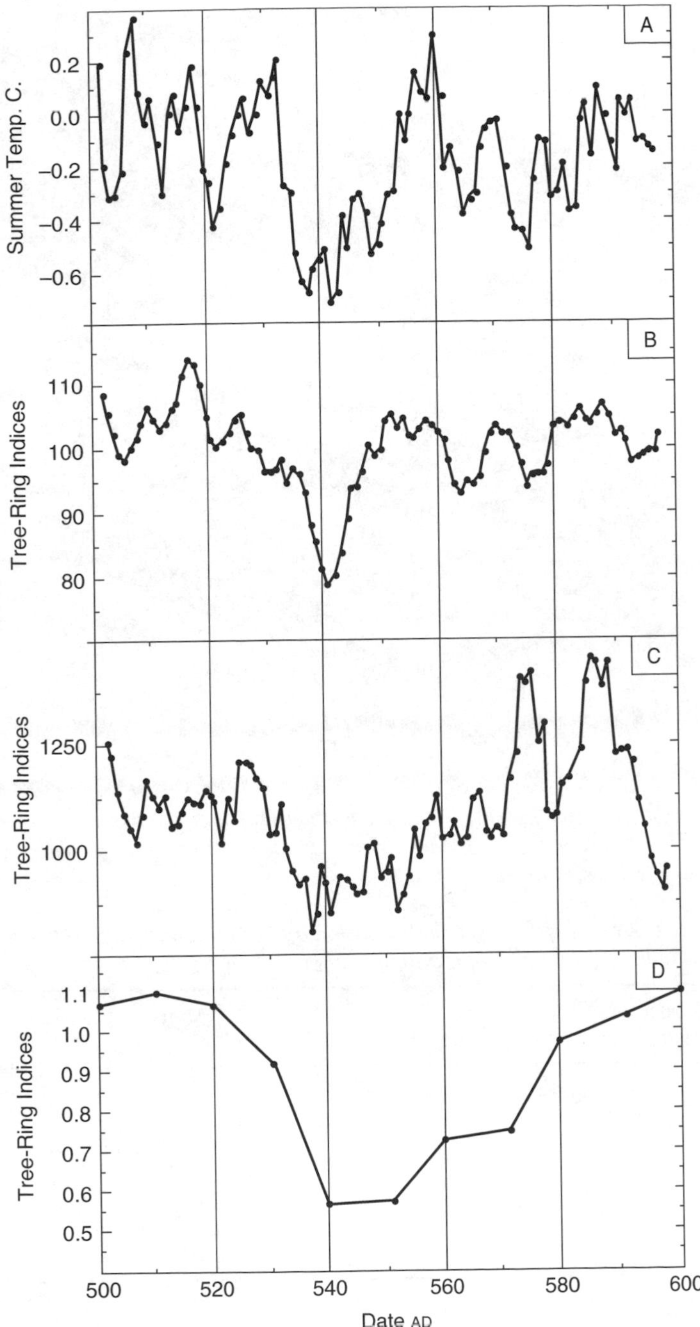

Figure 5.7 *Dendrochronological records from western Europe and North America showing the registration of the* AD *536–545 event. (A) Fennoscandian temperature record derived from tree-ring data. (B) European oak ring-width indices. (C) American bristlecone pine ring widths. (D) American foxtail pine ring widths (after Baillie, 1995). Reproduced by permission of Batsford, London*

Figure 5.8 *Goldcliff Building 1, South Wales: the well-preserved corner of an Iron Age rectangular building showing walls, flooring and internal subdivisions, possibly animal stalls. Split oak timbers forming the subdivisions are dendrochronologically dated 593–392* BC *scale 1 m (photo: J. Parkhouse)*

Table 5.1 *The dating of timber structures at Goldcliff, South Wales, as indicated by dendrochronology (after Hillam, 2000)*

Bronze Age Boat planks from Structure 1124	After 1017 BC
Iron Age	
Building 2	After 454 BC
Building 1	After 382 BC and possibly before 342 BC
Trackway 1108	336–318 BC
Building 6	Spring 273 BC
Building 6 (?) repairs	Spring 271 BC

5.3 Varve Chronology

A characteristic feature of many lake sediment sequences is the presence of regularly laminated sediments consisting of thin, horizontally bedded layers of different structure and texture. Often, these laminae are arranged in couplets with relatively coarse-grained layers alternating with finer-grained bands. In other situations, the couplets may consist of alternations between organic and inorganic laminae, or between sedimentary units rich in diatoms or iron oxides and thin sedimentary bands that are relatively deficient in these components. Such regular, or rhythmical, variations in sediment accumulation reflect seasonal variations in sediment supply to, or in chemical or biogenic processes within, the lake ecosystem. In many cases, the couplets will reflect the annual cycle of deposition, in which case they are termed **varves**.[2] Because they are deposited annually, varves form a basis for dating, for by counting varve sequences, time intervals can be established and a 'floating' chronology (section 5.2.1) developed. If varve series can be dated by radiometric or other means (by radiocarbon dating of included organic materials, for example), then the floating varve chronology can be linked to the calendar timescale.

The potential of varves as a basis for dating was initially recognised by the Swedish geologist Gerard de Geer (Figure 1.1) who, in 1884, made the first attempt to count and correlate varve sequences in the Stockholm area of Sweden. His aim was to use the varved sequences that had accumulated in front of the decaying Fennoscandian ice sheet to establish a timescale for deglaciation (section 5.3.3.1). Varve chronology has subsequently been applied in a range of other lacustrine contexts, and useful reviews (including applications) of the method are provided by Saarnisto and Kahra (1992), Kemp (1996) and Lamoreux (2001).

5.3.1 The Nature of Varved Sediments

There are several types of varved sediment. Perhaps the best known are **clastic varves**[3], which consist of alternations of coarse-grained layers (sands and silts) and finer-grained bands (silts and clays). Good examples of these are found in glaciolacustrine environments, in other words where sediments are accumulating in a lake in proximity to a glacier ice margin (Figure 5.9). During the spring and summer months, when the lake is ice-free, sediment is discharged into the lake and forms the coarser layer of the varve couplet. In winter, when the lake is frozen, deposition results solely from the settling out of fine silts and clays suspended in the water column which produces the finer component of the varve couplet. In some situations, **clastic-organic varves** develop. These reflect an annual cycle of erosion of mineral matter from the lake catchment and inwashing of this material during spring floods, followed by the accumulation of particulate organic material (e.g. plants and algae) in summer and autumn. The latter grades upwards into a fine-grained organic layer that accumulates during the winter months when the lake may be frozen (Ojala and Tiljander, 2003). In other lakes, purely **organic varves** may form. Here the varve couplet consists of a thin dark brown layer which is reflective of slow accumulation during the winter months, and a lighter-coloured organic horizon which forms as a result of higher levels of internal productivity within the lake ecosystem during the warmer summer months (Simola, 1992). Seasonal variations in precipitation of $CaCO_3$ (**calcareous varves**) may contribute to this colour distinction between the varve components, as also may summer blooms of aquatic organisms such as algae, chrysophytes[4] and diatoms (**diatom-rich varves**). In addition, seasonal variations in iron oxide precipitation may lead to the formation of **iron-rich varves** (Renberg, 1981).

Figure 5.9 Glaciolacustrine varves exposed in a section at Nummi-Pusula, southern Finland. These varved sediments were deposited in the Baltic Ice Lake which formed during the northward retreat of the last Fennoscandian Ice Sheet (photo: Jari Väätäinen, Geological Survey of Finland, GSF)

Varves can also form in ocean sediments. In glaciomarine environments, for example, where glaciers are calving into the sea, fine-grained laminated deposits are relatively common (Ó Cofaigh and Dowdeswell, 2001). These sometimes consist of couplets of laminated muds and diamict[5], the former reflecting meltwater discharge and subsequent settling of suspended sediment during the summer months, while the diamicts are deposited in winter when meltwater discharge is reduced and ice berg sedimentation dominates (Cowan *et al.*, 1997). These annual couplets can therefore be classified as **glaciomarine varves** (Cowan *et al.*, 1997). In deeper water, annual variations in ocean productivity related, for example, to upwelling are reflected in laminated sediment sequences, where lighter layers (which are rich in microorganisms such as diatoms) may be interbedded with darker, diatom-poor horizons (Hughen *et al.*, 1996). Together, these lighter and darker layers represent annual cycles of deposition and can therefore be considered as **marine varves**. Laminated sediments have also been found in some shallow-water marine contexts, such as estuaries, where individual couplets of finer and coarser sediments may reflect seasonal variations in sedimentation resulting, for example, from variations in river discharge, storminess or sea-water temperature. Here too an annual signal may be preserved within the sediment sequence (Allen, 2004).

As was the case in De Geer's time, counting of varved sediments can be carried out in the field, but most contemporary varve counting is now undertaken in the laboratory using either core samples or sediment monoliths taken from open sections. In the uppermost parts of lake sequences where sediments may be loosely compacted, undisturbed samples may be difficult to obtain using conventional coring techniques. In these situations, a specially adapted corer has been developed which allows the sediments to be frozen *in situ*, thereby preserving the fine structural features in the core (Saarnisto, 1986). In the laboratory varve couplets can often be identified by the eye, or with the aid of conventional light microscopy. Increasingly, however, varve counting involves more sophisticated approaches involving, for example, thin section analysis, scanning electron microscopy, photography and digital imaging techniques, and X-ray densitrometry (Lindeberg and Ringberg, 1999; Lotter and Lemcke, 1999; Ojala and Francus, 2002).

5.3.2 Sources of Error in Varve Chronologies

In a number of respects, problems encountered in varve chronology resemble those in dendrochronology. For example, in the same way that false rings can be produced in trees, more than one set of laminae (**sub-laminations**) can develop within an annual increment of lake sediment. This is especially likely in dimictic lakes[6], where a grey silt layer may mark either a spring or an autumn turnover, and organic bands may be deposited during both summer and winter. Iron precipitates may add to this doubled structure, with brown ferric hydroxides formed during spring and autumn turnover periods and black sulphides forming during intervening stagnation episodes (Simola, 1992). Sub-laminations may also form as a result of diurnal variations in sedimentation, coarser material being deposited by day and finer sediments at night (Ringberg, 1984), through episodic sedimentation from local wind-driven currents (Catto, 1987), or as a result of high-rainfall events or two or more periods of snowmelt during the course of a year (Lamoureux *et al.*, 2001). Alternatively, varves can be disturbed by bioturbation on the lake bed or, in some cases, may be removed entirely from lake floor sequences by slumping or by erosion during flood events (Leonard, 1986).

One way in which some of these problems can be addressed is by replication of varve series in multiple cores from a lake sequence. As in dendrochronology, correlation (**cross-dating**) between cores can be achieved using well-developed marker varves as pinning points for the chronology. This may help to resolve difficulties arising from sub-annual or missing varves. More problematical, however, are inter-site correlations (**teleconnections**) because matching varve thicknesses is usually only possible over limited geographical areas as local climatic and site-specific factors may result in quite different varved sequences, even in adjacent regions (Lowe and Walker, 1997).

5.3.3 Applications of Varve Chronologies

Varved or annually laminated sediment sequences are extremely valuable archives of palaeoenvironmental information because they enable events to be dated with a high degree of precision. In addition to providing a timescale for deglaciation, varve chronology has been employed in studies of climate change, vegetation change and land-use histories. Varved sequences have also been used to calculate rates of sediment influx into lakes and rates of erosion from their catchments, and they have also aided the monitoring of processes in the modern environment, such as heavy metal pollution, eutrophication and studies of the effects of 'acid rain' (O'Sullivan, 1983). In addition, varved sequences form a basis for calibrating the radiocarbon timescale (section 2.6). Some of these applications are considered in more detail in this section.

5.3.3.1 Dating regional patterns of deglaciation in Scandinavia. Since the pioneering work of De Geer (1912), numerous studies of varves have been undertaken in eastern Sweden and in other areas around the Baltic Sea. During the wastage of the great Fennoscandian ice sheet, a large freshwater lake existed in what is now the Baltic Sea. The eastern and southeastern margins of the ice sheet terminated in the 'Baltic Ice Lake' and glaciolacustrine clastic varves formed in close proximity to the ice margins. As the ice sheet wasted northwards and westwards additional varves were created, mainly as deltaic sediments in estuaries of large rivers, and by counting the varves and linking individual varve sequences, a 'master chronology' for deglaciation has been established. In De Geer's chronology, the starting point for the timescale was a distinctive varve in the Indalsälven Valley of north-central Sweden which he considered to reflect the sudden input of large quantities of meltwater, and which he took to mark the transition from glacial to postglacial conditions. De Geer termed this the 'zero' varve year, and older varves were given negative numbers and younger varves positive numbers in relation to this key varve horizon.

De Geer's master chronology has subsequently been repeatedly revised as more data have become available (e.g. Strömberg, 1994; Brunnberg, 1995), while connection of the upper part of the varve sequence to the present (AD 1978) has enabled a calendrical age of 9238 BP to be assigned to the 'zero year' (Cato, 1985). The Swedish Time Scale forms a more or less continuous record from ca. 13 300 varve years BP to the present, although recent work suggests that there may be gaps in the sequence, with more than 800 varve years missing from the postglacial chronology (Wohlfarth *et al.*, 1997; Andrén *et al.*, 1999). This problem can be addressed by employing varve data in combination with other lines of evidence, including pollen-stratigraphic records and calibrated radiocarbon dates from varved clays (Björck, 1999; Lundqvist and Wohlfarth, 2001), and correlation with independently dated records such as those from the Greenland ice cores (Andrén *et al.*, 1999). These approaches, which enable local varve chronologies to be developed for particular areas, each of which can be linked to calendar time via other dating methods (Wohlfarth *et al.*, 1998b), mean that clay varve chronology continues to form an important basis for establishing patterns of regional deglaciation in those areas of Scandinavia adjacent to the former Baltic Ice Lake (Figure 5.10).

5.3.3.2 Dating prehistoric land-use changes. As noted above (sections 2.7.1.4, 3.5.5.2), the depositional sequence on a lake bed contains a record of landscape changes that

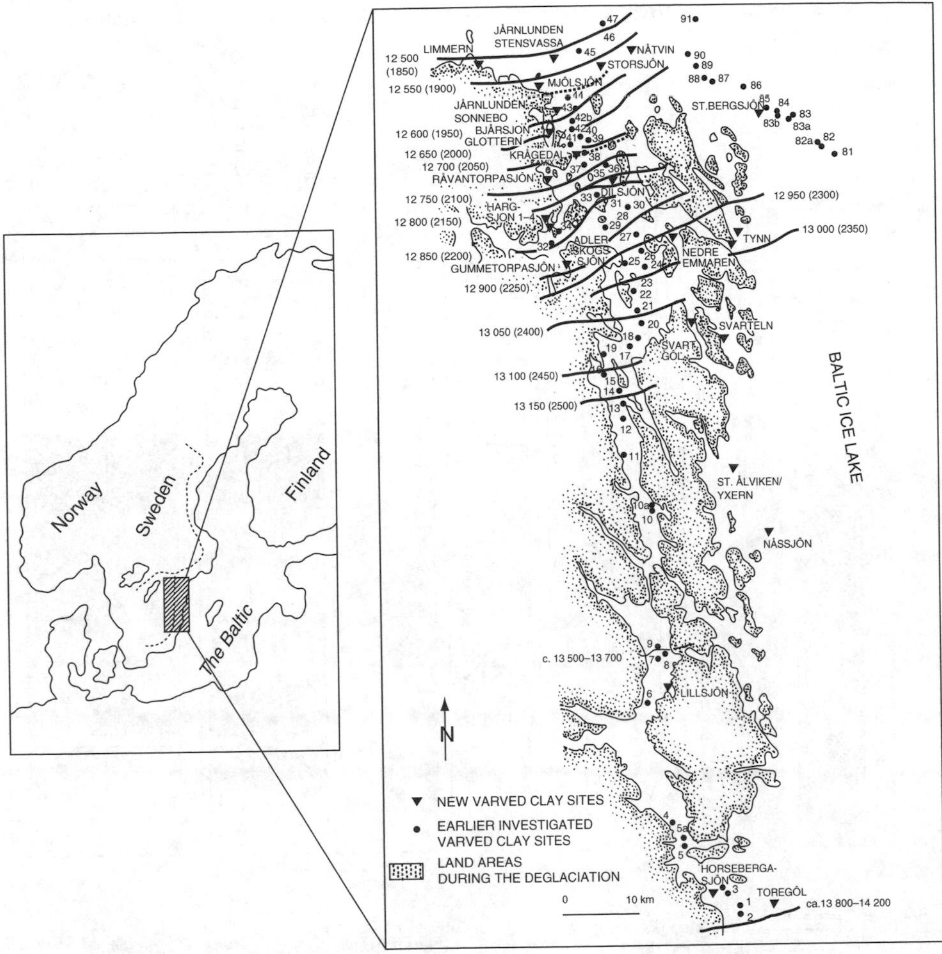

Figure 5.10 *Ice recession lines for part of southeast Sweden based on a combination of varve chronological and radiocarbon data. The dates are in calendar years* BP, *and the figures in brackets refer to the local varve years (after Wohlfarth* et al., *1988b). Reproduced by permission of Taylor & Francis AS, www.tandf.no/boreas*

have occurred around the lake catchment. These will be reflected in the physical and chemical composition of the sediments (which provide evidence, for example, of episodes of erosion slopes around the lake), and especially in the contained biological record, notably pollen and plant macrofossils, which reflect changes in vegetation cover and in land use in the vicinity of the lake. Varve chronology has proved to be especially valuable in providing a timescale for these local landscape changes.

One such record is that from Lake Jues in the southern Harz Mountains of central Germany (Zolitschka *et al.*, 2003). The lake contains an annually laminated sediment sequence which has been linked to calendar time by dendro-calibrated radiocarbon dates (Figure 5.11) and a detailed palynological record has been obtained from the

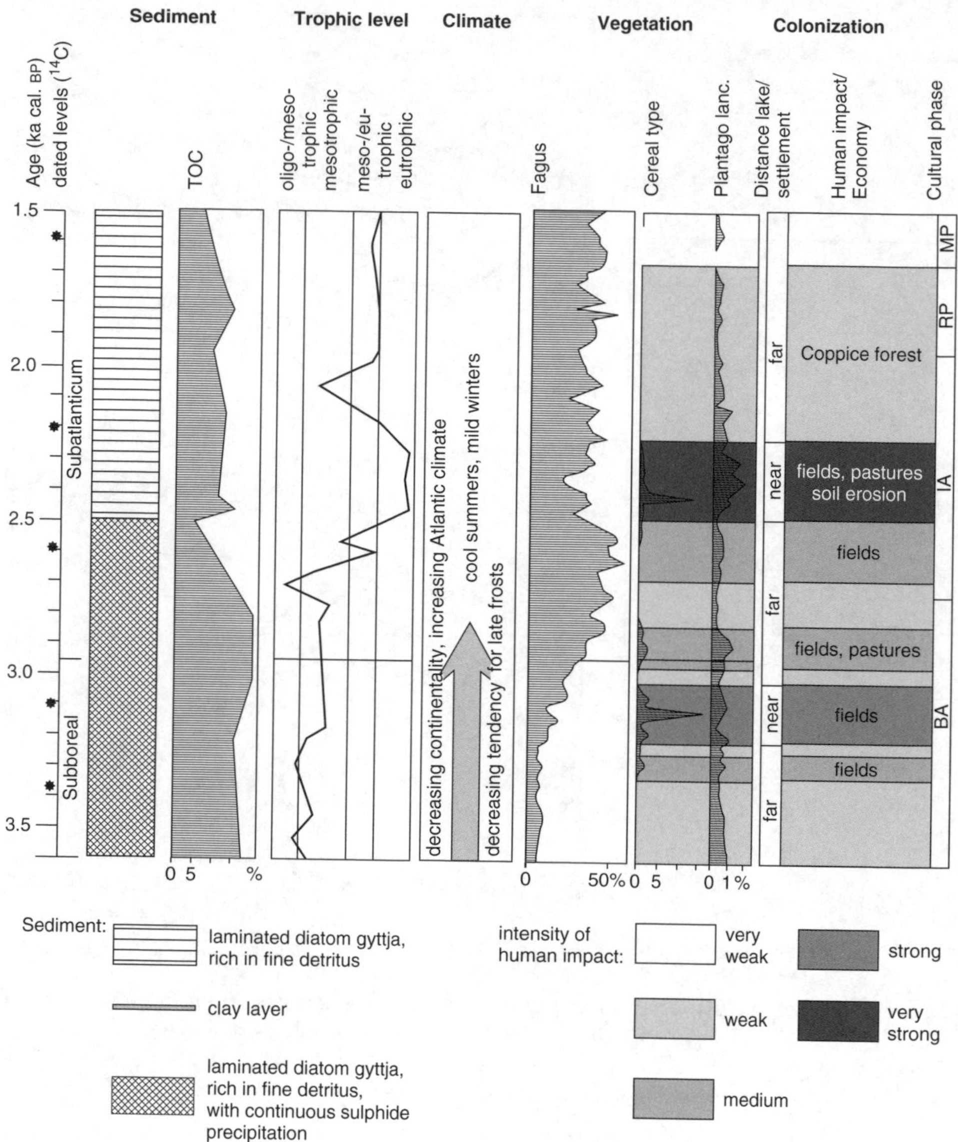

Figure 5.11 *Lacustrine sediment record from Lake Jues, Harz Mountains, Germany. Episodes of agricultural activity are reflected in increases in cereal pollen, and in pollen of* Plantago lanceolata. *The latter is an 'agricultural weed' and is often associated with pastoralism. BA: Bronze Age; IA: Iron Age; RP: Roman period; MP: Migration period (after* Zolitschka et al., *2003). Reproduced with permission of Elsevier*

deposits. This shows evidence of vegetation change and human activity in the hinterland of the site from the Bronze Age (around 3600 cal. years BP) through to the end of the Roman period. Human impact is evident from around 3350 cal. years BP, and during the Bronze Age three settlement phases can be clearly distinguished, each associated with an increase in agricultural activity. Woodland clearance for agricultural purposes enabled beech trees to colonise the cleared areas, as reflected in the rise in the pollen curve for *Fagus* between 3600 and 2950 cal. years BP. During the Iron Age, anthropogenic influence increased markedly with further deforestation leading to accelerated soil erosion. This intensive phase of agricultural activity, which appears to have been associated with a settlement in the vicinity of the lake, dates to between ca. 2500 and 2300 cal. years BP. Less intensive human impact is recorded during the later Iron Age and in the Roman period.

5.3.3.3 Dating long-term climatic and environmental changes. As well as providing records of local catchment histories, lake sediment sequences contain evidence of changes through time in the lacustrine ecosystem, which may reflect not only processes operating around the lake catchment (erosion, run-off, etc.), but also broader-scale environmental and climatic influences. In addition to biological proxies, such as pollen, climatic signals may be detectable in biochemical records, for example in carbon and oxygen isotope profiles (Street-Perrott *et al.*, 1997; Von Grafenstein *et al.*, 1999). These data are often available at high levels of resolution and over relatively long time periods, and varve chronology can be used to develop a high-precision timescale for such records.

Lücke *et al.* (2003) describe a composite varve-dated 11.4-m-long sediment sequence from Lake Holzmaar in the Eifel region of western Germany, which extends back over 14 000 years, and from which a detailed carbon isotope record has been obtained (Figure 5.12). Variations in $\delta^{13}C$ values reflect changes in the lake ecosystem (dissolved CO_2, nutrient availability, light, climate, water temperature, etc.) and constitute a proxy for environmental and climatic change. The mean time resolution for the isotope sequence is 14 years. The record shows clear evidence of a number of major ecosystem

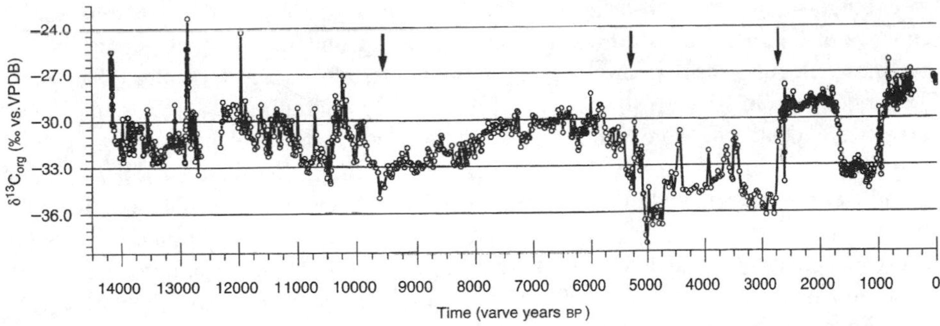

Figure 5.12 *The Lateglacial and Holocene carbon isotope record ($\delta^{13}C$) from Lake Holzmaar, western Germany. The shaded areas indicate distinctive periods in the isotopic record, while the arrows mark significant changes in $\delta^{13}C$ (after Lücke* et al., *2003). Reproduced with permission of Elsevier*

reorganisations, the major changes at around 14 200, 9600, 5500 and 2700 varve years being interpreted as reflecting major changes in climatic regime. The steady rise in $\delta^{13}C$ values during the Mid-Holocene is taken to represent a continuous climatic amelioration, reaching an optimum around 6500 varve years BP. Marked fluctuations in $\delta^{13}C$ over the past 2700 years appear to reflect human-induced changes in the lake catchment, relating either to deforestation, reforestation or run-off changes. Interestingly, a number of the major Holocene shifts in $\delta^{13}C$ correspond in time with major reorganisations in climate detected in other proxy data (e.g. Bond *et al.*, 1997), suggesting that the Holzmaar lacustrine record reflects climatic forcing at the hemispherical scale.

5.3.3.4 Varve sequences and the radiocarbon timescale. We have already seen how dendrochronology can be employed in the calibration of the radiocarbon timescale (section 2.6.1). However, as described above, continuous dendrochronological series currently only extend back to ca. 11 900 years BP, and hence alternative bases for calibration have to be found for the timescale prior to that date. Some of these were discussed in Chapter 2, where it was noted that one approach is to use varve chronology. Two varved sequences are considered here.

The INTCAL98 radiocarbon calibration (section 2.6.2) incorporates a varve series, namely the radiocarbon-dated marine varve sequence from the Cariaco Basin, Venezuela. There, the lighter layers in the varves, which contain large numbers of planktonic microfossils (mainly diatoms), reflect increased productivity over the basin during the winter–spring upwelling season, while the darker laminae are dominated by terrestrial material and are an indicator of run-off from the northern South American continent during the late summer–autumn rainy season (Hughen *et al.*, 1996). Radiocarbon dates from individual varves were obtained on species of planktonic foraminifers, and the 'floating' varved sequence was anchored to the calendar timescale by 'wiggle-matching' (section 2.6.5) the radiocarbon age versus calendar age variations to the tree-ring data (Hughen *et al.*, 1998). The calibration based on the Cariaco varve series (Figure 5.13A) covers the period from ca. 9000 to 14 500 cal. years BP.

An even longer radiocarbon calibration may be possible using the varve series from Lake Suigetsu in Japan (Kitagawa and van der Plicht, 1998; 2000). In that lake a sequence of sediments consisting of dark-coloured clays interbedded with lighter layers, the latter reflecting spring season diatom growth, extend to a depth of more than 75 m. The sequence from 10.42 to 30.45 m was scanned using digital image analysis and 29 100 varves were identified. More than 300 AMS radiocarbon dates on plant macrofossils have been obtained from the profile. The absolute age of the Lake Suigetsu floating varve chronology was determined by wiggle-matching 22 radiocarbon determinations from the younger part of the sequence with the German oak radiocarbon calibration curve. On the basis of this match, the Lake Suigetsu timescale covers the absolute age range from 8830 to 37 930 cal. years BP (Figure 5.13B). The accumulated counting error in the older part of the Lake Suigetsu varve chronology is estimated at 2000 cal. years, which means that the series cannot yet be regarded as a reliable calibration curve (van der Plicht, 2002). However, further refinement and revision of the chronology may, in due course, reduce the extent of the error. If so, then the Lake Suigetsu varve record could form the basis for a high-resolution calibration that potentially spans the complete radiocarbon-dating range (see also section 2.6.3).

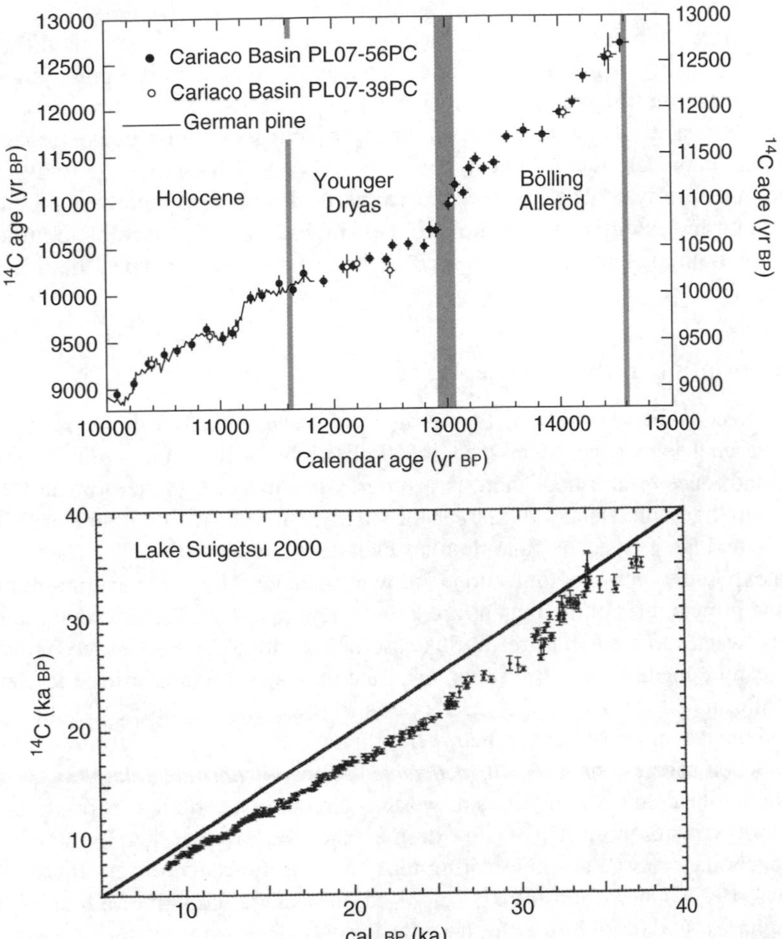

Figure 5.13 *Radiocarbon calibration curves from two varved sequences: (A) Radiocarbon versus calendar ages for the period 9000–15 000 cal. years BP based on evidence from the Cariaco Basin, Venezuela: solid line is German pine data (after Hughen* et al.*, 1998) Reproduced by permission of Radiocarbon (B) Radiocarbon versus calendar ages for the period 8830–37 390 cal. years BP based on data from Lake Suigetsu, Japan (van der Plicht, 2002). Reproduced by permission of Netherlands Journal of Geosciences*

5.4 Lichenometry

Lichens are composite plants consisting of algae and fungi. The association between the two is generally considered to be **symbiotic** in that both the fungus and the alga derived mutual benefits from the association. The alga furnishes food for the fungus which, in turn, supplies moisture and shelter for the alga. The body of a lichen is termed a *thallus*, and in some the alga and fungus are uniformly distributed through it. In others, the thallus is internally differentiated into zones with the green algae sandwiched between the fungal elements. Lichens can be divided into three broad groups based on the shape of the vegetative body: the **fruiticose** type consisting of small tubules and branches, the **foliose** type which have a leaf-like plant body, and the flattened **crustose** type. The last named are the most common members of the lichen family, and grow extensively on hard surfaces, including rock outcrops, boulders, tree bark, poles, buildings and gravestones. It is the crustose lichen types that are used in lichenometry.

5.4.1 Principles of Lichenometric Dating

That lichens could be used as basis for dating was first suggested by the Austrian scientist Roland Beschel in a paper published in 1950. Lichens are rapid colonisers of bare surfaces and, once established, there is a progressive increase in size of the thallus by slow marginal growth. Hence, the larger the thallus (in terms of its diameter), the older the lichen, and the greater the time that has elapsed since colonisation. Where a surface has been exposed to lichen colonisation (following glacier retreat, for example), provided that (a) the growth rates of the lichens are known, and (b) no significant time interval has elapsed between surface exposure (in this case deglaciation) and lichen colonisation, an estimate can be made for the timing of substrate exposure by measuring the size of the lichen thalli on that surface. It was on the basis of these two key principles that Beschel developed the technique of lichenometry (Beschel, 1973).

Some lichen species (such as *Rhizocarpum geographicum*) are relatively long-lived. Indeed, under the cold dry conditions of western Greenland, certain crustose lichens may live for 4500 years or more. This is an extreme example, however, and in the majority of cases, lichenometry as an absolute dating tool has a useful range of less than 500 years (Matthews, 1992). Once a substrate is colonised, the general growth of a lichen proceeds in three phases: (1) a rapid growth phase during which growth occurs at a logarithmic rate; (2) a linear growth phase during which growth is nearly uniform; and (3) a slow growth phase where lichen growth gradually declines until death (O'Neal and Schoenenberger, 2003). By measuring lichen thallus diameters (usually the largest lichen) on surfaces of known age, lichen growth rates can be established and used to construct a **growth curve** showing the relationship between lichen size and age (Figure 5.14). Surfaces of known age (referred to as **fixed points**) might include humanly constructed features such as building walls or gravestones, or natural features such as rock surfaces whose ages might be inferred from old photographs or historical records. Once the growth curve has been developed, surfaces of unknown age can be dated by relating lichen diameters on those surfaces to the growth-rate curve, thereby deriving a calendar age.

5.4.2 Problems Associated with Lichenometric Dating

Not all lichens are suitable for lichenometry and only those that have a pattern of gradual and progressive growth can be employed. Moreover, growth curve must be based on

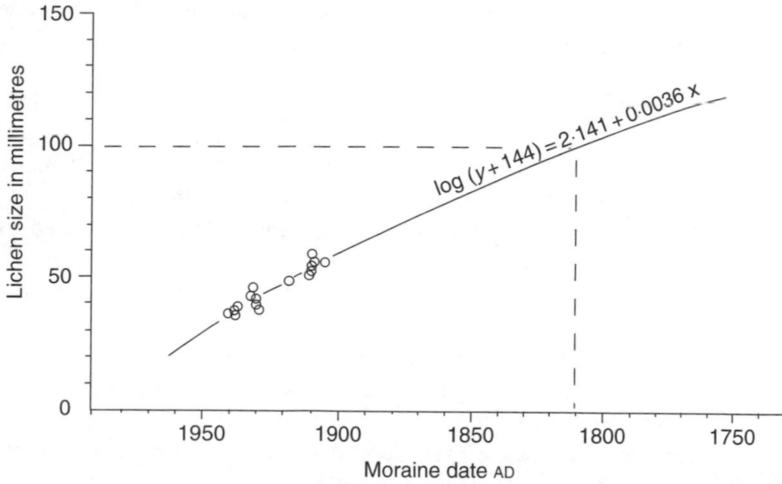

Figure 5.14 *Lichenometric growth curve based on the five largest lichens per moraine on four moraine ridges of known age at Nigardsbreen, western Norway (after Bickerton and Matthews, 1993). Using this curve, a moraine of previously unknown age with a largest lichen size of 100 mm would date to ca. AD 1810. Reproduced by permission of John Wiley & Sons Ltd*

single lichen types as differences in predicted ages have been found where different lichen datasets have been employed (Bickerton and Matthews, 1992). In addition, because the growth rate and the duration of the different growth phases described above are influenced by a range of local and regional environmental factors, such as air temperature, day length and snow cover (Benedict, 1990), lichenometric dating is effectively area-specific. Growth-rate curves that have been developed for one region will not necessarily be applicable elsewhere. Field sampling and lichen measurement are not always straightforward, and considerable effort may be required in order to ensure reproducibility of results within any one region (Innes, 1985). Moreover, in some remote areas, such as the mountains of Greenland or Norway, there may be no documentary or other evidence for establishing the fixed points on the lichen growth curves (Winkler *et al.*, 2003). Above all, there is the fundamental assumption that there has been no significant time interval between surface exposure and lichen colonisation. This, of course, can never be proven and must always remain a source of uncertainty in lichenometrical dating.

5.4.3 Lichenometry and Late Holocene Environments

The above caveats notwithstanding, there is no doubt that lichenometry has proved to be a useful technique for dating relatively recent events. While it is perhaps best known as a basis for establishing rates and patterns of deglaciation, particularly following the Little Ice Age glacier maximum in the eighteenth and nineteenth centuries, lichenometry has also been used in a number of other palaeoenvironmental contexts. Some examples of these are discussed in the following sections.

5.4.3.1 Dating post-Little Ice Age glacier recession in Norway. Jostedalsbreen is a plateau ice cap in the mountains of western Norway with a surface area of 487 km^2, and is the largest single ice mass in northern Europe. It formed during the Mid-Holocene and reached its maximal extent during the Little Ice Age in the mid-eighteenth century. The glacier variations of Jostedalsbreen have been more intensively studied than any other ice body in Scandinavia, particularly the pattern of retreat from the Little Ice Age maximum, where the principal chronological technique employed has been lichenometry (Matthews, 1994).

Detailed lichenometrical investigations at Nigardsbreen, a major outlet glacier on the eastern flanks of Jostedalsbreen, have generated a growth curve (Figure 5.14) that can be used as a basis for lichenometric dating of recessional moraines left by other receding glaciers (Bickerton and Matthews, 1993). One such glacier is Bergsetbreen which has retreated by ca. 4 km from its maximum downvalley position in the eighteenth century. At least 14 separate moraines were formed by short-lived readvances during this overall retreat phase (Figure 5.15). Three of these moraines, L, M and N, can be precisely dated to 1937, 1931 and 1910, respectively, on the basis of observation and measurement, and the predicted ages for these moraines from the lichenometrical data are in excellent agreement with the independent historical dates (±4 years). Moreover, the predicted lichenometrical date for moraine H is consistent with the position of the glacier in 1851 as observed by J.D. Forbes (1853). The dating of the older moraines (A–D) is less secure, largely because of a shortage of suitable sites for lichen growth. Of these, moraine B with a predicted age of 1779, appears the most reliably dated, although it is possible that it could have formed slightly earlier, between 1755 and 1762. Establishing a detailed chronology for deglaciation is not only important for understanding glacier behaviour and rates of operation of glacial geomorphic processes, but it is also important in other investigations, such as studies of plant colonisation of recently deglaciated terrain (Matthews, 1992).

5.4.3.2 Dating rock glaciers and Little Ice Age moraines in the Sierra Nevada, western USA. One of the problems with the glacial geomorphological evidence described above is that it relates only to the last phase of glacier activity within an area, as glacier readvances destroy any moraines that have formed during an earlier glacial phase. In the case of Jostedalsbreen, therefore, it is much more difficult to reconstruct the history of glaciation from landform evidence prior to the Little Ice Age maximum. An example of where this can be achieved, however, and where the chronology of Little Ice Age and earlier glacier activity can be constrained by lichenometry, is provided by Konrad and Clark (1998) in their investigation of rock glacier and glacial moraine evidence in the Sierra Nevada, western USA.

Rock glaciers are accumulations of rocky debris that are lobate in shape, contain varying amounts of interstitial ice and, in form and position, resemble true glaciers. Some, which have small relatively clean ice field at their heads, are effectively debris-covered glaciers, and are referred to as glacigenic rock glaciers. Such landforms are common throughout the Sierra Nevada (Clark *et al.*, 1994). In the upper Owens Valley (Figure 5.16), active glacigenic rock glaciers occur in close proximity to true glaciers with clearly defined

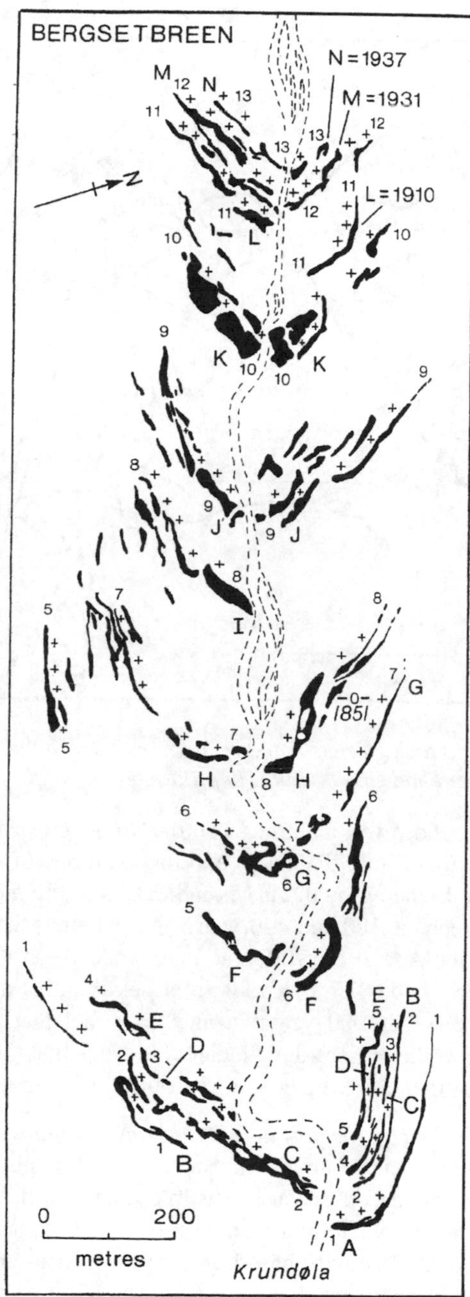

Figure 5.15 *The Bergsetbreen moraine-ridge sequence. The approximate position of the glacier front as described by J.D. Forbes (1853) when he visited the area in 1851 is shown by (-o-). The historical dates for the moraines are also shown as are the lichen-measurement plots which were used for dating purposes (+). The individual moraine ridges are lettered and the inferred ages for these, based on lichenometrical measurements, are shown in brackets (after Bickerton and Matthews, 1993). Reproduced by permission of John Wiley & Sons Ltd*

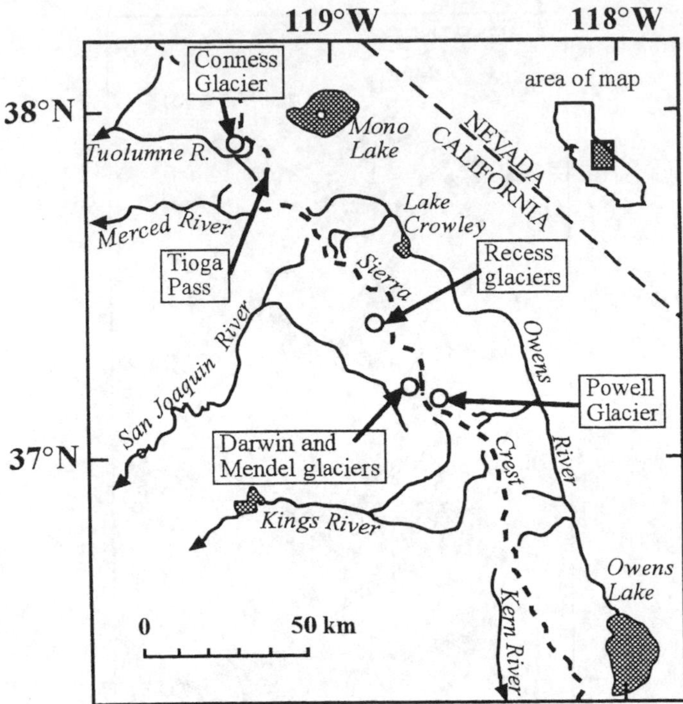

Figure 5.16 Glaciers and rock glaciers dated by lichenometry in the Sierra Nevada, western United States (after Konrad and Clark, 1998). Reproduced by permission of INSTAAR

moraine ridges. There is, however, a marked contrast in lichen size between those on the glacier moraine ridges (30–45 mm diameter) and those on one of the large rock glaciers (78–105 mm diameter). Lichen growth curves indicate that the moraines formed during the Little Ice Age between ca. 100 and 430 years ago, whereas the rock glacier predates the onset of the Little Ice Age at ca. 700 years ago. Indeed, on the basis of lichen size measurements elsewhere in the Sierra Nevada, it is possible that the large lichens on the rock glacier are in excess of 1000 years in age. The fact that no pre-Little Ice Age moraines are found below the present-day glaciers indicates that this glacial episode must have been less extensive than that which occurred during the subsequent Little Ice Age.

5.4.3.3 Dating Late Holocene rockfall activity on a Norwegian talus slope. Obtaining long-term records of rapid mass movement such as rockfall is difficult, for rates of rockfall are likely to be influenced by variations in climate, particularly by the frequency and intensity of cold intervals when frost-weathering is likely to be enhanced. This means that extrapolations of rockfall activity based on present measurements and observations may be misleading. Talus accumulation surfaces and debris-strewn hillslopes are essentially diachronous, with episodic rockfall activity leading to the deposition of boulders at different times. However, if the age–frequency distribution of the surface boulders can be established, then it may be possible to reconstruct a history of spatial and temporal patterns of rockfall onto these surfaces. One way in which this can be achieved is by means of lichenometry.

An example of this approach is described by McCarroll *et al.* (1998) for a talus slope in Jotunheimen, western Norway, where an accumulation of boulders has formed in an approximately triangular shape beneath a 100-m-high cliff. The largest lichen was measured at 28 different sites across the talus slope, with 100 boulders being measured at each site. Sites with similar lichen-size frequency distributions were then grouped together and used to interpret the spatial and temporal patterns of rockfall supply. Absolute ages were obtained by relating lichen size to a growth curve based on a sequence of dated moraine ridges downvalley from the snout of the nearby Storbreen glacier. Simulation modelling was carried out in order to reconstruct possible temporal patterns of debris supply to different parts of the talus slope. The results suggest that the normal rate of rockfall during the Late Holocene would lead to the burial of about 4% of the talus surface every 25 years. Rates of rockfall supply during the eighteenth century, the coldest part of the Little Ice Age, are estimated to have been almost five times the normal Holocene rate. The predicted increase in rockfall activity during the Little Ice Age is in good agreement with historical records, and it coincides with the maximum Late Holocene advance of most Norwegian mountain glaciers.

5.4.3.4 Dating archaeological features on raised shorelines in northern Sweden. At the last glacial maximum, the Fennoscandian ice sheet was centred over the Gulf of Bothnia in the northern Baltic Sea (Siegert, 2001). As the ice sheet decayed, isostatic recovery[7] began and this process is continuing at the present day. Tide-gauge records and detailed survey show that the current rate of uplift in the northern Baltic is in excess of 7 mm year^{-1}, although modelling results suggest that the uplift rate may be as high as 9 mm year^{-1} (Fjeldskaar *et al.*, 2000). This means that the coastline of the Gulf of Bothnia has risen by 10 m or more since Viking times!

The process of uplift is reflected in a series of raised rock shorelines that form a staircase around parts of the present-day coast. Numerous archaeological sites, dating from the Late Iron Age through to the Medieval periods (ca. 1500–500 years ago), as well as more recent historical sites have been found on these shorelines. Most occur on rocky islands and shores and consist of boulder and cobble constructions. As such they are difficult to date using conventional means, but they may be dated by means of lichenometry. Broadbent and Bergqvist (1986) measured lichen diameters on a number of raised shorelines and developed a lichen growth curve for the period ca. 200–1400 years using ages inferred from shoreline displacement rates to provide the fixed points for the lichen growth curve. The shore date gives a *maximum* age for a surface, whereas the lichen date provides a *minimum* age for a particular feature. The combination thus frames a given archaeological feature between an upper and a lower time limit. Application of this approach to a series of small rocky cairns, which had at one time supported posts for drying fishnets, produced a lichen-based date for the cairns of AD 1660. The elevation above sea level of these features is 3.8 m which, on the shoreline elevation curve, provides a maximum date of AD 1545. This suggests that the first net-drying posts stood about 1 m above the contemporary mean sea level, something that can still be observed in the northern Gulf of Bothnia at the present day. Moreover, historical records show that this type of net system first appeared in the region in the seventeenth century, which is fully in accord with the results of lichenometric dating.

5.5 Annual Layers in Glacier Ice

Alongside deep-ocean coring, one of the other great achievements in Quaternary science during the course of the twentieth century was the development of a technology for drilling through the great ice sheets of the world and recovering cores of glacier ice (Figure 5.17A). The first deep ice cores were obtained from Camp Century in Greenland (1966) and since then drilling has been carried out at a number of sites in Greenland and Antarctica, as well as on smaller ice caps on Devon Island in the Canadian Arctic, and on low-latitude ice caps in Tibet and South America (Thompson, 2000). Perhaps the most important of these drilling programmes, in terms of scientific data that have been generated and methodological refinements in coring technology, were two cores obtained from the summit of the Greenland Ice Sheet in the early 1990s: the Greenland Ice Core

(A)

Figure 5.17 *(A) An ice core from NorthGRIP, Greenland, being split lengthways using a horizontal band saw prior to sampling. (B) An automatic line-scanner used to determine the stratigraphy of the ice core. It employs a dark field indirect light source to detect cloudy bands in the ice (reflecting impurities) as the camera moves along the core. (C) A 1.65 m section of the NorthGRIP core from a depth of 1837 m (ca. 25 000 years ago) showing the visual stratigraphy revealed by the line scanner. Clear ice shows up black, whereas the cloudy bands, which contain relatively large quantities of impurities, in particular micrometre-sized dust particles from dry areas in eastern Asia, appear white. The visual stratigraphy is essentially a seasonal signal. This record shows how well the stratigraphy in an ice core can be preserved, even at depths approaching 2 km (photos: Jørgen-Peder Steffensen, Sigfus, Johnsen and Anders Svensson)*

Figure 5.17 (Continued)

Project (GRIP) co-ordinated through the European Science Foundation, which reached bedrock at 3029 m in 1992, and the American-led Greenland Ice Sheet Project (GISP2), located only 30 km away to the west of GRIP and which hit bedrock at a depth of 3053 m in 1993 (Hammer *et al.*, 1997). These two cores spanned more than 100 000 years of Late Quaternary time. More recently, a new Greenland site has been drilled by a European team, NorthGRIP (NGRIP). This reached bedrock at a depth of 3085 m in July 2003, and the record extends back to the last interglacial at ca. 123 000 years (North Greenland Ice Core Project Members, 2004). Important coring sites in Antarctica are at Vostok Station, a Russian–French collaboration drilled during the 1980s and which contains a record of ice accumulation going back more than 400 000 years (Petit *et al.*, 1999), and the new European Project for Ice Coring in Antarctica (EPICA) at Concordia Station, Dome C. The latter, which involves more than ten European countries, promises to be the most exciting ice-core drilling yet undertaken, as the record already extends back to 740 000 years, and it is possible that, in due course, it may reach almost 1 million years (EPICA Community Members, 2004).

5.5.1 Ice-Core Chronologies

Because glacier ice accumulates in a sequence of annual increments, it is possible to establish a chronology based on the visible stratigraphy and other properties of glacier ice. In the upper levels of polar ice sheets, for example, the annual additions of snow to the ice mass are revealed in clearly defined layers, and in an ice core these can be counted and a chronology established for the depth of the core (Alley *et al.*, 1997a). Other constituents of the ice, such as seasonal variations in stable isotope content, in major chemical elements or in electrical conductivity[8], can also help to identify annual layers within the ice. Such a 'multi-parameter' continuous count approach was used to establish a chronology for the GISP2 ice core down to a depth of 2800 m where the ice was dated at 110 000 years BP (Meese *et al.*, 1997). Recent developments using digital scanners, computers and large storage media (Svensson *et al.*, 2005) enable the visual stratigraphy of ice cores to be determined with remarkable clarity (Figure 5.17B and C).

At deeper levels, deformation within the ice means that the annual layers become more diffuse and consequently may be more difficult to recognise. This problem is exacerbated at low accumulation sites where the seasonal layers or seasonally varying tracers tend to be smoothed by diffusion during the process of conversion of snow to ice (firnification). In these circumstances, recourse has to be made to other methods. One technique is to use ice-flow models based on a knowledge of ice dynamics in order to derive age estimates. This approach has been employed to develop a timescale for the deeper sections of the GRIP ice core from Greenland, and for the EPICA and Vostok cores from Antarctica. Other means of dating include the tuning of proxy-climate records obtained from the ice cores, for example atmospheric trace gases such as methane (section 5.5.3), to variations in the earth's orbital parameters (Chapter 1: Note 4) which have an independent chronology based on astronomical calculations (Ruddiman and Raymo, 2003). Indeed, an orbitally tuned MOI curve has been employed to develop an age–depth model for the EPICA core (EPICA Community Members, 2004). An alternative approach is to correlate the isotopically defined events in the ice cores with independently dated (by ^{14}C or $^{230/234}U$) events in marine cores or speleothem records, and then to 'calibrate' the ice-core timescale using these age measurements. This method has recently been used to 'calibrate' the Greenland timescale based on the GRIP and GISP2 ice cores (Shackleton *et al.*, 2004).

5.5.2 Errors in Ice-Core Chronologies

Errors in ice-core chronologies arise from imperfections in the ice-core record, from variations in core quality, or from human error in interpreting the ice-core sequence. In terms of the nature of the record, it is possible that layers of snow will be disturbed by wind scouring and, in some instances, snowfall from individual years may be removed entirely leaving gaps in the ice-core stratigraphy. This is especially problematic where annual snowfall has always been low (parts of Antarctica, for example), or where the core records date from times in the past where colder temperatures have resulted in reduced snowfall, such periods often being associated with increased wind strength. A further complication arises from deformation within the ice which causes folding and possible loss of stratigraphic continuity (Alley *et al.*, 1997b). At the base of both the GRIP and GISP2 cores, for example, there is clear evidence of severe structural disturbance in the lower 200–300 m, and this has almost certainly resulted in discontinuities or breaks in the stratigraphic sequence (Meese *et al.*, 1997). Chronological problems may also arise as a result of incomplete core recovery, in which case interpolations have to be made based on annual layers above and below the missing levels. Such interpolations must inevitably have some associated errors. Counting of visible layers within the ice is normally carried out by teams of scientists, and while they are highly experienced in ice-core research, there will inevitably be a 'personal factor' in the interpretation of ice layers as annual or sub-annual horizons (Alley *et al.*, 1997a). Although replicate counts are the norm and intercomparisons between different parameters ensure that the highest possible levels of agreement are obtained, some degree of error cannot be avoided. These various factors mean that ice-core chronologies can never be regarded as absolute, and all contain an element of uncertainty. In the GISP2 timescale, errors associated with multi-parameter layer counting are of the order of 1–2% back to 12 000 years, ±2% back to ca. 40 000 years, ±5% at ca. 45 000 years and up to 10% at ca. 55 000 years. The estimated error at 110 000 years is of the order of 20% (Meese *et al.*, 1997). In the EPICA core, where the chronology is based largely on ice-flow modelling, the timescale has an estimated error of ±10 years for the period 0–700 years, of up to ±200 years back to 10 000 years, and of up to ±2000 years back to 41 000 years (Schwander *et al.*, 2001). In the deeper part of the core, the preliminary timescale suggests an error of ±10 000 years at 807 000 years, and when this is extended to the base of the core (960 000 years), the error is estimated to be of the order of ±20 000 years (EPICA Community Members, 2004). In the 400 000-year Vostok record, Petit *et al.* (1999) conclude that the error associated with the timescale is always better than 15 000 years, better than 10 000 years over most of the record, and better than 5000 years for the last 110 000 years.

5.5.3 Ice Cores and the Quaternary Palaeoenvironmental Record

Glacier ice has proved to be an archive of palaeoenvironmental information of enormous significance to Quaternary scientists. It contains, *inter alia*, a record of changes in atmospheric trace gases (such as carbon dioxide and methane) that become trapped in tiny bubbles within the ice; of past variations in atmospheric particulate matter, such as soot, dust or volcanic ash; of variations in stable isotopes (particularly isotopes of oxygen); and of a range of other trace substances (Oeschger and Langway, 1989). These various lines of evidence can be used to reconstruct changes in atmospheric circulation, former temperature and precipitation regimes, histories of volcanic activity, and human impact on the global climate system (Alley, 2000a; Mayewski and White, 2002). Some examples of the types of environmental record that can be obtained from ice cores, and which demonstrate the value of ice-core chronologies, are considered in the following sections.

5.5.3.1 Dating climatic instability as revealed in the Greenland ice cores. One of the most remarkable discoveries to have emerged from the study of polar ice cores relates to the instability of past climate. Until the 1990s, the prevailing view had been that during the 100 000 years or so prior to the Holocene, the North Atlantic region had experienced a climatic regime of unremitting cold (glacial conditions) interrupted only occasionally by short-lived interstadial episodes. Data from the Greenland ice cores, however, show that this is far too simplified a view of climatic history, for between 20 000 and 115 000 years ago, no fewer than 25 major climatic fluctuations appear to have occurred at regular or quasi-regular intervals, and with an amplitude of up to 15 °C (Johnsen *et al.*, 2001; North Greenland Ice Core Project members, 2004). These climatic reconstructions, based on changes in oxygen isotope ratios in the ice ($\delta^{18}O$), are at a remarkably high degree of resolution and the records are underpinned by timescales based on ice-layer counting and on other parameters (Figure 5.18A). The evidence suggests that changes in climate were cyclic, with each cycle lasting between 500 and 2000 years. Moreover, there is structure in the data, in that climate warming was very rapid, and in some cases may have been completed within 50 years, whereas the subsequent cooling trend was a more gradual process (Figure 5.18B). These major climatic fluctuations, which appear to reflect a combination of feedback mechanisms involving ice-sheet fluctuations, oceanographic changes and atmospheric circulation variations, are known as **Dansgaard–Oeschger cycles**, after the Danish glaciologist Willi Dansgaard and his Swiss colleague Hans Oeschger, two of the pioneers of ice-core science.

5.5.3.2 Dating rapid climate change: the end of the Younger Dryas in Greenland. The discovery of the Dansgaard–Oeschger cycles not only challenged the conventional view of general climatic stability during the Last Cold Stage, it also demonstrated that climatic change could occur extremely rapidly. Received geological wisdom has it that climatic

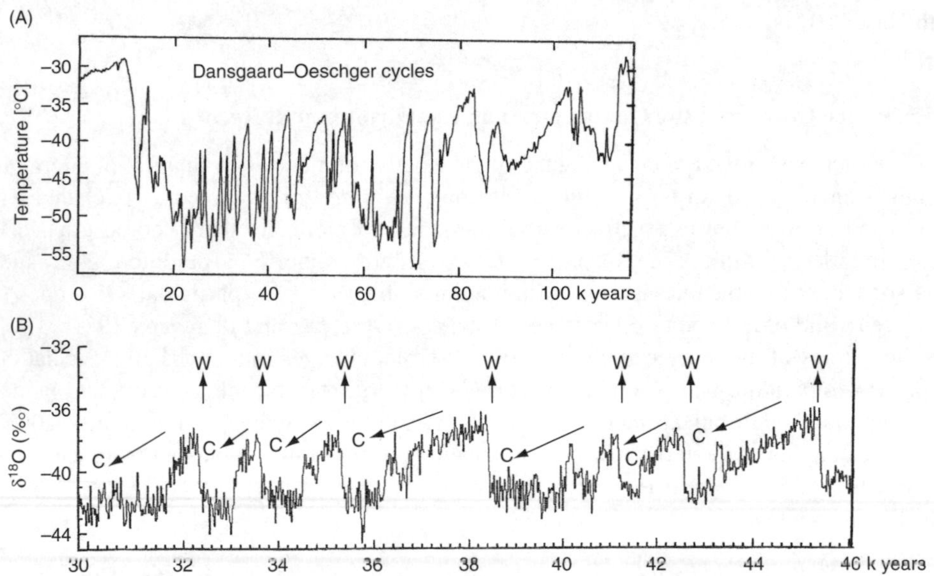

Figure 5.18 *(A) Past temperature changes for summit, Greenland, based on data from the GRIP ice core. The temperature at the Last Glacial Maximum (ca. 25 000 years BP) were about 20 °C colder than today, and the temperature shifts of the Dansgaard–Oeschger cycles are up to 15 °C (after Johnsen* et al.*, 2001). Reproduced by permission of John Wiley & Sons Ltd. (B) Part of the GISP2 $\delta^{18}O$ record showing the rapid warming and slower cooling in successive Dansgaard–Oeschger cycles between 30 000 and 45 000 years BP (after Stuiver and Grootes, 2000). Reproduced with permission of Elsevier*

change is a slow process that is measured on timescales of hundreds or, more likely, thousands of years. Ice cores, however, with their multi-parameter records and near-annually resolved incremental chronologies, show that significant shifts in climate can take place within decades or even less. A good example is provided by the end of the Younger Dryas cold stage (Figure 1.5) as recorded in the GISP2 ice core. The Younger Dryas, which has a measured duration of 1300±70 year in GISP2 (Alley *et al.*, 1993), was an abrupt return to near-glacial conditions at ca. 12 900 years when temperatures fell to 15 °C ± 3 °C colder than today (Severinghaus *et al.*, 1998). Evidence from the GISP2 core shows that at the end of the Younger Dryas there was a 5–10 °C warming and a doubling of snow accumulation in central Greenland, a large drop in wind-blown materials (Figure 5.19), indicating reduced wind speed and other changes in distant source regions or between source regions and Greenland, and a large increase in methane, indicating expansion of global

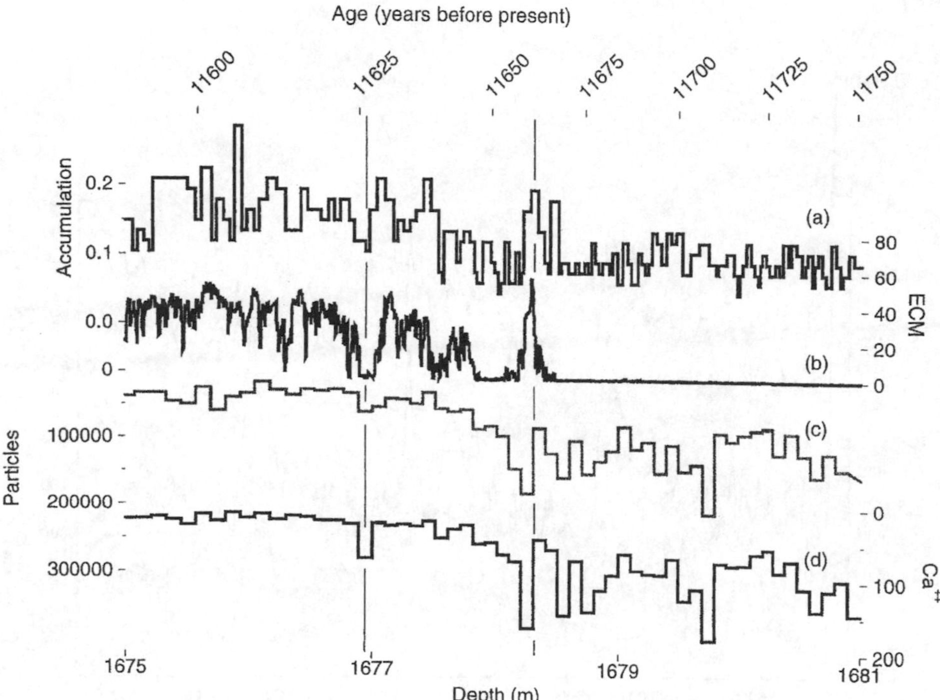

Figure 5.19 *The end of the Younger Dryas in the Greenland GISP2 ice core, as reflected in changes in (a) snow accumulation in m/ice per year, (b) electrical conductivity in microamps (ECM), (c) insoluble particulate concentrations in number per millilitre, and (d) soluble calcium concentrations in ppb. Note the increase in snow accumulation indicating rise in temperature, the increase in ECM values indicating a reduction of dust content within the ice, and a decrease also in other particulates and Ca. The last three all reflect reduced wind strength. The main climatic step at the end of the Younger Dryas falls between 1677 and 1678 m depth in the core (between the two vertical lines), corresponding to an age of ca. 11 625–11 655 years* BP *(after Alley, 2000b). Reproduced with permission of Elsevier*

wetlands (Alley, 2000b). What is so remarkable is that the GISP2 ice-core chronology shows that these major climatic and environmental changes occurred within a few decades, and possibly within a few years, in other words in less than a modern Western human lifetime!

5.5.3.3 Dating long-term variations in atmospheric Greenhouse Trace Gases. The world's ice sheets are unique archives of past changes in atmospheric Greenhouse Trace Gases (GTGs), and the longest record so far obtained is that from the Vostok Station in Antarctica (Petit *et al.*, 1999). In Figure 5.20, a 420 000 history of carbon dioxide (CO_2) and methane (CH_4) is shown plotted against the ice isotopic trace (δD), the latter being a proxy for temperature. Peaks in the δD curve reflect interglacial periods, and troughs cold or glacial periods. The trace gas signals mirror this temperature record very closely, with the highest CO_2 and CH_4 mixing ratios during the interglacials and the lowest during the glacial maxima. This close correspondence between temperature variations and past

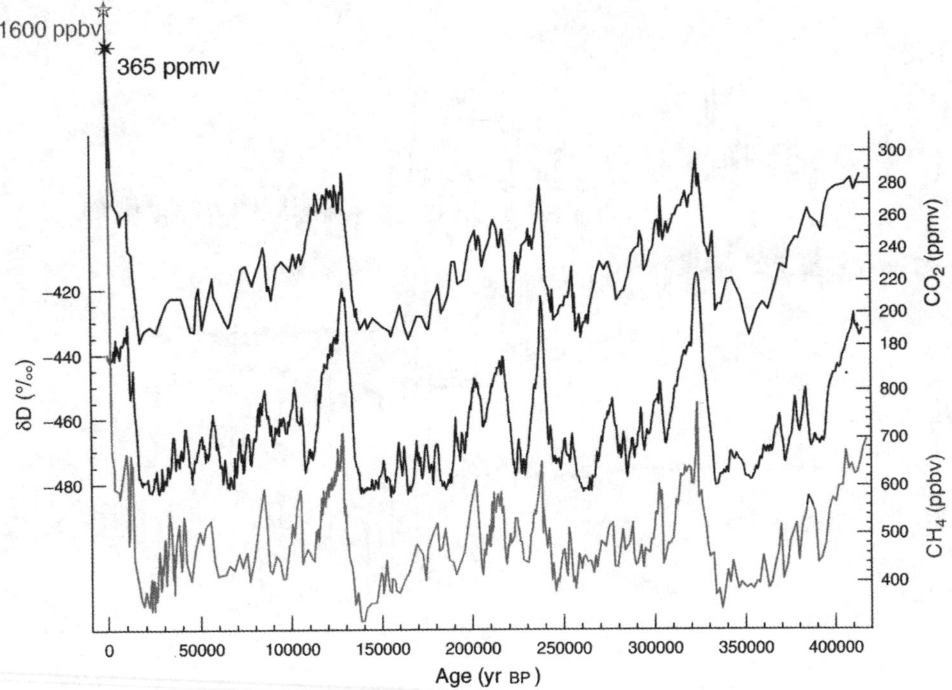

Figure 5.20 *Greenhouse trace gases (GTGs) over the last four glacial–interglacial cycles in the Vostok ice core. The peaks in the three curves, CO_2, CH_4 and δD (a proxy for temperature), reflect interglacial periods, while the troughs represent glacial episodes. Note that the present levels of CO_2 at 365 ppmv (parts per million by volume) and CH_4 at 1600 ppbv (parts per billion by volume) are significantly higher than at any time over the past 420 000 years. This reflects the impact on the earth's atmosphere of recent industrial activity (after Petit* et al., *1999; Raynaud* et al., *2000). Reproduced with permission of Elsevier*

changes in atmospheric trace gas content implies that GTGs have been an important forcing factor in long-term climate change, with climate simulation modelling suggesting that they may contribute as much as half (2–3 °C) to the globally averaged glacial–interglacial temperature change (Raynaud *et al.*, 2000). What makes these records so valuable is that they are not only continuous, and therefore provide time-series data, but they are underpinned by an independent timescale based on the unique properties of glacier ice.

5.5.3.4 Dating human impact on climate as reflected in ice-core records. The polar ice sheets not only contain evidence of natural atmospheric changes, they also preserve a record of human impact on the atmosphere. The well-documented 'Greenhouse Effect', in other words the significant increase in CO_2, other trace gases and particulate matter, following the industrial revolution in Europe during the course of the eighteenth century (Houghton *et al.*, 2001), is clearly manifest in ice cores from the polar ice sheets. Ice-layer counting enables the timing of the increase of these atmospheric GTGs to be dated precisely. The evidence shows that atmospheric concentration of CO_2 has increased by 31% since 1750, and that of CH_4 by 151% over the same period, both gases rising to levels that have not been exceeded over the past 420 000 years (Figure 5.20).

It is possible, however, that human activity has been affecting the atmospheric environment over a much longer timescale. Figure 5.21 shows the record for atmospheric CH_4 in the Vostok ice core over the course of the Holocene. During the Early

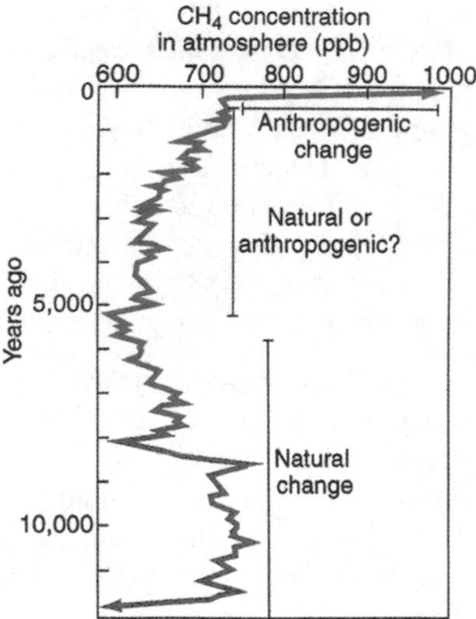

Figure 5.21 *The Holocene CH_4 record in the Vostok ice core showing naturally induced change during the Early and Mid-Holocene, and the increasing anthropogenic influence during the later Holocene. Note, in particular, the abrupt increase in CH_4 at ca. 5000 years ago which may reflect the expansion of rice farming (after Ruddiman and Thompson, 2001). Reproduced with permission of Elsevier*

Mid-Holocene, changes in CH_4 appear to reflect naturally induced variations (in areas of peatland, for example), while the abrupt Late Holocene increase is clearly indicative of recent industrial activity (see preceding text). There is, however, a sharp increase in atmospheric methane in the Mid-Holocene (ca. 5000 years on the Vostok timescale), and it has been suggested that this reflects the expansion of rice farming across southeast Asia (Ruddiman and Thompson, 2001). Early attempts at rice farming would have required relatively extensive flooding of weed-infested fields which would have provided a major source for methane emissions. If this is so, then the first major human impact on the earth's atmospheric environment may predate the industrial revolution by more than 4500 years!

5.6 Other Media Dated by Annual Banding

5.6.1 Speleothems

As we saw in Chapter 3, speleothems are valuable sources of palaeoclimatic data, and can be dated by U-series. Speleothem structures form through successive growth layers of calcite and, in certain circumstances, it may prove possible not only to identify these individual growth bands, but also to recognise them as *annual* increments of material (Baker *et al.*, 1993). In these instances, a chronology can be established by counting the individual laminae (as in dendrochronology or varve chronology), and a timescale established which is either 'floating' or anchored to calendar years either by counting back from present-day levels (Baker *et al.*, 1999) or by using other dating methods to calibrate the sequence (Paulsen *et al.*, 2003). The thickness of the laminae is controlled by a range of factors, including the drip rate and carbonate deposition processes, and individual lamina can be identified visually using both UV and visible light. However, they are also detectable by other techniques such as the variations in fluorescence (or luminescence). The latter derives from organic acids that have been trapped within the speleothem calcite having been introduced from the overlying soil and vegetation by the groundwater feeding the speleothem (McGarry and Baker, 2000). Two examples of speleothem records dated partly or entirely by lamination counting are now considered.

5.6.1.1 Dating a proxy record for twentieth-century precipitation from Poole's Cavern, England. Annual luminescence variations in three stalagmites from Poole's Cavern, near Buxton, England, were compared with rainfall data from a nearby meteorological station to determine whether the speleothem records could be used as a palaeoprecipitation proxy (Baker *et al.*, 1999). Counting of annual laminae from the top down showed that the speleothems accumulated over the period AD 1910–1996. Ten of the laminae exhibit a double luminescent band structure and these coincide with years with either high monthly or daily mean precipitation. This suggests that periods of high-intensity (>60 mm per day) and high-quantity (>250 mm per month) precipitation may flush luminescent organic material onto the stalagmites from either the soil or the groundwater zones, thereby generating double laminae. If so, then there may be considerable potential in the analysis of stalagmite lamina structure for deriving a palaeoprecipitation signal, especially one of precipitation intensity which may be detectable at the sub-annual scale.

5.6.1.2 Dating climate variability in central China over the last 1270 years. A high-resolution speleothem record extending back to 1270 years has been obtained from a stalagmite from Buddha Cave, which is located some 900 km southwest of Beijing, China (Paulsen *et al.*, 2003). Oxygen isotope (δ^{18}O) and carbon isotope (δ^{13}C) variations were measured throughout the speleothem which has been dated by a combination of U-series, ^{210}Pb and lamination counting to a time resolution as fine as 1–3 years. The δ^{18}O profile provides a record of temperature change (heavier δ^{18}O values reflecting warmer cave temperatures), while carbon isotope ratios reflect changes in precipitation (Figure 5.22, lower part). The latter is based on the principle that the carbon isotope composition of cave speleothem reflects that of the overlying soil (and hence of vegetation). Plants adapted to a cold and wet climate (C_3 type) typically have δ^{13}C values lighter than those for C_4-type plants which tend to grow in a warmer and drier climate (see also section 2.7.1.4). The resulting palaeoclimatic reconstruction (Figure 5.22, upper part) shows very clearly the Medieval Warm Period, the Little Ice Age and the twentieth-century warming trend, lending support to the global extent of these events. The isotopic records also show clearly defined cycles of 33, 22, 11, 9.6 and 7.2 years. Several of these have been recorded in other proxy records, including another lamina-based speleothem record from

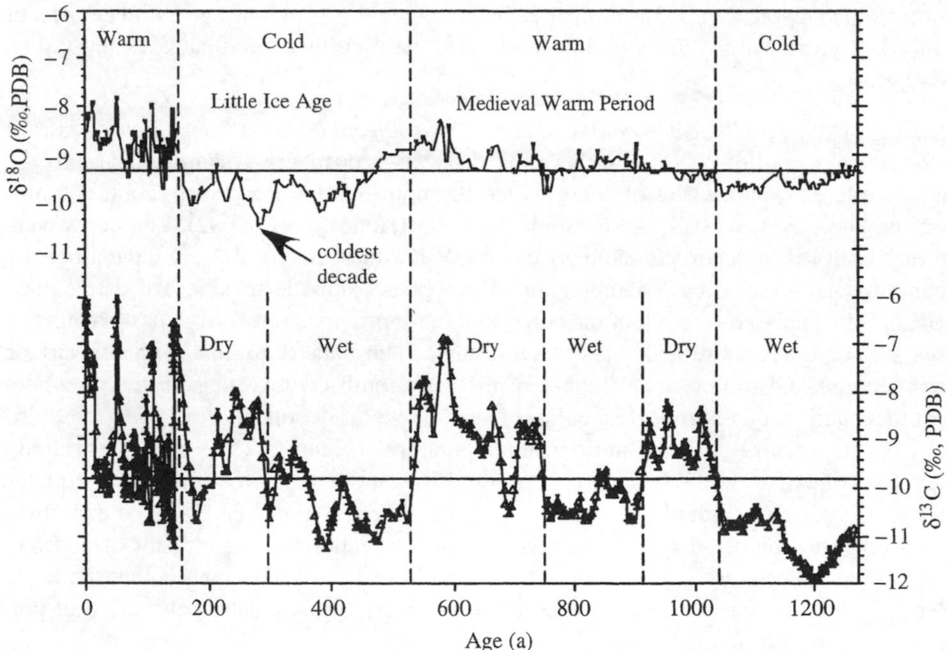

Figure 5.22 *Oxygen (δ^{18}O) and carbon (δ^{13}C) records from the 1270-year stalagmite from Buddha Cave, China. The oxygen isotope trace can be read as a qualitative temperature record, the peaks in the curve (heavier δ^{18}O values) reflecting higher temperatures, while the carbon isotope trace indicates precipitation variations (after Paulsen* et al., *2003). Reproduced with permission of Elsevier*

China (Qin *et al.*, 1999), and point to a strong external forcing factor (e.g. solar irradiance variations) in short-term climate change in eastern China.

5.6.2 Corals

Corals are valuable sources of high-resolution palaeoclimatic data for shallow-water regions of the tropical oceans. Stable isotope and trace-element data from coral can be used to reconstruct sea-surface environmental conditions, including variations in temperature and salinity, as well as precipitation, evaporation and freshwater dilution from river run-off (Beck *et al.*, 1992; Charles *et al.*, 1997; Hendy *et al.*, 2002). Moreover, because coral aragonite is deposited at rapid rates of several millimetre to several centimetre per year, variations in environmental conditions over timescales of months, or even weeks, can be reconstructed from coral skeletons. Skeletal growth rate is primarily a function of coral species, water temperature and depth, and it varies seasonally, producing growth bands of differing densities that are visible by X-radiography (Quinn *et al.*, 1993). More recently, luminescent banding within the coral structure, which is evident under long-wave UV light, has been used to identify annual aragonite increments. These reflect either the incorporation into the coral skeleton of terrestrial humic substances during river flood events (as in speleothems: see section 5.6.1.1), or they result from episodes of reduced calcification as the coral responds to lower salinity conditions during coastal run-off (Hendy *et al.*, 2003). Typical errors associated with coral age estimates are of only 1–2 years per century and, in some cases, the records extend back for more than 400 years.

5.6.2.1 Dating a 420-year-coral-based palaeoenvironmental record from the southwestern Pacific. A record of changes in the tropical ocean-atmosphere system since 1565 was reconstructed from a series of cores drilled through massive *Porites* sp. coral colonies on the Great Barrier Reef, northeastern Australia (Hendy *et al.*, 2002). The cores were cross-dated (as in dendrochronology) using UV-luminescent bands, and annual density banding was revealed by X-radiography. Three palaeoclimatic tracers, $\delta^{18}O$, Sr/Ca and U/Ca, were analysed for each of the cores, and these provided a record of past changes in sea-surface temperature and sea-surface salinity. The data show that both sea-surface temperature and salinity were higher in the eighteenth century than in the twentieth century, and that an abrupt freshening of surface water occurred after 1870 and coincided with a marked cooling in tropical temperatures (Figure 5.23). The higher salinity levels in the southwestern Pacific between 1565 and 1870, a period which equates with the later part of the Little Ice Age in the northern hemisphere, may be explained by a combination of advection[9] and wind-induced evaporation as a result of a strong latitudinal temperature gradient and intensified circulation. This suggests that the Little Ice Age glacial expansion may have been driven, in part, by a greater poleward transport of water vapour from the tropical Pacific.

5.6.2.2 Dating a 240-year palaeoprecipitation record from Florida, USA. A core through a sample of coral of *Montrastraea faveolata* taken from an offshore reef in Biscayne Bay to the south of Miami, Florida, was shown by X-radiography to contain a growth record extending back 240 years (Swart *et al.*, 1996). The density banding of the coral was regular and distinct, with an average growth rate of 7.87 mm $year^{-1}$. Analysis was carried

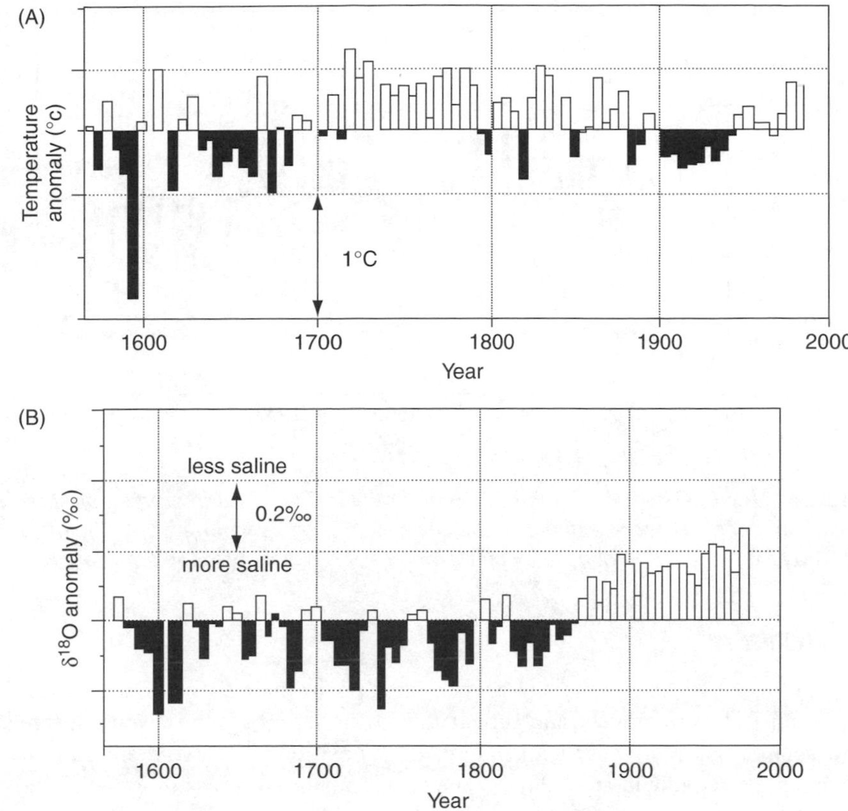

Figure 5.23 *Proxy climate records from the Great Barrier Reef for the past 420 years. (A) The Sr/Ca record, which is a proxy for temperature. (B) The $\delta^{18}O$ trace from which the temperature signal has been removed thereby creating a proxy palaeosalinity record. Note the marked decrease in sea-surface salinity after 1870 and the complementary fall in temperature (from Hendy* et al.*, 2002). Reprinted with permission from AAAS. © AAAS*

out on the $\delta^{18}O$ content of the coral skeleton in order to establish variations in former levels of precipitation. The latter can be inferred from the $\delta^{18}O$ trace, because in years when precipitation was heavier than the average, precipitation tends to be inversely correlated with $\delta^{18}O$ (heavier precipitation is indicated by 'lighter' $\delta^{18}O$ values), reflecting the greater input of isotopically depleted precipitation. In contrast, when precipitation is below normal, there is a positive relationship between precipitation levels and $\delta^{18}O$. At these times, there is a greater input of isotopically 'heavy' water flowing out of the nearby Everglades swamplands. The $\delta^{18}O$ record, which extends back to 1745 (Figure 5.24), suggests that the late eighteenth and nineteenth centuries were dry compared with the early part of the twentieth century when precipitation levels appear to have increased. Again, the fact that these are annually dated records means that specific years of heavier or lighter precipitation can be readily identified in the dataset.

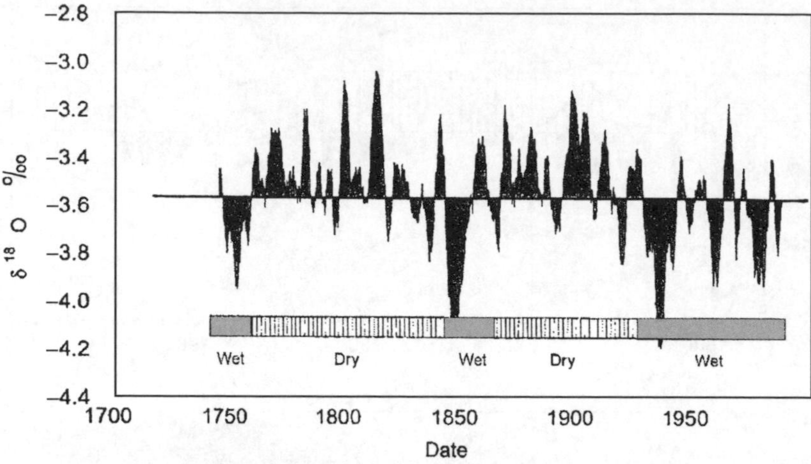

Figure 5.24 *The $\delta^{18}O$ signal from the Biscayne Bay, Florida, coral interpreted as a palaeo-precipitation record. Wetter weather is indicated by $\delta^{18}O$ values falling below the mean line (after Swart* et al.*, 1996). Reproduced by permission of SEPM Society for Sedimentary Geology*

5.6.3 Molluscs

Marine molluscs also exhibit annual growth banding in their shells. These range from semi-diurnal (tidal) to annual, and reflect the influence of a range of environmental factors, such as salinity, water temperature, tidal position and, above all, nutrient and food availability (Goodwin *et al.*, 2001). The variations in thickness of growth bands are therefore potential recorders of a range of environmental parameters, as well as providing temporal frameworks for geochemical environmental proxies such as $\delta^{18}O$ and $\delta^{13}C$. To date, fewer palaeoenvironmental studies have been carried out using molluscs as opposed to corals, but the results that have been obtained suggest that marine molluscs could prove to be a valuable source of information about changes in marine environments over the course of recent centuries (see below). In addition, certain species of freshwater mollusc may also provide palaeoenvironmental records, including changes in air temperature (Schöne *et al.*, 2004).

5.6.3.1 The development of a sclerochronology using the long-lived bivalve Arctica islandica. The ocean quahog, *A. islandica*, is an ideal species to use in palaeo-environmental reconstructions. It has a wide range in the North Atlantic (35–70°N and 10–280 m depth), it forms easily identified growth rings (Figure 5.25), its growth is continuous throughout life, and it has great longevity, ages of 100 years and more being common in most populations (Witbaard *et al.*, 1997). Moreover, studies of modern specimens have shown that the $\delta^{18}O$ record in *Arctica* is in phase with the growth banding, is in isotopic equilibrium with the ambient seawater, and can record ambient bottom water temperature with a precision of ±1.2 °C (Weidman *et al.*, 1994). As such, it could prove to be a valuable source of information on recent changes in ocean water temperature.

A. islandica is also an ideal species to use in sclerochronology. Given that growth-band characteristics are strongly influenced by the immediate environment,

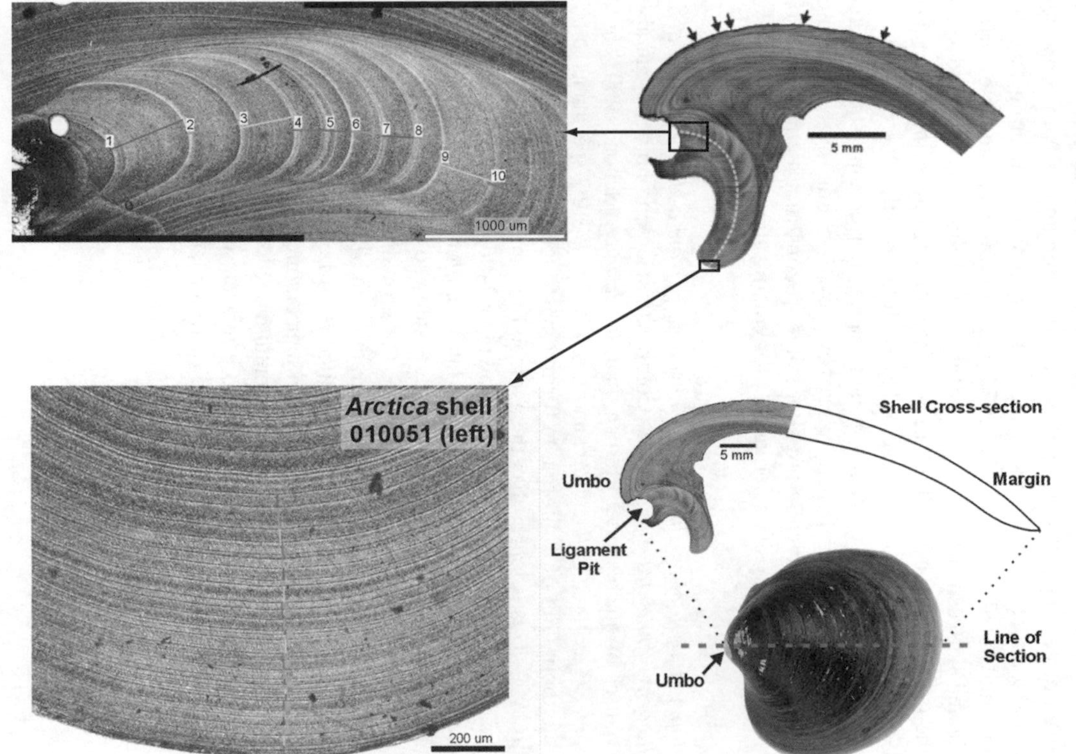

Figure 5.25 *Annual growth rings in a shell of A.* islandica. *The upper left plate shows the juvenile annual growth rings in the ligament pit below the umbo, while the lower left plate shows the older annual growth rings taken from below the umbo. The mollusc grows relatively quickly during the juvenile stage and hence adds thicker increments of aragonite than in old age, just as more vigorous growth in young trees is followed by slower growth in later life and a reduction in tree-ring width with age. Hence, sclerochronologies based on mollusc shells require the use of standardisation procedures for ring-width variations, similar to those used in dendrochronology (Figure 5.2) (photo: James Scourse)*

individuals from within a population might be expected to share similar growth records. Hence, each member of a population might form a thick annual band during a 'good' year, and thinner bands during subsequent 'bad' years. If so, then records of these band widths in different individuals, both living and dead, can be correlated and spliced together to produce long sclerochronologies using the same principles as in dendrochronology. An example of this approach is described by Marchitto *et al.* (2000) using samples of *A. islandica* from Georges Bank off the coast of New England, eastern USA. They employed three live-collected and four dead-collected shells to construct a 154-year composite sclerochronology, the band width matching indicating that the dead-collected individuals died in 1950, 1971, 1978 and 1989. A weak inverse correlation could be detected between measured sea-surface temperature records and shell ring width, with the period of relatively high sea-surface temperatures during the late 1940s and early 1950s coinciding with a phase of lower shell growth. One explanation for this is that warm surface temperatures led to increased water-column stratification and colder bottom water which resulted in narrower growth bands. Marchitto and co-workers believe that it will eventually prove possible to construct a continuous, or near-continuous, 1000-year composite sclerochronology using this particular mollusc species.

5.6.3.2 The development of a 'clam-ring' master chronology from a short-lived bivalve mollusc and its palaeoenvironmental significance. Composite sclerochronologies can also be derived from molluscs of much shorter lifespan than *A. islandica*. Schöne (2003) used three species of *Chione* (*C. cortezi, C. fluctifraga* and *C. californiensis*: 67 specimens in all) collected from the northern Gulf of California to form a master chronology covering the period from 1982 to 1999. This is longer than the average lifespan (typically 6–10 years) of these *Chione* species. A total of 63–76% of the growth variation in these molluscs can be explained by summer sea-surface temperature variation and/or fluctuations of annual river inflow into the Gulf of California. What makes this study so important is that in one species, *C. cortezi*, it has proved possible to resolve the growth ring record into daily increments, from which daily changes in water temperature can be inferred (Schöne *et al.*, 2002). The mollusc, therefore, contains a highly detailed record of how long a particular temperature regime existed, and just how quickly water temperatures increased or declined. This is one of the highest resolution palaeoenvironmental records to be derived from any organism, and by linking living and dead shells into a master sclerochronology it may eventually prove possible to extend this daily record of water temperature change back over several decades.

Notes

1. **Subfossils** are those fossil remains in which there has been little structural change since death. Examples include wood and mollusc shell.
2. The term derives from the Swedish word '**varv**' which means something that begins and ends in the same place, such as 0° and 360° in a circle, a turn, a lap, or yearly cycle.
3. **Clastic sediments** are those formed from fragments of weathered bedrock (sands, silts and clays) and have usually been transported to a point of deposition.

4. **Chrysophytes** are a group of planktonic algae that are covered with silicious scales. Many are sensitive to pH variations in lake water and, as their remains preserve well in lake sediments, they may provide a useful record of changes in the lake ecosystem over time.
5. The term **diamict** or **diamicton** refers to non-sorted terrestrial sediments or rock containing a wide range of particle size, regardless of mode of origin.
6. **Dimictic lakes** are those where there are two phases of overturn within the water column, one in the spring and one in the autumn. These are interspersed with periods of relative stagnation during the summer and winter months. Dimictic lakes contrast with **monomictic lakes** where stagnation occurs only in one season.
7. **Glacial isostasy** refers to the crustal deformation that results from the buildup of ice sheets. As the ice accumulates, the earth's crust becomes depressed by a value proportional to the weight of the overlying ice. Once deglaciation begins, the crust gradually returns to its pre-depression state, a process known as **isostatic recovery**.
8. **Electrical conductivity measurements (ECM)** provide a continuous high-resolution record of low-frequency electrical conductivity of glacier ice which is related to the acidity of the ice. When placed in an electrical current, strong acids, such as sulphuric acid from volcanic activity and nitric acid controlled by atmospheric chemistry, cause an increase in current flow. By contrast, when the acids are neutralised due, for example, to an increase in alkaline dust from continental sources, the current is reduced. For further details refer Taylor *et al.* (1993) and Meese *et al.* (1997).
9. **Advection** is the process of transmission by horizontal movement, and is usually applied to the transfer of heat by horizontal movement of air. The most familiar example is the transfer of heat through the movement of tropical air from low to high latitudes.

6

Relative Dating Methods

All things began in order, so shall they end, and so shall they begin again...
Sir Thomas Browne

6.1 Introduction

Relative dating methods determine whether something is older than or younger than something else; in other words, they group sedimentary horizons, fossils, artefacts, etc., and rank them in terms of **relative order of age**. This approach has its roots in the geological sciences, and particularly in **stratigraphy**. The great British geologist William Smith, whose classic map of the British Isles was published in 1815 (Winchester, 2002), formulated the two basic principles of stratigraphy, namely that (1) if one series of rocks lies above another then the upper series was formed after the lower series (unless it can be demonstrated that the beds have been inverted, for example by tectonic activity); and (2) each bed (or group of beds) contains a characteristic assemblage of fossils. The first of these principles is known as the **Law of Superposition**, and it is on this basis that the various stratigraphic units within a geological sequence can be placed in relative order of age. By the same token, it also enables the fossil assemblages within those strata to be classified in terms of relative age.

In Quaternary science, however, there are ways other than stratigraphy whereby phenomena can be ranked in order of antiquity. Many of these involve the operation of physical or chemical processes that are wholly or partially time-dependent. For example,

Quaternary Dating Methods M. Walker

sub-aerial weathering[1] of rock surfaces and pedogenesis will bring about gradual changes on rock and ground surfaces; the outer layers of certain minerals will slowly be affected by hydration;[2] chemical precipitates from groundwater will progressively accumulate in buried bone; while following the death of an organism, biochemical changes will lead to the breakdown of tissues to produce compounds of a more simple chemical structure. In each of these cases, the extent to which change has occurred will increase with time, and this therefore forms a basis for relative dating. In this chapter, we look at a number of dating methods that are based on these various processes.

6.2 Rock Surface Weathering

As we saw in section 3.4, radiometric ages for the timing of initial exposure of older rock surfaces can be obtained using cosmogenic nuclides (^{10}Be, ^{36}Cl, etc.), while younger surfaces may be dated using lichenometry (section 5.4.3). However, it is also possible to use the degree of rock weathering as a basis for establishing a chronology. Unlike radiometric dating or lichenometry, however, these weathering characteristics do not, in themselves, allow the timing of exposure to be established in years. Rather, they enable one rock surface to be shown to be similar to, or different from, another in terms of age. If weathering rates can be calibrated in one locality by independent means (cosmogenic nuclide dating, for example), then it may prove possible to use that information to establish a timescale for surface weathering in adjacent areas.

6.2.1 Surface Weathering Features

All landscape surfaces are subject to weathering processes that lead to physical and chemical alteration of surficial deposits and of the underlying bedrock, and both weathering and soil formation (see below) begin immediately upon exposure of the geomorphic surface. Physical weathering processes, such as frost wedging and exfoliation,[3] lead to disintegration of the surface of rocks and boulders, and operate mainly at the ground surface. Chemical weathering processes, such as oxidation, hydrolysis and dissolution, occur where water is in contact with the rock, and can lead to the breakdown of rock both at the ground surface and at depth. Typical weathering features include decomposed boulders, the crumbly nature of certain rocks and clasts (of sandstone, for example), and the relative concentrations of more durable materials, such as quartz and chert, on the ground surface. Most rocks tend to weather inward from their surfaces, irrespective of whether they are exposed at the surface or buried within shallower parts of the soil. The result is a **weathering rind** within the rock that approximately parallels the outer surface, and variations of thickness of these rinds can be used to establish relative order of antiquity of rock surfaces. Data from New Zealand and the western United States, where weathering rind variations have been frequently used to establish relative chronologies, suggest that surface rinds are best suited for application to shorter time intervals, such as the Holocene, whereas subsurface rinds can be utilised more successfully for longer time intervals, perhaps several hundred thousand years (Knuepfer, 1994).

A number of techniques have been employed to determine variations in weathering on rock surfaces and on boulders. Some rock types, such as andesite, basalt and fine-grained sandstones, develop clear rinds with sharp inner edges, and these can be measured often very precisely (to within 0.1 mm). Other rock types, such as granite and quartzite, often have more diffuse rind inner edges, while some rocks, such as schists and limestone, tend to display very poor rind formation (Knuepfer, 1994). In these instances, alternative approaches have to be employed. A qualitative estimate of the degree of surface weathering can be obtained by striking rocks or boulders with a hammer. Weathered stone or rock surfaces produce a dull sound and weaker recoil than fresh boulders which generate a sharp ring and strong hammer rebound (Mahaney *et al.*, 1984). More quantitative estimates of degree of weathering may be obtained using instruments such as the **Schmidt hammer** (McCarroll, 1994). This measures the distance of rebound when a spring-loaded mass strikes a surface. The amount of rebound is a reflection of the surface hardness of the rock surface and of the degree of weathering; unweathered surfaces thus give higher values while more weathered surfaces result in lower rebound values. Another approach is to use a **microseismic timer** which determines the compressional wave (P-wave) velocity through stones and boulders, the velocity of the P-wave being determined by the soundness of the rock material (Crook, 1986). In arid areas, coatings of clay minerals cemented to the underlying bedrock by oxides of manganese and/or iron are found on many surfaces. Known as **rock varnish**, this thin layer of material (typically less than 200 μm in thickness) can provide a relative measure of age of the surface. This is because some cations, such as calcium and potassium, are easily mobilised, whereas others, notably titanium, are more stable. Hence the ratio of potassium plus sodium to titanium in rock varnish decreases over time, and provides an indication of relative age. This technique is known as **cation-ratio dating** (Dorn, 1994). In addition, rock varnish may contain minute quantities of organic material that can be dated by AMS ^{14}C, thereby providing numerical ages for exposure of the underlying rock surface (Dorn *et al.*, 1992). Where well-jointed rock crops out at the ground surface, variations in depths of joints can also be employed to determine relative order of age, the deeper and more expanded joints indicating a longer exposure to weathering agencies, for example under periglacial conditions (Ballantyne *et al.*, 1998b).

6.2.2 Problems in Using Surface Weathering Features to Establish Relative Chronologies

Although relative-age dating using variations in surface weathering provides a useful and often relatively straightforward means of deriving age information from bedrock surfaces or boulders, difficulties can easily arise. While degree of weathering may indeed be time-dependent, it is also influenced by a range of other variables including local climate, altitude, aspect and, above all, bedrock composition. Hence local variations in lithology, for example, may be as responsible for variations in Schmidt hammer data as time (McCarroll, 1991). In glaciated areas, the incorporation of older materials into younger glacial landforms can also lead to errors in Schmidt hammer data (Evans *et al.*, 1999), while in high Arctic regions, wind polishing and case-hardening of rind surfaces may result in anomalously high Schmidt hammer rebound values (Lilliesköld and Sundqvist, 1994). Particular difficulties have arisen in cation-ratio dating where questions have been raised about the precise nature of the mechanisms that lead to cation-ratio variations (Reneau and Raymond, 1991) and about the long-term stability of the ultra-thin varnishes on rock surfaces (Bierman and Gillespie, 1994). There have also been differences of opinion over the reliability of radiocarbon dates obtained from organic residues preserved within rock varnish (Beck *et al.*, 1998; Dorn, 1998).

6.2.3 Applications of Surface Weathering Dating

Although surface dating techniques are not as precise as numerical-age methods due to the greater variability in the controlling physical and chemical parameters, they have proved valuable in Quaternary studies because they can be applied to a wide range of surface contexts. They have perhaps been most effectively employed in glacial geomorphological investigations to group exposed glacial land surfaces of similar age, and to use weathering contrasts on clasts and boulders in moraines to distinguish landforms of different ages (Bursik and Gillespie, 1993). In some areas it has proved possible to calibrate weathering rind development by radiometric dating to generate ages in years for glacial moraines and other landforms. Surface weathering dating has also been employed in other areas of geomorphological research, for example, to date rock avalanche events, river terrace development, patterns of river incision, and rates of fault displacement of river terrace surfaces (Knuepfer, 1994). Rock varnish dating has been most widely applied in the western United States where, in addition to providing a basis for deglacial chronologies, it has been used to constrain the ages of desert landforms such as alluvial fans (Dorn, 1994). In the majority of cases, however, these relative dating methods have been most effectively used in conjunction with other techniques such as lichenometry or radiometric methods, most notably cosmogenic nuclide dating.

6.2.3.1 Relative dating of Holocene glacier fluctuations in the Nepal Himalaya. Shiraiwa and Watanabe's (1991) work in the Nepal Himalaya provides a good example of how relative dating methods can be combined with the results of radiometric dating to develop a Holocene glacial chronology for a high mountain region. In the Langtang Valley, about 60 km north of Kathmandu, numerous moraines occur downvalley from the present-day glaciers in the region (Figure 6.1). These were grouped into distinctive moraine complexes on the basis of geographical position, and a range of relative age data, including Schmidt hammer rebound values (R-values), weathering rind thickness and soil development characteristics (section 6.4), were obtained on each of the complexes. The relative age evidence was supported by radiocarbon dates from organic materials either beneath or within the glacial deposits. This combination of relative and radiometric dating enabled the moraines to be assigned to five glacial stages: the Gora Tabela Stage (probably pre-Holocene); the Langtang Stage (3650–3000 ^{14}C years BP) which appears to have been the most extensive Holocene advance in the valley; the Lirung Stage (2800–550 ^{14}C years BP); and two smaller Little Ice Age advances (Yala I and Yala II Stages) between ca. AD 1400 and 1910. The differences in degree of weathering between the moraine complexes supported the initial classification of the landforms based on geographical position, and proved to be a key line of evidence in the development of a Holocene glacial chronology.

6.2.3.2 Relative dating of periglacial trimlines in northwest Scotland. In recent years, considerable attention has been directed towards reconstructing not only the spatial extent of glaciers and ice sheets, but also ice-sheet thickness. In upland regions, an important geomorphological indicator of the altitude of former ice masses is the **periglacial trimline**,

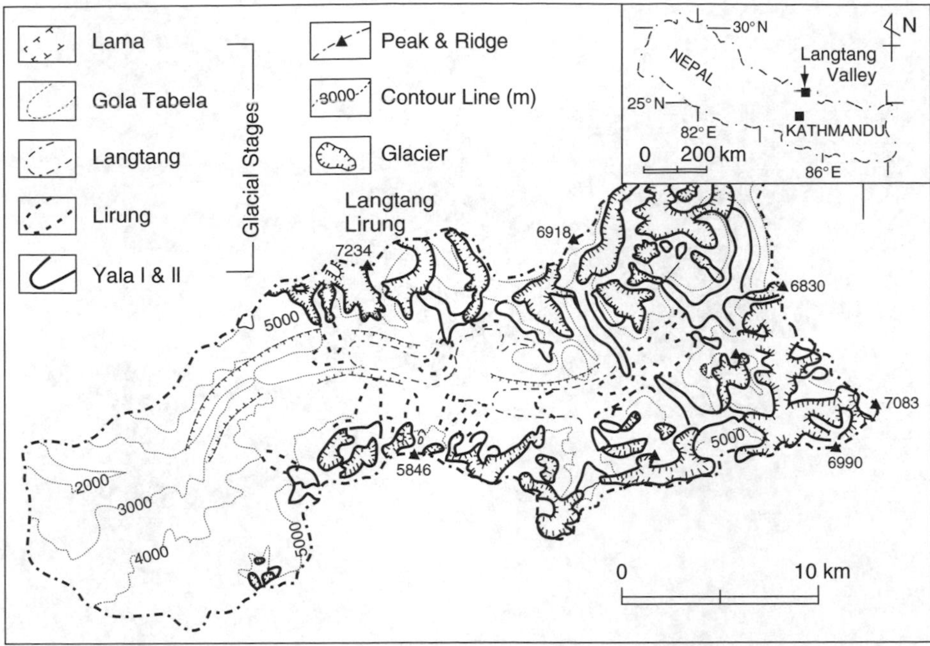

Figure 6.1 *The extent of glaciers in the Langtang Valley, Nepal Himalaya, during the Late Quaternary (after Shiraiwa and Watanabe, 1991). Reproduced by permission of INSTAAR*

i.e. the maximum level to which glacier ice has eroded or 'trimmed' frost-weathered debris on a hillslope (Ballantyne and Harris, 1994). The periglacial trimline effectively marks the uppermost limit of glacial erosion on a hillside; below the trimline will be a zone of glacial erosion or deposition, while above the trimline will be a zone of frost weathering (Figure 6.2). In practice, however, the periglacial trimline is not always easy to identify in the field, and relative dating methods are frequently employed to confirm its existence, working on the principle that a greater degree of weathering will be recorded above the trimline than below it because of more prolonged exposure to surface (periglacial) weathering processes.

In their reconstruction of the last ice sheet in northwest Scotland, Ballantyne *et al.* (1998b) employed the depth of dilation joints and Schmidt hammer measurements on rocks above and below the mapped trimline to derive weathering indices. The depth of joints on both Torridonian Sandstone and Lewisian Gneiss lithologies (Figure 6.3A) showed a statistically significant difference above and below the trimline, the joint depth being markedly greater on bedrock outcrops above the trimline. Schmidt hammer R-values on Torridonian Sandstone (Figure 6.3B) formed a similar pattern, with a statistically significant difference apparent between R-values above and below the trimline. Collectively these data pointed to a much greater degree of rock weathering above the hypothesised trimline, confirming the view that this did indeed reflect the upper limit of glacial erosion on these mountains, and enabling the vertical extent of the last ice sheet to be located

Figure 6.2 A trimline (T) cut by a glacier advance during the Little Ice Age (14th–18th century AD) on Mt Elbrus, Caucasus (photo: Colin Ballantyne)

with a relatively high degree of precision. The results show that the upper ice limit ranged from 425–450 m in the Outer Hebrides to >950 m on the mainland of Scotland.

6.2.3.3 Relative dating of archaeological features by Lake Superior, Canada. Surface exposure dating may offer considerable potential to the archaeologist, either in terms of the physical and/or chemical weathering characteristics of rock surfaces (McCarroll, 1994), or in terms of the cosmogenic nuclide dating of large rock or stone structures (Stuart, 2001). Thus far, however, there are relatively few studies that have employed either of these approaches to resolving an archaeological problem. One that has is Betts and Latta's (2000) investigation of a series of stone structures in eastern Canada.

Along the northern shore of Lake Superior, a number of archaeological features, known as the Pukaskwa pits, occur on the cobble beach terraces that formed through progressive glacio-isostatic emergence of Lake Superior following deglaciation. The pits typically appear as excavated ovoid depressions on the surface of the cobble terraces, with a ring or wall of stones erected around their circumference. Their function is unclear and their age is also unknown, estimates ranging from the prehistoric Late Archaic period (ca. 3000 years ago) to the historic Terminal Woodland period (1250–300 years ago). Dating had hitherto proved impossible, but Betts and Latta outline a methodology using the Schmidt hammer which has shed new light on the age of these unique structures. Because the beach terraces are emergent features, the highest is clearly the oldest and the lowermost

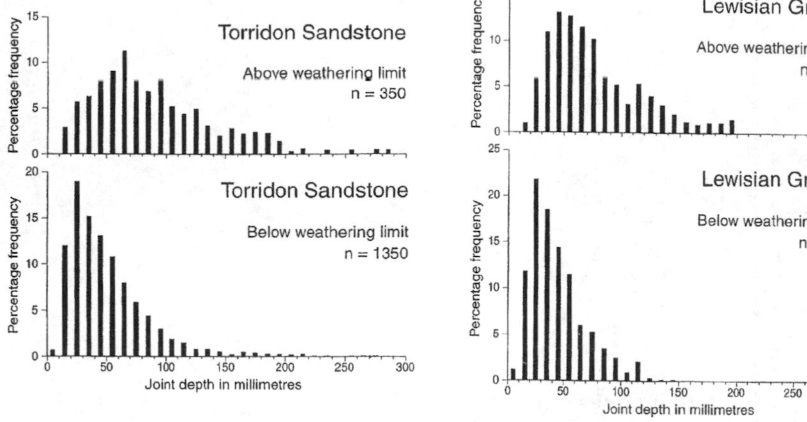

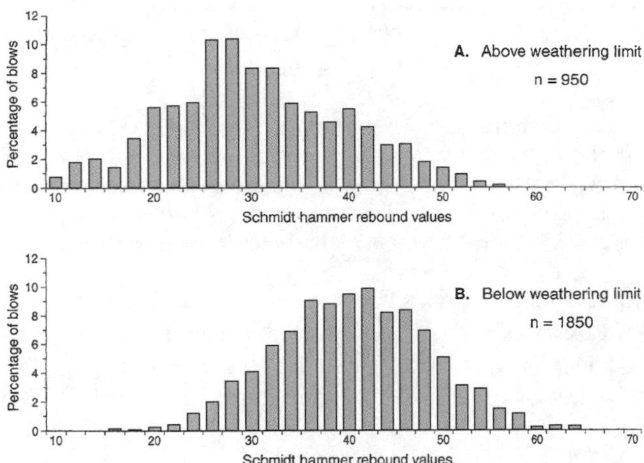

Figure 6.3 *(A) Depths of horizontal joints above and below the weathering limit (trimline) on Torridonian Sandstone (mainland Scotland) and Lewisian Gneiss (Harris, Outer Hebrides). (B) Schmidt hammer rebound values (R-values) from above and below the weathering limit on Torridonian Sandstone mountains on the Scottish mainland (after Ballantyne* et al.*, 1998b). Reproduced with permission of Elsevier*

the youngest. This was confirmed by Schmidt hammer readings which showed a measurable difference in surface hardness (R-values) of granitic cobbles of different age. A significant difference was also detected at most sample sites between rock hardness of cobbles in the pits and that on the adjacent terrace surface. For a number of beach terraces in the Pukaskwa region, the exposure ages are known from other evidence, and these data form the basis for a regression model of R-values against age (Figure 6.4). This relationship between rock hardness and age can then be used to infer a date for the structures,

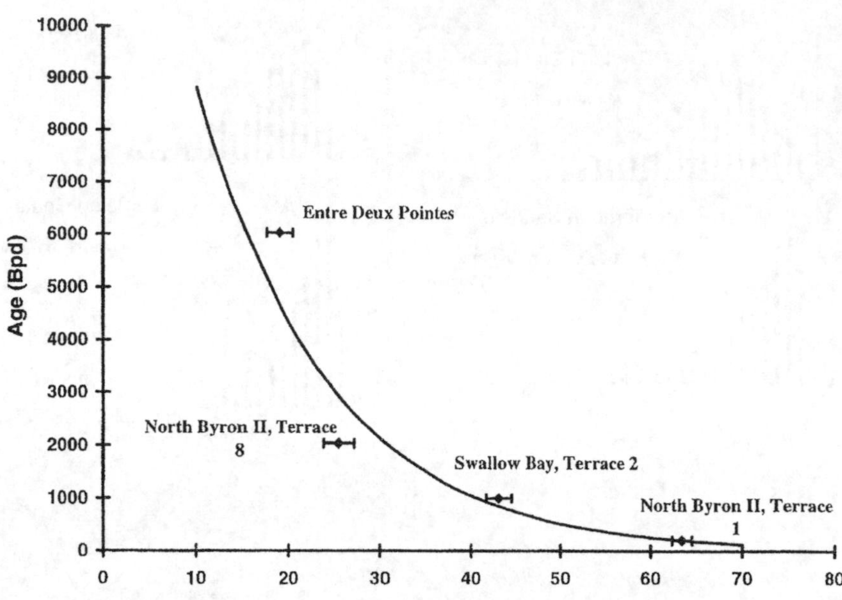

Figure 6.4 *Age versus R-value plot for selected terraces from the Pukaskwa region of Lake Superior, Canada. The curve shows that the surface hardness of a granite cobble rapidly decreases in the early stages of exposure, but this decrease becomes smaller with age (after Betts and Latta, 2000). Reproduced by permission of Blackwell Publishers*

based on the R-values obtained from stones within the pits. The evidence shows that one group of Pukaskwa pits was probably constructed around 900 years ago, and a second group between 300 and 500 years later. These ages are consistent with the known period of occupation of this area by the Blackduck people during the Terminal Woodland period.

6.3 Obsidian Hydration Dating

Obsidian is a dark coloured (usually black) glassy volcanic rock. It breaks with a conchoidal fracture and may be fashioned into a sharp cutting edge; consequently it has been widely used for making primitive cutting tools. Freshly broken surfaces of obsidian absorb moisture from their surroundings to form a hydration layer known as **perlite**. One factor governing the extent of perlite development is time, and hence by measuring the depth of the hydration layer, it may prove possible to estimate the relative age of exposure or breaking of the obsidian surface, working on the principle that the thicker the perlite layer, the older the exposed surface. Obsidian hydration dating (OHD) therefore enables archaeological artefacts, which are often manufactured from obsidian, to be ranked in terms of relative order of age. If the hydration rate can be quantified or calibrated, then it may be possible (at least in theory) to use the thickness of the perlite layer to determine the age of the artefact in years.

6.3.1 The Hydration Layer

It is important to emphasise that the hydration layer (or rim) on obsidian differs from the chemical weathering rinds developed on other rock surfaces that were discussed earlier, in that it forms entirely through the process of moisture absorption. The progress of this absorption is marked by a diffusion front that separates the hydrated glass from the unaltered glass below. A number of factors determine the rate at which hydration proceeds, and hence the thickness of the hydration layer. These include the chemical composition of the obsidian, the temperature of the hydrating environment, and the relative humidity of the hydrating environment (Beck and Jones, 1994), and each of these needs to be quantified in order for the method to work (see below). The hydration layer can be seen in thin section under a microscope when illuminated by polarising light. However, because hydration rims are very thin (sometimes less than 1.0 μm), optical measuring techniques are not always successful and computer-assisted imaging technology also is now used to determine the thickness of the hydration layer (Ambrose, 1994). More recently, secondary ion mass spectrometry (SIMS) has been employed to measure the concentration of hydrogen as a function of depth in obsidian samples, a technique known as **Obsidian Diffusion Dating** (Riciputi *et al.*, 2002). Estimates of the rates of hydration have been made on the basis of correlation of perlite layer thickness with associated radiocarbon dates, or through experimental work involving the exposure of fresh obsidian samples to steam or liquid at high temperature and measurement of the hydration layer that developed (Stevenson *et al.*, 1996). Dates have been reported in the age range from 200 to 100 000 years (Aitken, 1990).

6.3.2 Problems with Obsidian Hydration Dating

Of all the relative dating methods discussed in this chapter, there is no doubt that OHD has proved to be the most controversial. When first developed more than 40 years ago, it promised to be a rapid, inexpensive, relatively straightforward technique for dating obsidian artefacts. As such it was eagerly adopted by archaeologists, and there is now a substantial body of literature describing the results of this work. Although methodological problems have always been a concern, over the last ten years or so fundamental questions have been raised about the validity of the technique (Anovitz *et al.*, 1999), with one critic even posing the question, 'Where in the world does obsidian hydration dating work?' (Ridings, 1996). Other scientists, however, have mounted a robust defence of the method, acknowledging that while OHD is not without difficulties, it can deliver useful results (Hull, 2001).

There are several problems with the technique. Optical measurement, which has traditionally been the basis for determining hydration layer thickness, has been shown to be relatively imprecise. This means that many obsidian hydration ages that were obtained prior to the adoption of digital imaging techniques are likely to be in error. More serious, however, are problems relating to the determination of the hydration rate. We have already seen that hydration rate is influenced by a number of factors, including chemistry of the obsidian, temperature and relative humidity. All three may be highly variable, and hence the correct modelling of hydration rate, which is fundamental even if only a relative time inference is to be drawn, is far from straightforward. The calibration of obsidian hydration rates to an absolute timescale is even more problematical. The so-called 'empirical-rate dating', which involves the correlation of the width of optically measured hydration rims with independent chronometric data (e.g. radiocarbon dates), has frequently generated conflicting results, while the more complex and widely applied approach of 'intrinsic dating', which requires experimentally determined rate constants and a measure of site temperature, also appears to be compromised. The mathematical model that is used in intrinsic dating rests on the assumptions that the hydration rim grows at a rate proportional to the square root of time, and that the diffusion coefficient is constant. However, recent work involving SIMS analysis (section 6.3.1) indicates that both of these assumptions are incorrect and, if so, the rate equation which has long been used in traditional obsidian hydration dating must also be incorrect (Riciputi *et al.*, 2002).

These shortcomings would seem to imply that OHD simply does not work. However, even the most trenchant critics of the technique acknowledge that hydrated obsidian artefacts remain a potentially valuable source of chronometric data, but that in order for this potential to be realised, a more rigorous analytical approach is required (Anovitz *et al.*, 1999). Moreover, it may well be that the problem is not so much with obsidian hydration dating *per se*, but rather with unreasonable expectations of the temporal resolution that is possible with the technique. If OHD is accepted as a relative dating method that can provide rapid answers to site-specific questions, then it may still have a valuable role to play, particularly in establishing the order of antiquity of artefacts. Indeed, in many cases it may be the only practical dating method that is capable of resolving research problems relating to artefact – as opposed to site or component – dating for the foreseeable future (Hull, 2001).

6.3.3 Some Applications of Obsidian Hydration Dating

Although OHD has been most widely applied by archaeologists to date obsidian artefacts, the technique has also been used in geological and geomorphological contexts, for instance in the dating of volcanic activity (Friedman and Obradovich, 1981) and glacial events (Pierce *et al.*, 1976). Globally, the OHD has been used most extensively in western North America and in New Zealand, where outcrops of obsidian are relatively common. Two examples are considered here, one archaeological and one geomorphological, although both of these studies should be viewed in the context of the foregoing discussion on the limitations of the technique.

6.3.3.1 Dating of a Pleistocene age site, Manus Island, Papua New Guinea. At the base of a Miocene limestone cliff on the southern side of Manus Island, Papua New Guinea, is the Pamwack rock shelter, a site that has yielded a rich faunal and artefactual record and a history of occupation believed to span several thousand years (Ambrose, 1994). The site is important because it probably represents the oldest evidence for long-distance oceanic settlement yet recorded, being more than 200 km from the nearest islands of New Ireland across the Bismark Sea to the east. Radiocarbon dates are available from the site, but there are a number of anomalies of age with depth, probably due to site disturbance during prehistoric use of the site.

Several thousand flakes of obsidian were recovered from the uppermost 1.8 m of deposits in the rock shelter. The protected nature of the site, the deep profile with abundant obsidian, and a stable equatorial mean temperature regime make it a particularly suitable location for obsidian hydration dating. The value of applying OHD lies in the comparison with the dating series provided by radiocarbon, particularly where organic material is sparse. To counteract possible effects of surface weathering on the outer perlite layer, OHD measurements were made on internal crack surfaces of the obsidian flakes. The results show a consistent increase in hydration thickness on the crack surfaces with depth through the site (Figure 6.5A). An age–depth chronology was developed by comparing these results with available radiocarbon dates (Figure 6.5B). While this timescale must be considered as provisional, the increase in hydration layer thickness with depth can be regarded as a working model for a relative chronology for the site, and confirms the view that the Pamwak rock shelter has a long history of human occupation.

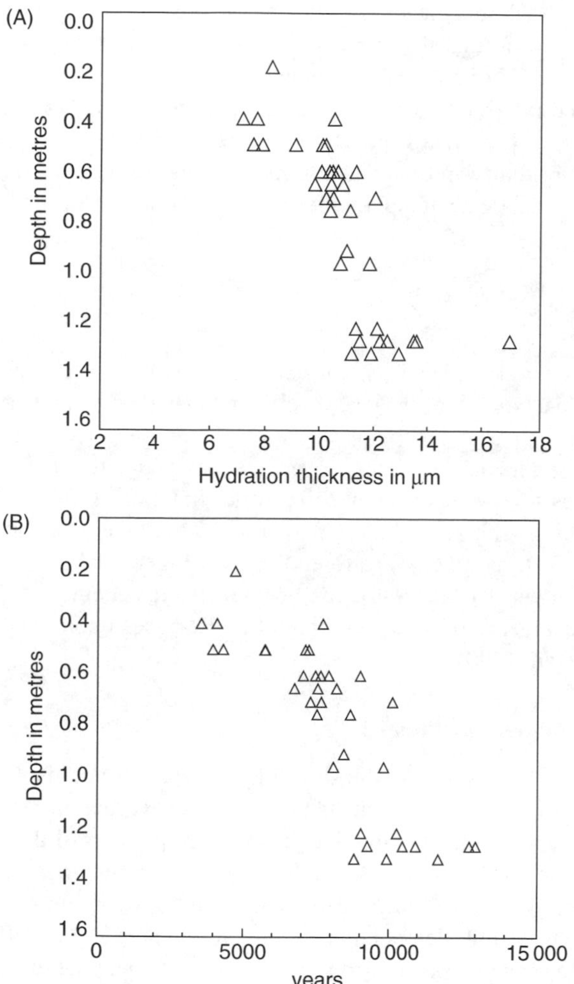

Figure 6.5 *(A) Relation of hydration thickness on internal crack surfaces with depth in the Pamwak site, Manus Island, Papua New Guinea. (B) Provisional age–depth model for hydration thicknesses for the Pamwak site. The ages are in calibrated years BP (after Ambrose, 1994). Reproduced with permission of Elsevier*

6.3.3.2 Dating of fluvially reworked sediment in Montana, USA. An unusual application of obsidian hydration dating is described by Adams *et al.* (1992) in their study of fluvial terraces cut into an outwash plain near West Yellowstone, Montana, USA. The floor of the West Yellowstone basin is covered by several tens of metres of obsidian-rich gravel (the Obsidian Sand Plain), the age of which is controversial, with estimates ranging from a glaciation predating the last interglacial (Bull Lake) to the Last Glacial Maximum (Pinedale Glaciation). The gravel surface has been dissected by the Madison River following the last deglaciation leaving at least five well-developed terraces. Working on the principle that fluvial transport will fracture obsidian grains, OHD dating was applied to samples from each of the terraces. The results showed that there was no significant difference between hydration layer thicknesses on obsidian samples from the different terrace levels, suggesting that the terraces were cut in a shorter period of time than the technique can discern. Numerical age estimates of the terraces, based on two previously published hydration rate estimates for the area, gave ages of 19 000 ± 1000 years BP and 24 400 ± 1100 years. These dates are entirely consistent with the hypothesis that final reworking of the Obsidian Sand Plain and terrace formation occurred during initial glacial retreat from the Pinedale glacial maximum (ca. 30 000 years ago).

6.4 Pedogenesis

Another index that can provide a basis for relative chronology is the degree of soil development. This approach involves the use of **soil chronosequences**, i.e. a series of soils that have formed within a particular region in which all the factors of soil formation except time (climate, parent material, vegetation cover, etc.) are held more or less constant (Huggett, 1998). Thus contrasts between different soil profiles in terms of, for example, particle size variations, soil micromorphological characteristics, depth of soil development, or indeed of other parameters such as magnetic susceptibility (Singer *et al.*, 1992) or clay fraction mineralogy (Ballantyne *et al.*, 1998b), can be interpreted as being a function of time and hence can provide a means of relative dating of the surfaces upon which those soils have developed (Birkeland, 1999).

6.4.1 Soil Development Indices

Although qualitative estimates of soil development have made useful contributions to the establishment of local relative chronologies, recent research has tended to adopt a more quantitative approach. This involves the generation of **soil development indices** which aim to reduce soil property data from a soil profile to a single number that reflects the degree of pedogenic development. This allows site to site comparisons, making it possible to identify temporal and regional trends in soil development (Goodman *et al.*, 2001) and to rank soil profiles in relative order of age. One example is the **profile development index (PDI)** which quantifies soil development at a site by comparing properties of a soil to those of the parent material (Harden and Taylor, 1983). This index is calculated on the basis of a range of soil properties (texture, structure, rubification, etc.) and points are allocated for stepwise departures of these properties from the soil parent material values. The profile values are then divided by a standard profile thickness to produce an index ranging from 0 to 1, with the higher values denoting greater departure from those of the parent material. Another index reflects degree of clast disintegration in a soil profile. The **clast disintegration index (CDI)** is calculated on the basis of the variation in degree of clast weathering in different soil horizons. Again, points are allocated up to a value of 90 (maximum weathering), and a weighted mean value of clast disintegration can be calculated for each soil profile using a standard profile thickness (Berry, 1994). Other soil indices, based on soil field and laboratory data, are described by Birkeland (1999).

While soil development indices can be used to determine relative order of age, they also constitute a basis for dating in years through the development of **soil chronofunctions**. Chronofunctions, which have been established for a number of relative dating criteria in addition to soils, are a means whereby independently dated sequences (based on radiocarbon dating or lichenometry, for example) are used to assign numerical ages to correlated deposits, or extrapolated to estimate the age of undated elements within or beyond the area of the dated sequence (McCarroll, 1991). Chronofunctions are usually derived by means of regression analysis in which the dependent variable is a particular landscape property (e.g. degree of weathering) and the independent variable is time. Hence, in the case of soil chronofunctions, the soil development indices are modelled as a function of time (Figure 6.7), and this enables the ages of the underlying surface to be inferred from the degree of pedogenesis that has occurred on those land surfaces.

6.4.2 Problems in Using Pedogenesis as a Basis for Dating

Several difficulties arise in using degree of pedogenesis as a basis for dating. Probably the most serious of these is the requirement that all soil-forming factors apart from time be held constant. Other influences, notably topography and vegetation, have been shown to be as important as time in explaining soil genesis, while the character of soils in an area may also reflect past conditions (Reider, 1983). Equally, the properties of a soil profile may have been affected as much by variations in climate as by time (Birkeland *et al.*, 1989). Questions also arise about the integrity of the soil profile, particularly where this may have been affected by erosion and/or redeposition. Studies in New Zealand, for example, have shown that in some older (pre-Holocene) soils, episodic loess erosion and redeposition have resulted in soils that are an order of magnitude younger than the underlying glacial deposits (Rodbell, 1990). Erosion has also been noted as a problem in soil chronosequences in the western United States, and may be the reason for a lack of age-related trends in clast weathering and soil development on moraines of manifestly different age (Berry, 1994). That particular investigation also highlighted a further difficulty with this form of dating, namely the inability of the method to provide finely tuned age resolution in regions where rates of soil development are relatively low. All of these problems suggest that while rates of pedogenesis (reflected in soil chronofunctions) may, in certain areas, be relatively reliable indicators of subsurface age (Goodman *et al.*, 2001), this particular form of dating is again perhaps best employed in combination with other chronometric techniques. Indeed, while there are numerous instances of relative ages being deduced from chronosequences, there are very few examples where soil chronofunctions alone have been employed to date surfaces of unknown age.

6.4.3 Some Applications of Dating Based on Pedogenesis

Pedogenesis has been most widely used in the dating of glacial deposits, especially moraines. In many mountain regions, older soils have been found to be characterised by deeper profiles, thicker horizons, a greater depth of oxidation and increased clay accumulation, and these and other characteristics have enabled relative ages to be assigned to moraine sequences, sometimes extending back over timescales of more than 100 000 years (Colman and Pierce, 1986). In a number of areas, including the Himalayas (Shiraiwa and Watanabe, 1991) and the Peruvian mountains (Rodbell, 1993), soil development has been employed in conjunction with surface weathering data in the development of glacial chronologies. Where independent chronometric control is available, relative glacial histories have been placed within a temporal framework on the basis of soil chronosequences. Two examples of these approaches are discussed here.

6.4.3.1 Relative dating of moraines in the Sierra Nevada, California. Along the eastern slopes of the Sierra Nevada, California, there is evidence in the form of moraine sequences for multiple glaciations. Geomorphological mapping has differentiated these into, from youngest to oldest, the Tioga, Tenaya, Tahoe, pre-Tahoe/post-Sherwin, Sherwin and McGee glaciations. Berry (1994) describes the results of a soil-geomorphic investigation which aimed to resolve the moraines into distinct relative age groups on the basis of soil development, weathering of surface and subsurface clasts, and geomorphic criteria. Various properties of the soils on the moraine crests and slopes, including carbonate content and pedogenic clay content (Figure 6.6), were quantified in the laboratory, and these data were combined with indices of soil development (PDI) and clast weathering (CDI: section 6.4.1). The results showed that it was possible to differentiate moraines of the Tioga and Tahoe Glaciations in the northernmost valley studied, while in two other valleys Tioga, Tahoe and pre-Tahoe/post-Sherwin moraines could be differentiated. In none of the three valleys could moraines believed to be from a Tenaya glaciation be distinguished. Berry suggested that this was probably because the soil-geomorphic methods are insufficiently sensitive to resolve what may be small age differences between the Tioga and Tenaya moraines. This conclusion has subsequently been confirmed by cosmogenic nuclide dating of the moraines which gave ages of 15 500–17 000 years for the Tioga and 18 000–19 500 years for the Tenaya moraines in the eastern Sierra Nevada (Phillips, F.M., 1995, personal communication). Hence although soil geomorphic methods do not allow finely tuned age resolution, they do provide important data on the relative age of the glacial landforms, and they also constitute a basis for evaluating dates generated by chronometric techniques.

6.4.3.2 Dating glacial events in southeastern Peru. In the Cordillera Vilcanota and Quelcayya Ice-Cap regions of southwestern Peru, there is a detailed record in the form of moraines and associated landforms of a complex history of Late Quaternary glaciation. Goodman *et al.* (2001) show how variations in degree of soil development, in association with a programme of radiometric dating, can be employed to reconstruct the glacial history of this mountain region. They generated soil development indices for profiles on 19 separate moraine crests, and these were calibrated by reference to radiocarbon dates on buried soils or on peats, and cosmogenic isotope dates on boulders on the moraine surfaces (Figure 6.7). Soil age estimates from extrapolation of field and laboratory data suggest that the most extensive Late Quaternary glaciation occurred >70 000 years BP. This is the first

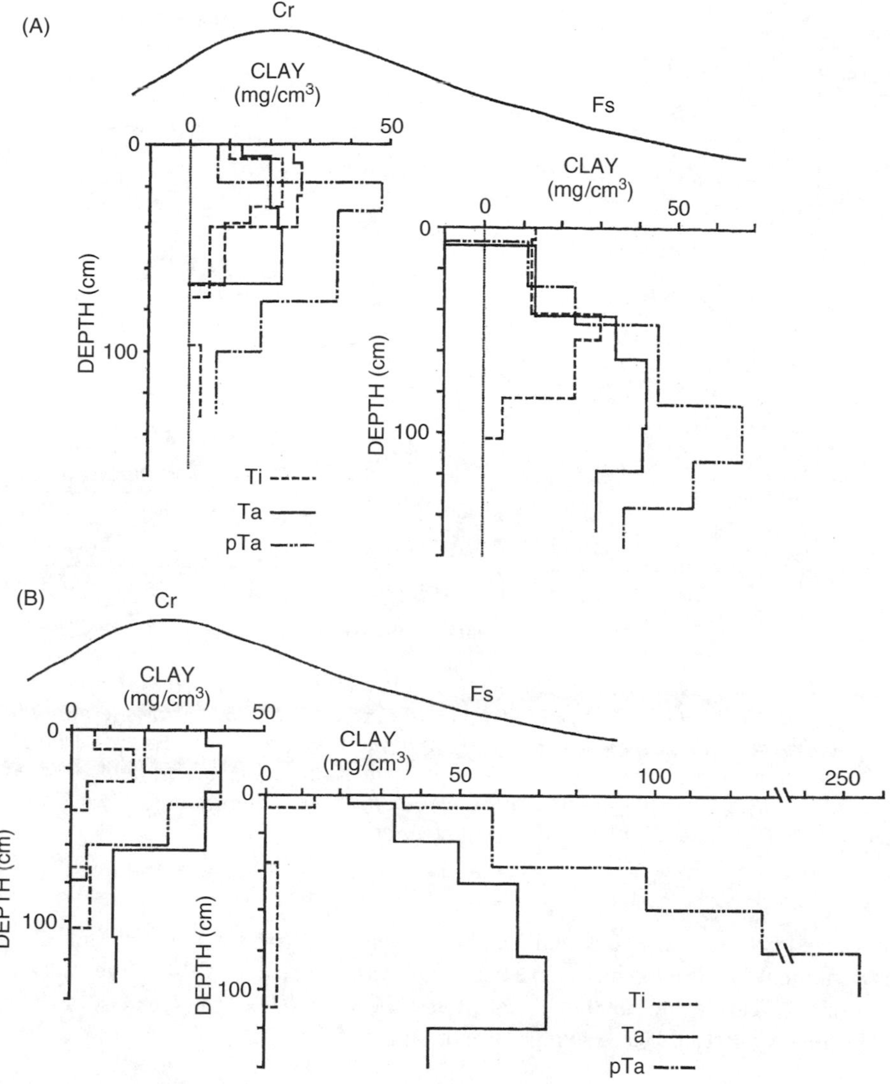

Figure 6.6 *Distribution of pedogenic clay content with depth in soils on Tioga (Ti), Tahoe (Ta) and pre-Tahoe (pTa) moraines in (A) Pine Creek and (B) Bishop Creek in the eastern Sierra Nevada, California. The data are taken from the moraine ridge crests (Cr) and moraine footslopes (Fs) as shown in the schematic cross-profiles above. Note how moraines of different age display different distributions of pedogenic clay down-profile (after Berry, 1994). Reproduced with permission of Elsevier*

semi-quantitative age estimate for maximum ice extent in southern Peru, and is supported by a limiting radiocarbon date of ca. 41 500 ^{14}C years BP. Younger moraines occurring upvalley from this glacial maximum appear to be associated with a glacial advance that culminated at ca. 16 650 cal. years BP in the Cordillera Vilcanota. Following a phase of rapid deglaciation, outlet glaciers on the northern margin of the Quelcayya Ice-Cap

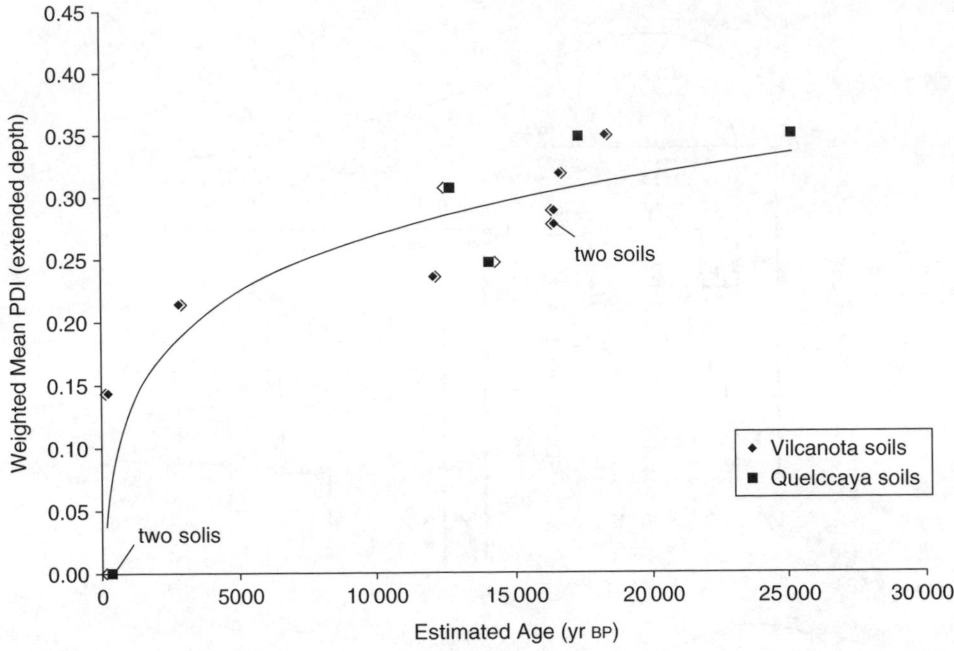

Figure 6.7 *Variations in PDI with soil age in the Vilcanota and Quelcayya regions of Peru. The regression line in the soil chronofunction has an* r^2 *value of 0.87. Arrows to the right (>) indicate that the age assignment is based on minimum-limiting* ^{14}C *date(s) and true age is older than that plotted; arrows to the left (<) indicate that the age assignment is based on maximum-limiting* ^{14}C *date(s) and true age is younger than that plotted (after Goodman* et al., *2001). Reproduced with permission of Elsevier*

readvanced between ca. 13 090 and 12 800 cal. years BP. The youngest moraines, which show minimal soil development, have been radiocarbon-dated to <394 cal. years BP in the Cordillera Vilcanota and <300 cal. years BP on the west side of the Quelcayya Ice-Cap, and correlate with Little Ice Age moraines of other regions.

6.5 Relative Dating of Fossil Bone

A number of elements not present (or having only a limited presence) in living bone are found in buried bones. These include fluorine and uranium from groundwater that become absorbed into the hydroxyapatite, the phosphatic mineral of which bone is composed. Secondary minerals, such as calcite and pyrite, can form in voids in archaeological bone, and in some cases replace the bone mineral itself. Internal changes in bone structure, involving changes in crystallinity, have also been detected (Millard, 2001). Because time is a factor in all of these **diagenetic changes**,[4] analysis of the chemical and structural content of fossil bone material can provide a basis for relative dating.

6.5.1 Post-Burial Changes in Fossil Bone

In terms of relative dating, the most widely used indicator of chemical change in bone is the fluorine content. Because hydroxyapatite absorbs fluorine over time, the higher the fluorine content of a piece of fossil bone, the longer the time that has elapsed since burial. Older bone assemblages will therefore be distinguishable from younger assemblages on the basis of their fluorine content. Moreover, because bones that have been buried in a particular context for the same length of time will have a broadly comparable fluorine content, contamination of the assemblage by older or younger bones will be readily revealed by anomalously higher (older) or lower (younger) fluorine contents (Demetsopoulos *et al.*, 1983). This not only provides valuable information about the taphonomy[5] of a bone assemblage, but it is also a valuable screening mechanism in the selection of bones from that assemblage for radiocarbon or U-series dating.

An alternative approach to the relative dating of bone material is by comparison of **fluorine profiles** through individual bones. In younger bones, the fluorine concentration falls rapidly from the outer part of the bone towards the centre, whereas in older bone material, the profile is relatively flat. The shape of the profile is therefore age-dependent and, as such, it provides an approximate dating tool (Aitken, 1990). Again, this technique can be used to establish relative ages in a mixed bone assemblage, and also as a guide to selection of samples for radiocarbon dating. Uranium incorporated into fossil bone material from groundwater can be used in a way similar to the fluorine method, and to some extent has an advantage in that the counting of uranium emissions does not involve the destruction of the bone material. The relative ages of fossil bones in assemblages can also be established by analysing the nitrogen content of the bone collagen, for as the collagen breaks down after burial, the nitrogen levels will also fall. Hence, decreasing nitrogen content will be an indicator of increasing age.

Post-burial crystallinity change in bone may also provide a basis for relative dating. This can be measured by a 'crystallinity index' based on XRD or IR spectra (Bartsiokas and Middleton, 1992; Sillen and Parkington, 1996). Studies have shown that bone crystallinity increases with time, and that the method can be used to establish the relative age of bone material up to around 20 000 years. In addition, it can distinguish between younger bone (under 3000 years) and more ancient material (>40 000 years), and hence may prove to be a useful means of determining antiquity where the provenance of the bone has been questioned. It also has the advantage of minimal destructiveness (Sillen and Parkington, 1996).

6.5.2 Problems in the Relative Dating of Bone

In common with a number of the chemical methods considered in this chapter, the use of diagenetic changes in fossil bone as a basis for relative dating suffers from problems of non-uniform alteration and low levels of accuracy and reliability. Questions arise, in particular, over the nature of the processes involved in post-burial diagenesis, and while some, such as uranium uptake and carbonate exchange, might be described as being partly understood, others cannot and remain the focus of much recent research (Millard, 2001). Similarly, while progress has been made in understanding crystallinity changes (Hedges *et al.*, 1995), more work is needed on the variation of crystallinity and solubility characteristics in bone mineral material before this can be regarded as a completely reliable basis for establishing relative order of age.

6.5.3 Some Applications of the Relative Dating of Bone

Perhaps the most famous application of the relative dating of fossil bone material was the use of fluorine analysis to show that the skull of 'Piltdown Man' was a hoax, an assertion that was subsequently confirmed by radiocarbon dating (Oakley, 1969). In subsequent years, however, with the development of a range of radiometric dating techniques and growing concerns about diagenesis, interest in relative dating of bone tended to wane. More recently, partly because of the research that has been undertaken into the complexities of bone chemistry, and partly because of advances in instrumentation and analytical methodologies, attention has been focussed again on the potential of relative dating, albeit perhaps as an adjunct to other dating techniques. Two examples of these applications are considered here.

6.5.3.1 Fluoride dating of mastodon bone from an early palaeoindian site, eastern USA. Tankersley *et al.* (1998) analysed the fluoride content of a large number of mastodon bones from the Hiscock site, Genesee County, New York, an early palaeoindian site that had also yielded artefacts (six bifaces and an end scraper), worked mastodon bone and ivory, and a stone bead. The stratum from which these artefacts were recovered also contained the bones of caribou, stag-moose, giant beaver, Californian condor and mastodon (*Mammut americanum*). The mastodon bed was believed to be terminal Pleistocene in age and, like other early palaeoindian bone beds, was thought to reflect a dynamic and evolving system of biological, cultural and sedimentary processes. Fluoride analysis (Figure 6.8A) showed that this was indeed the case, as the results display a distinctive trend in relative time, indicating that the bones accumulated through time rather than being amassed from a single event. The trend was further confirmed by radiocarbon dating (Figure 6.8B) as this showed that the accumulation phase lasted for more than 1000 years. This study re-emphasises the potential of fluoride dating for providing a rapid assessment of the taphonomy of animal bone assemblages.

6.5.3.2 Chemical dating of animal bones from Sweden. A second example of the use of fluorine dating is Johnsson's (1997) study of animal bones found buried in soil at a number of different localities in Skåne, southern Sweden. Eighty-seven bones were analysed for fluorine content, and 61 of these were also analysed for 13 other major and minor chemical elements. The results showed that not only fluorine but also silicon is incorporated into the bone mineral material over time, while silicate progressively replaces phosphate in the bone apatite crystals. These findings were subsequently corroborated by the results of radiocarbon dating (Figure 6.9). Several old specimens of bone from a large quantity of stored and undated material were analysed in a similar manner, and some of those with high fluorine and silicon content were radiocarbon-dated. As a result, the oldest finds of wild horse in Scandinavia (five specimens dated between 11 180±95 and 10 495±95 ^{14}C years BP) and one of the oldest finds of domestic cattle in Scandinavia (4690±80 ^{14}C years BP) were confirmed. As was the case with the Hiscock analysis, these results show that valuable information on relative order of antiquity can be obtained from the chemical analysis of ancient bone material.

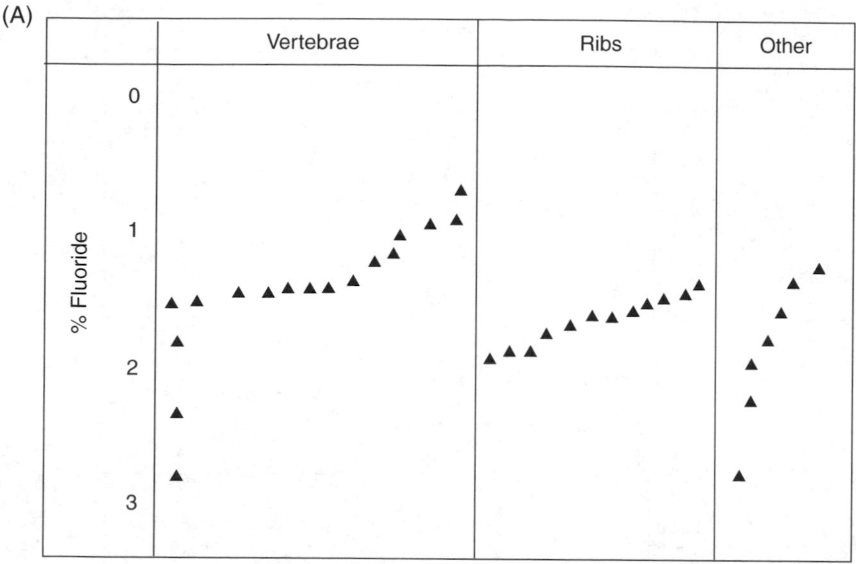

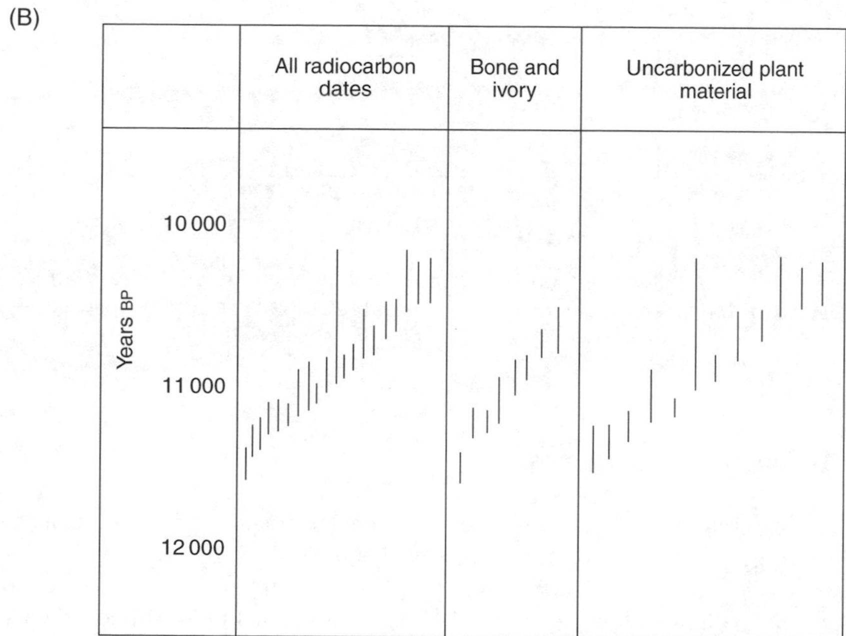

Figure 6.8 *(A) The distribution of fluoride content of individual mastodon skeletal elements from the Hiscock site, New York. The horizontal scale reflects relative time. (B) The distribution of radiocarbon dates in bone, ivory and uncarbonised plant material from the Hiscock site. The vertical bars represent 1σ (after Tankersley* et al.*, 1998). Reproduced with permission of Elsevier*

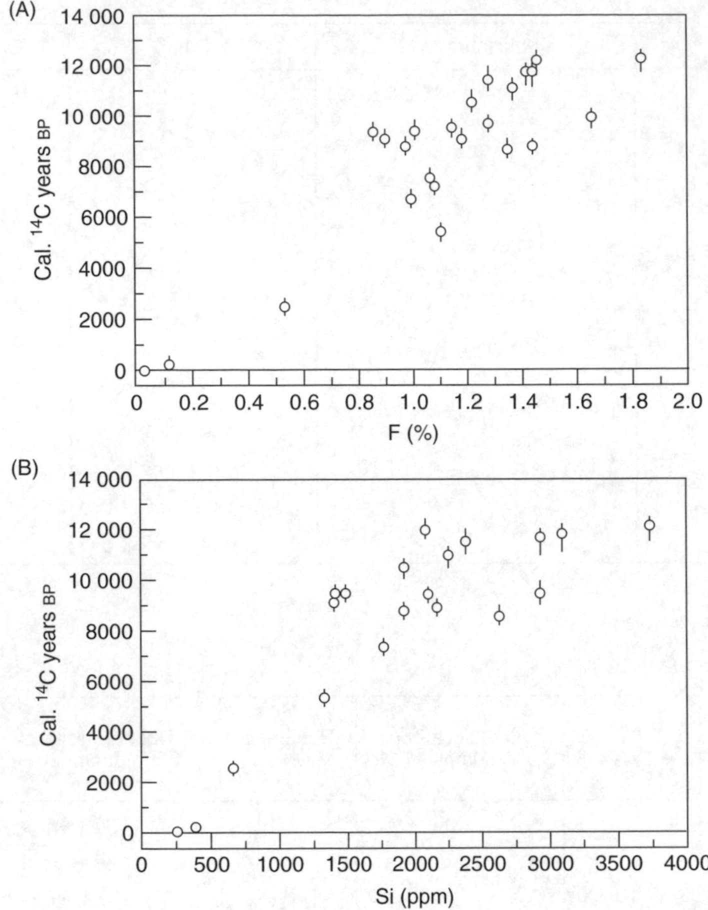

Figure 6.9 *Plots of (A) fluorine and (B) silicon content of mammal bones from sites in Skåne, Sweden. The timescale on the* y*-axis is in calibrated years* BP *(after Johnsson, 1997)*

6.6 Amino Acid Geochronology

All living organisms contain **proteins**. These are complex organic compounds that contain atoms of oxygen, carbon, nitrogen, hydrogen and sulphur. Proteins survive in carbonate structures, such as bone or shell, for long periods of time and have even been recovered from Cretaceous dinosaur bones (Wykoff, 1980). In the 1950s, Philip Abelson, then at the Carnegie Institute in the United States, found that proteinaceous residues (amino acids) manufactured by an organism during life do not disappear at death; rather they become locked in the mineral matrix of molluscs or egg shells, for example, where molecular changes occur, some of which are time-dependent. Hence, the analysis of certain protein residues from the Quaternary fossil record can provide the basis for a chronology. Unlike the decay of radiocarbon, however, the precise rates at which chemical change occurs are not easy to establish, and protein diagenesis can therefore only be used to rank fossil materials in relative order of age. As with all of the other techniques discussed in this chapter, it does not, at least in the first instance, provide a means for establishing age in years.

6.6.1 Proteins and Amino Acids

Proteins are some of the largest molecules that exist in living organisms. They are composed of **amino acids**, the basic chemical units from which proteins are synthesised by the body. Amino acids are linked together to form compound structures known as **peptides** or **peptide chains**, each amino acid being connected to the next by **peptide bonds**. The patterns formed in this way are specific to each protein type, and the various amino acid arrangements create thousands of different proteins, including enzymes and antibodies. Approximately 80 amino acids occur in nature, of which about 20 are commonly found in proteins. Human beings need 20 amino acids for metabolism or growth of body tissues; the human body can synthesise 9 of these, the remainder (essential amino acids) are obtained from animal or plant tissues in the diet. A list of the more common amino acids includes alanine, glycine, isoleucine, leucine, serine and aspartic acid.

All amino acids (with the exception of glycine) exist in two molecular forms, known as **isomeric forms**. In terms of chemical and biochemical properties, these two forms are very similar. When examined under a microscope, however, there is a very clear difference, for they rotate plane polarised light in opposite directions. In effect, they form two non-superimposable mirror images (Figure 6.10) like left and right hands. Indeed, these optical isomers are termed **L-amino acids** (after the Latin *laeva* – left hand) and **D-amino acids** (*dextra* – right hand). Biologically, there is an important distinction between the L- and D-configurations because living proteins contain only L-isomers. The D-isomeric form can occur in a free state or as a component of non-proteinaceous structures. It also exists, however, in fossil material as a result of the breakdown of proteins, and it is this molecular

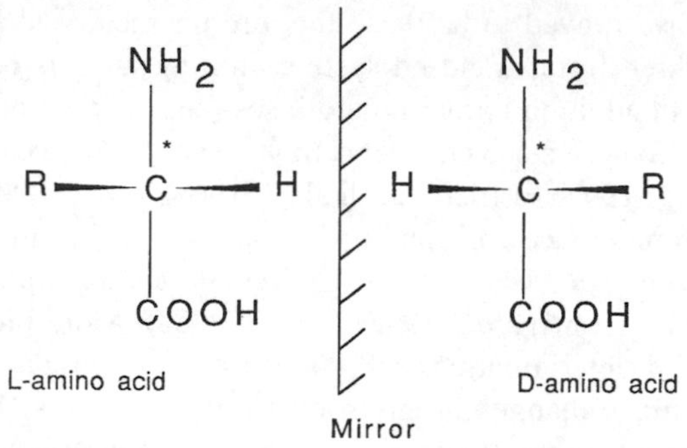

Figure 6.10 *Generalised representation of D- and L-amino acids. The side chain (R) is different for each amino acid. The carbon atom at the centre of the isomers (the chiral carbon atom) forms the point of asymmetry which allows the formation of the two optical isomers (after Sykes, 1991). Reproduced by permission of the Quaternary Research Association*

change that forms the basis for relative dating. Further details on amino acid geochemistry and its applications in geochronology are provided by Goodfriend *et al.* (2000), Wehmiller and Miller (2000) and Blackwell (2001b).

6.6.2 Amino Acid Diagenesis

Following the death of an organism, chemical changes occur that lead to the disruption of the peptide chains and the release of free amino acids. Degradation will be rapid in situations where the proteinaceous residues are exposed to the atmosphere, but the process will be retarded where the proteins are contained within harder skeletal materials. In such media, where effectively 'closed-system conditions' obtain, reaction times may span thousands or even millions of years.

In terms of Quaternary chronology, the most significant diagenetic change to occur in fossilised material is the inversion (intraconversion) of the L-amino acids to their respective D-configuration, a reaction that proceeds until an equilibrium position has been achieved. This process is known as **racemisation**. If the rate of inversion can be calculated, the time that has elapsed since the death of an organism can be determined from the ratio of D/L-amino acids. In Quaternary science, the most commonly used reaction is that of the amino acid L-isoleucine (L-Ile) to its non-protein diastereoisomer, D-alloisoleucine (D-aIle). This is a more complex process than racemisation and is referred to as **epimerisation**. By far the most widely used media for amino acid epimerisation measurements have been marine molluscan shells. In these, the ratio of D- to L-isomers increases from near zero in modern material to an equilibrium ratio of 1.30 ± 0.05 (Miller and Mangerud, 1985). Thus when amino acid ratios are measured, the results are expressed along the continuum from 0 to 1.30 (e.g. 0.26, 0.50, 0.75); the higher the ratio, the older the material, reflecting the fact that as time since death increases, the greater will be the number of L-isomers that have been inverted to the D-configuration. Measurements of amino acid ratios (AARs) are made using either ion exchange chromatography or gas–liquid chromatography, and kinetic models are used to determine the trend of protein diagenesis (Sykes, 1991).

As amino acid diagenesis occurs in all proteinaceous materials, in theory at least, a range of fossils could be used as dating media. In practice, however, the only really suitable materials have proved to be those that are characterised by a 'tight' skeletal carbonate matrix which forms a closed system and, as such, is relatively unaffected by external factors. In addition to marine molluscs (see above), successful amino acid measurements have also been made on other carbonaceous materials, including non-marine molluscs (Goodfriend, 1992), ostrich eggshells (Brooks *et al.*, 1990), organic residues derived originally from surface soils and now contained within the carbonates of cave speleothems (Lauritzen *et al.*, 1994), and organic materials incorporated into carbonate rocks in coastal areas (Hearty *et al.*, 1992; Hearty, 2003). More problematic, however, have been amino acid determinations on wood, as such non-carbonaceous materials are not buffered against changes in environmental pH (Sykes, 1991). Bone is also difficult, for although numerous attempts have been made to use amino acid diagenesis (especially of aspartic acid) to date archaeological materials, bone appears to be particularly unsuited for this purpose (Gernaey *et al.*, 2001). This is principally because bone does not constitute a closed system, so that once buried the original amino acids can be lost and younger ones added. Indeed, studies have shown that leaching of the more soluble proteins from bone and retention and degradation of the insoluble proteins can, potentially, give any value for a racemisation ratio (Collins *et al.*, 1999).

6.6.3 Problems with Amino Acid Geochronology

Amino acid ratios in fossil material are affected by a number of factors, including temperature, taxonomy and processes within the depositional environment. Temperature is a major influence, for rates of epimerisation increase exponentially with temperature, with an approximate doubling of the rate for every 4 °C temperature rise. This means that the shell of a mollusc that died during the last interglacial will have experienced significant variations in rates of epimerisation over the course of the past 130 000 years (Figure 6.11). If meaningful comparisons are to be made between amino acid ratios from different sites, then those sites have to have experienced similar climatic, and especially thermal, histories. A further complication is that proteins in different structural layers of a shell may often differ in terms of their amino acid composition and rate of epimerisation. Ideally, therefore, the same portion of each shell should be used for analysis, but in practice this is extremely difficult (indeed seldom possible) when samples are taken from fragmented molluscan assemblages. For some media, contamination and exchange of amino acids with the depositional environment may lead to aberrant amino acid ratios, although this may be less of a problem for molluscs than for other fossil materials such as bone (see above). More serious, however, can be the degradation of proteins due to microbiological activity during the early stages of diagenesis, as this may result in significant differences in amino acid

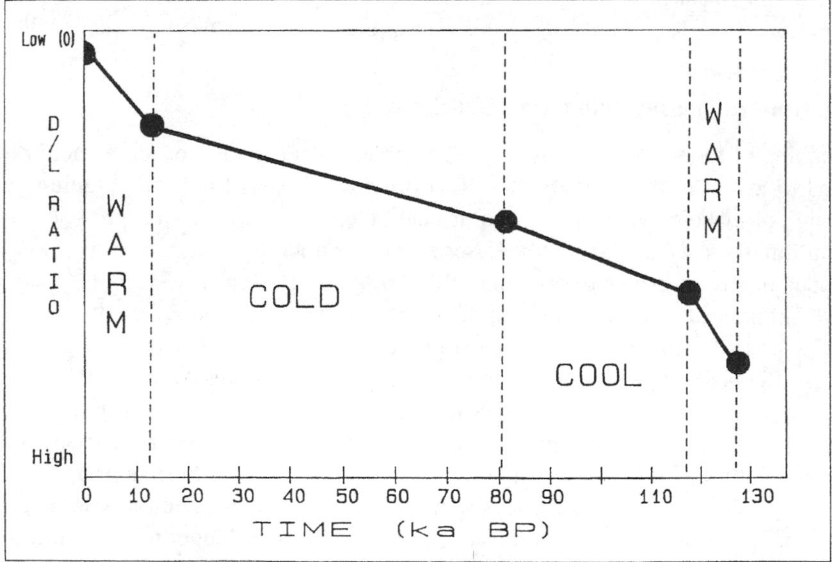

Figure 6.11 *Simplified diagram showing the increase in D/L ratio in a molluscan fossil from northwest Europe over the past 130 000 years. The most rapid epimerisation rate occurs during the warmer interglacials, whereas during the intervening cold stage the rate was markedly reduced (after Miller and Mangerud, 1985). Reproduced with permission of Elsevier*

ratios in molluscs from sites of the same age with similar thermal histories (Sejrup and Haugen, 1994).

Although perhaps less contentious than OHD dating (section 6.3), amino acid geochronology has still proved to be controversial. The method has not always had a 'good press' and, in some quarters, continues to be treated with a degree of scepticism. Its image was undoubtedly damaged in the 1970s by the alleged dating of some Californian palaeoindian remains to 50 000–60 000 years ago (Bada *et al.*, 1974) leading to claims of very early colonisation of America. The bones were subsequently redated by AMS radiocarbon to 5000–6000 years BP (Bada, 1985), and the error therefore corrected. Nevertheless, inconsistencies between amino acid evidence and results from other dating techniques continued to raise questions about the validity of the technique (Marshall, 1990). More recently, the interpretation and application of molluscan AARs has been challenged, especially where these have been employed to characterise individual stratigraphic units, or to differentiate between fossil beach sequences from successive interglacial episodes (McCarroll, 2002). AARs are naturally variable, and it is often the case that a range of values will be obtained from any one stratigraphic context. This means that the technique needs to be applied with due regard to the natural variability of AARs and hence age estimates must be based on samples that are sufficiently large to characterise this natural variability.

The foregoing critique does not mean that amino acid geochronology cannot be used to good effect. On the contrary, because it can be applied to both marine and non-marine mollusca, over a wide range of ages it is, potentially, one of the most useful techniques available in Quaternary dating (McCarroll, 2002). However, AARs should perhaps be considered more as a guide to relative age, to be assessed with other evidence, rather than as a definitive stratigraphical or dating tool. Indeed, many studies in Britain, Europe and elsewhere have used amino acid geochronology in this way, as the examples in the next section will show.

6.6.4 Applications of Amino Acid Geochronology

The ranking of fossils and their associated sediments on the basis of amino acid ratios is referred to as **aminostratigraphy**, and this provides a basis for relative dating and for inter-site correlation. When the AARs are calibrated by methods that provide an independent timeframe (^{14}C, OSL, ESR, U-series, etc.), an amino acid geochronology can be developed that is applicable beyond the calibration site (Bowen, 1999), and in some cases relative amino acid ages may range up to 1 million years (Bowen *et al.*, 1998). We have already seen that a major controlling factor on amino acid diagenesis is temperature, and that this affects the time required to attain an equilibrium state. At mid-latitude sites, it may take ~2 million years to reach equilibrium, whereas at Arctic sites it could require 20 million years or more to reach the same state (Miller and Mangerud, 1985). In view of the generally low temperatures that existed at many mid-latitude regions during the Quaternary cold stages, and their likely retarding effect on rates of amino acid diagenesis, it might be anticipated that it would have taken considerably longer than 2 million years to attain the equilibrium state. This has been confirmed by amino acid ratios from East Anglia in England, where D-alle/L-Ile ratios have been obtained on molluscs from the Norwich Crag (0.80) and from the earlier Red Crag (1.09), some of the earliest Quaternary deposits in the British Isles (Bowen, 1991). Four examples of the applications of aminostratigraphy are considered in the following section.

6.6.4.1 Dating and correlation of the last interglacial shoreline (~MOI substage 5e) in Australia using aminostratigraphy. One of the most widely used applications of aminostratigraphy has been in the identification of, and correlation between, interglacial shorelines of different age. This approach involves the measurement of D/L ratios on marine molluscs associated with those shorelines, and on the basis of these measurements, regionally applicable **aminozones** can be developed which are characterised by a particular range of AARs and reflect individual high sea-level (interglacial) events. Often these aminozones are constrained by other evidence, such as geomorphological and stratigraphical relationships between shorelines, and by radiometric dating (radiocarbon, U-series, etc.), which is important given the potential limitations of amino acid geochronology described in section 6.6.3. High sea-level events in many coastal regions have been differentiated on the basis of aminostratigraphy including, for example, the English Channel where raised beaches have been linked with high sea-level events in MOI stages 5e, 7, 9, 11 and 13 (Keen, 1995; Bates *et al.*, 2003).

One of the problems with this approach is that it assumes that the sites that are being compared have experienced similar thermal histories (section 6.6.3). While this may be the case in a relatively limited geographical region, such as the coasts of southern Britain and northern France, it may not necessarily be so at the continental scale. However, providing that the effects of temperature variability can be corrected for, aminostratigraphy can still prove a powerful correlative and chronometric tool over relatively large areas. Murray-Wallace *et al.* (1991) provide an example from southern Australia. There, amino acid racemisation data on marine molluscs associated with the last interglacial shoreline from sites ranging from the western coast of Australia to the eastern coast of New South Wales show how racemisation rates increase exponentially with contemporary mean annual temperature (CMAT), the latter being a function of latitude (Figure 6.12). Once this relationship has been established, however, then D/L ratios on shorelines of hitherto unknown age but which fall within the envelope of the amino acid data shown on Figure 6.12 with respect to the CMAT for a given sample site, are likely to be of last interglacial age. Figure 6.12, therefore, represents a predictive model that may be used in subsequent studies to assign preliminary ages to coastal marine deposits in southern Australia on the basis of the amino acid ratios in associated marine molluscs.

6.6.4.2 Quaternary aminostratigraphy in northwestern France based on non-marine molluscs. The classification of Quaternary deposits in northwestern France has conventionally been based on the study of loess profiles, slope deposits, palaeosols and interbedded fluvial sediments. In the major river valleys, such as the Seine and the Somme, the fluvial sequences have been grouped into geomorphological 'terraces'. However, correlation between the various deposits has often proved problematical and an overall time-stratigraphic framework has been difficult to develop.

Bates (1993) has proposed an alternative approach using amino acid geochronology. Amino acid ratios on non-marine molluscs from a number of sites in northwestern France and from the Seine and Somme Valleys formed the basis for a regional aminostratigraphic scheme in which seven amino acid groups could be identified on the basis of mean standardised D/L ratios. These are consistent with previous stratigraphic schemes and confirm the relative order of the Somme terrace sequence (Figure 6.13A). The D/L ratios were subsequently calibrated using TL dates and magnetostratigraphy, and

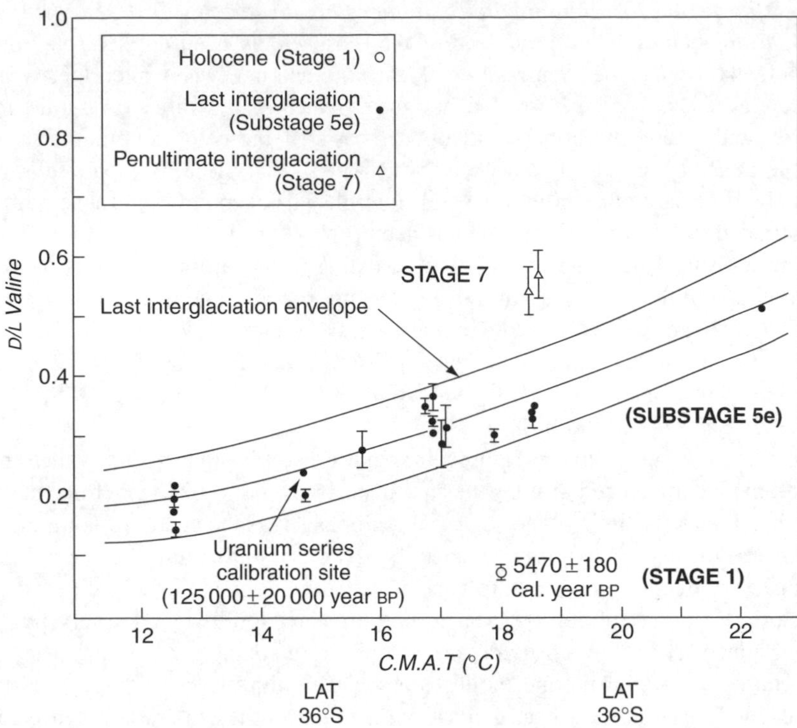

Figure 6.12 *Valine racemisation in last interglacial (MOI substage 5e) molluscan fossils plotted against contemporary mean annual temperature (CMAT °C) for coastal sites in southern Australia within the latitudinal range 24°S–43°S. Note how the amino acid data for both MOI stage 7 and MOI stage 1 (radiocarbon-dated) sites plot well outside the envelope for MOI substage 5e sites. A single uranium series date calibrates the amino acid data to the last interglacial (after Murray-Wallace* et al.*, 1991). Reproduced with permission of Elsevier*

compared with the MOI stratigraphic framework (Figure 6.13B). This suggests that amino group 1 is of Weichselian age (MOI stages 2–4), that group 2 is of last interglacial (Eemian) age (MOI substage 5e), and that group 3 is either late MOI stage 7 or early stage 6 in age. Amino acid groups 4–6 are less easily correlated with the marine sequence, and may span more than one marine isotopic stage. A more secure ascription may be made in terms of amino group 7, however, as these samples were taken from sediments that display reversed geomagnetic polarity (Matuyama reversed: section 7.4.3), and hence must be earlier than MOI stage 20. Despite the difficulties in correlating groups 4–6 with marine isotope stages, this scheme offers an independent means of subdividing, correlating and dating the complex sequence of deposits in northern France and offers the basis for a new regional time-stratigraphic framework. Similar approaches, using amino acid geochronology, have also been employed in the subdivision and correlation of river terrace sequences in central and southern Britain (Bridgland *et al.*, 2004).

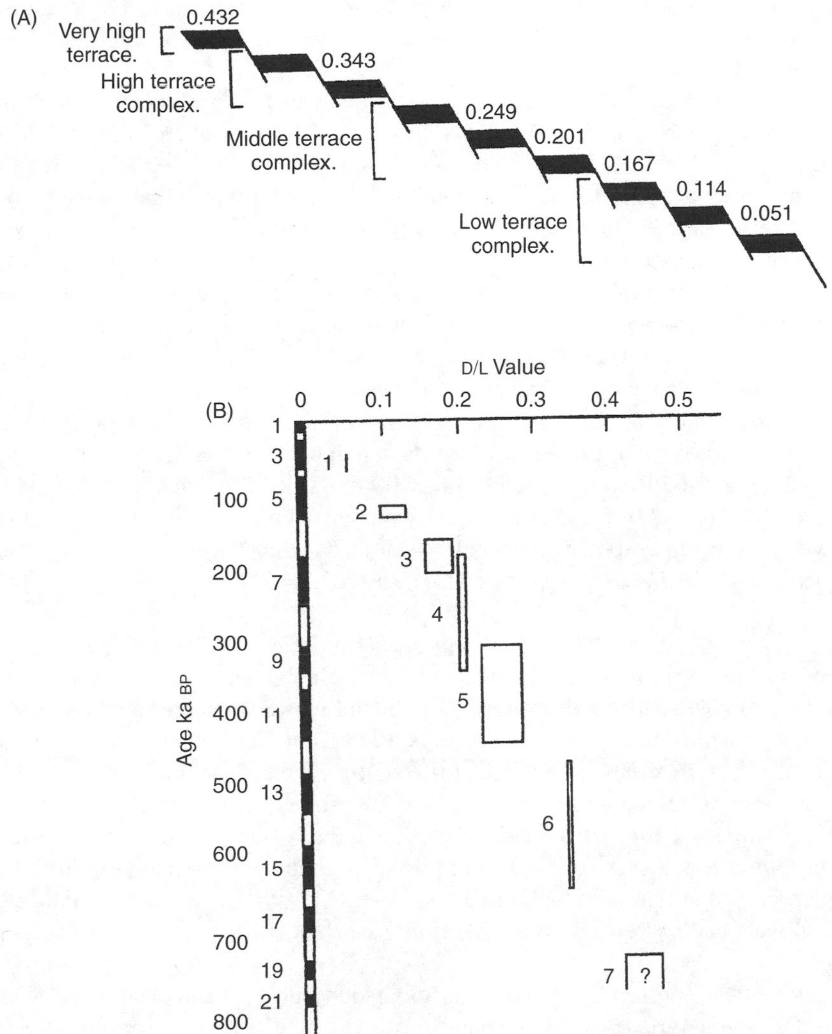

Figure 6.13 (A) Composite diagram showing the terrace sequence and amino acid groups in the Somme Valley, northwest France. (B) Correlation of amino acid groups in northwest France with the MOI sequence. Note how groups 1–3 are readily differentiated from each other and are temporally well constrained, by contrast with groups 4–6 which are less well differentiated and appear to span much greater time ranges (after Bates, 1993). Reproduced with permission of Elsevier

6.6.4.3 Dating the earliest modern humans in southern Africa using amino acid ratios in ostrich eggshell. The oldest anatomically modern human remains are beyond the range of radiocarbon dating, and the depositional contexts in which these bones have been found often lack material suitable for other dating methods. As a consequence, age estimates for early human skeletal material and associated sedimentary horizons in southern

Africa, a key area for human evolutionary studies, are often based on long-distance correlations with the deep-ocean record, a procedure that can only provide a general indication of age.

Miller *et al.* (1999) offer an alternative approach to this problem by using amino acid ratios from ostrich shell. They focus on a particular substage of the Middle Stone Age in southern Africa, the Howiesons Poort industry, which is a distinctive cultural-stratigraphic marker in sequences south of the Zambesi River. Anatomically modern human material has been found associated with, or in horizons older than, the Howiesons Poort layer in stratified deposits in Border Cave and at Klasies River (Figure 6.14A). Miller and co-workers took samples of ostrich shell material, which had been brought into the cave for food and was interstratified with lithic material and other evidence of human occupation, from the Howiesons Poort horizons in Border Cave, Boomplaas Cave and Apollo 11 cave. Isoleucine epimerisation data from the ostrich shells were calibrated to a timescale using radiocarbon dates on organic materials found in the upper parts of the sequences (Figure 6.14B). Collectively the data suggest that the Howiesons Poort lithic industry is bracketed by limiting dates of 56 000 and 80 000 years, and is most likely centred on 66 000 ± 5000 years.[6] Anatomically modern human remains in deeper levels in the caves are more than 100 000 years old, which lends strong support to the hypothesis of an African origin for *Homo sapiens* (see also section 3.2.4.2).

6.6.4.4 Dating sea-level change in the Bahamas over the last half million years. Developing long-term records of global sea-level change (known as eustatic change) is often frustrated by the fact that few areas of the world are tectonically stable. This means that it is difficult to separate the eustatic effect from changes in levels of the land resulting, for example, from glacio-isostatic changes or other tectonic effects. One area that is tectonically stable is the Bahamas, which makes these islands especially important for global sea-level studies (Carew and Mylroie, 1995). Although there is abundant evidence for long-term sea-level change on the islands, direct dating has often proved difficult as much of the evidence lies beyond the range of radiocarbon, while only a small percentage of the older marine deposits contain corals for U-series dating.

An alternative approach is to use amino acid geochronology, and Hearty and Kaufman (2000) describe a variant on this technique known as 'whole-rock aminostratigraphy'. The coasts of the Bahamas are characterised by sequences of limestones, which are marine deposits formed during interglacial high stands of sea level, and interbedded palaeosols that reflect low stands of sea level during glacial events. Each 'couplet' of limestone and overlying soil therefore reflects a full interglacial–glacial cycle, with six separate couplets being identified in the Bahamas sequence (I–VI: Figure 6.15A). The marine limestones include coated grains and aragonite muds, and their organic residues contain concentrations of amino acids similar to those in marine and terrestrial molluscs. D-aIle/L-Ile ratios from these residues enabled the sea-level events to be ranked in terms of relative order of age. Seven aminozones could be identified (A–I in Figure 6.15A) and these were calibrated by a kinetic model of epimerisation rate and U-series dating (Table 6.1). The aminozones were then correlated with the marine isotope sequence, aminozones A, C, E, F/G, and I with MOI stages 1, 5e, 7/9, 11 and ≥13, respectively (Figure 6.15B). This study provides a detailed record of global eustatic change over the course of the last

(A)

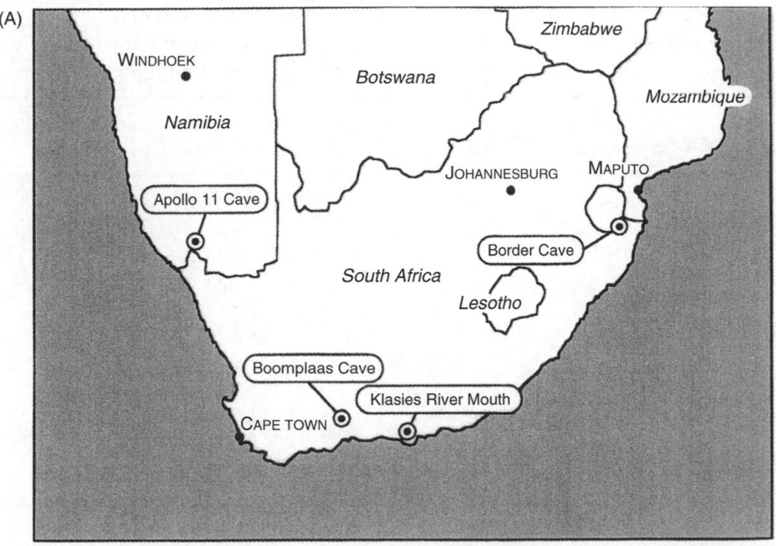

(B)

Border Cave

Lithostratigraphy			Industry	^{14}C age (ka)	aIle/Ile	AAR age (ka)
1	1BS	UP				
		LR.A				
		LR.B	ELSA	38 ± 1	0.271 ± 0.018 (25)	*
		LR.C				
	1WA			38–40	0.255 ± 0.022 (9)	*
2	2BS	UP	MSA 3b	> 49	0.328 ± 0.022 (2)	47
		LR.A				
		LR.B				
		LR.C	MSA 3a		0.388 ± 0.012 (18)	56
	2WA				0.468 ± 0.018 (4)	69
3	3BS		MSA 2 (H.P.)			
	3WA					
4	4BS/4WA		MSA 1		0.87 ± 0.11 (4)	>100
5	5BS				0.97 ± 0.07 (7)	
	5WA					
6	6BS					

Figure 6.14 *(A) Southern Africa showing the locations of Border Cave, Apollo 11 Cave, Boomplaas Cave and Klasies Mouth Cave. (B) Summary of the lithostratigraphy, cultural industries, radiocarbon dates and amino acid ratios for Border Cave. MSLA – Early Middle Stone Age; MSA – Middle Stone Age; HP – Howiesons Poort horizon. Amino acid ratios are the mean and standard deviation based on the number of separate preparations shown in parentheses. * are the radiocarbon calibration levels. Uncertainty in the AAR dates is estimated to be 10% (after Miller* et al.*, 1999). Reproduced with permission of Elsevier*

Table 6.1 *Summary of sea-level data for the Bahamas over the past half million years. The right-hand column shows amino acid ratios with the number of separate analyses in parentheses, and also uranium-series dates (after Hearty and Kaufman, 2000)*

Marine isotope (sub) stage	Estimated sea level ($m\pm$present datum)	Limestone couplet	Regional whole-rock A/I ratio (Table 1and [U-series])
Late 1	0 Datum	VI	0.09 ± 0.01 (6)
Mid 1	−5		0.11 ± 0.01 (7)
5a	0 to −5	V	0.31 ± 0.02 (19)
Latest 5e	+6 to +10	IV	0.40 ± 0.03 (90) [128 − 117 10^3 year]
Early 5e	+3		0.48 ± 0.04 (28) [130 − 119 10^3 year]
7	0 to −5	III	0.55 ± 0.03 (17)
9	0 to +3		[300 10^3 year]
Late 11	+18 to +20	II	0.68 ± 0.03 (36)
Mid 11	+7.5		[≤550 10^3 year]
Early 11	+2		
13	−5 to −10	I	0.76 ± 0.04 (2) [≥550 10^3 year]

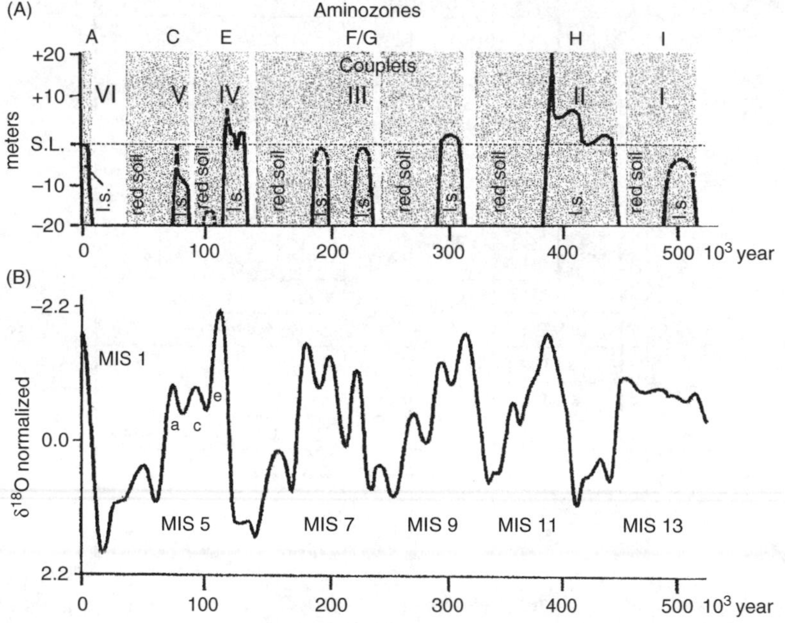

Figure 6.15 *(A) Sea-level high stands in the Bahamas over the course of the last 500 000 years and (B) comparison with the SPECMAP $\delta^{18}O$ curve. Aminozones (A–I) and limestone red soil couplets (I–VI) are shown. Note how the high stands of sea level occur early in each 'interglacial' as reflected in peaks in the SPECMAP curve (after Hearty and Kaufman, 2000). Reproduced with permission of Elsevier*

half million years, and clearly demonstrates the value of the whole-rock aminostratigraphy method in dating and correlating widespread emergent marine deposits.

Notes

1. **Sub-aerial weathering** refers to the combination of weathering processes that occur on a land surface.
2. **Hydration** is the process through which an anhydrous mineral (one that does not contain water within its crystal structure) takes up water to form a new and crystallographically distinct mineral.
3. **Exfoliation** refers to the weathering of a rock by the peeling off of the surface layers. It is sometimes referred to as 'onion-weathering'.
4. **Diagenesis** refers to the alteration of minerals and sediments by a combination of processes. These include the influences of oxidation and reduction, hydrolysis, solution, biological changes, compaction, cementation, recrystallisation, and changes to the lattice structure of clays through the expulsion of water and ion exchange.
5. **Taphonomy** is the study of the processes that lead to the formation of a fossil assemblage.
6. These age estimates are in broad agreement with a more recent date from Border Cave of 74 500 ± 5000 BP obtained by ESR dating on a human mandible (Grün *et al.*, 2003).

7

Techniques for Establishing Age Equivalence

For time is the longest distance between two places.
Tennessee Williams

7.1 Introduction

A long-established approach to dating in geology is to employ geological boundaries between the principal stratigraphic units. Working on the assumption that these reflect major environmental changes, for example between the tropical forest environments of the Carboniferous and the more arid landscapes of the Permian, and that these environmental shifts were global or, at least, hemispherical, in terms of their effects, then the geological boundaries could be considered to be broadly synchronous. If so, they can be regarded as **time-planes**, and hence a geological boundary in one area will be of a broadly equivalent age to the same boundary elsewhere. As such, these geological boundaries form the basis for **time-stratigraphic correlation** (correlation between stratigraphic horizons of comparable age) between often widely dispersed sites. Moreover, by using radiometric or other techniques to assign ages to the boundaries, there is a basis for dating, working on the principle that if the boundary has been dated radiometrically in one locality, that age can be applied to another locality where the boundary may not have been dated directly.

In Quaternary deposits, however, especially those that have accumulated during later Quaternary time, these approaches are more difficult to apply. This is because of the phenomenon known as **time-transgression**. In older geological strata, the sedimentological and, more particularly, the fossil record which forms the cornerstone of stratigraphical subdivision is a reflection of major climatic changes, and while these may well have been synchronous, the geological and biological response to those changes will not necessarily be so. It takes time, for example, for glaciers to waste away, for rivers to adjust to new rainfall regimes, and for plants and animals to react to a change in climate. As a consequence the response will occur earlier in some areas than in others, and this will be reflected in the geological record. Because of the vast spans of time that are involved, however, the geological boundaries *appear* to be synchronous, and indeed we treat them as such, even though we know that, strictly speaking, they are time-transgressive. In Quaternary deposits, by contrast, where ultra-rapid climatic shifts are often reflected in stratigraphic sequences, and where the sedimentary record can be examined at a far higher level of resolution than that for earlier geological periods, the time-transgressive nature of the physical (i.e. sedimentological) and biological responses to climatic change is all too apparent. Hence, lithostratigraphic and biostratigraphic boundaries cannot be regarded as synchronous time-planes, and cannot therefore be used as a basis for correlation and for dating.

In many Quaternary depositional sequences, however, there may be marker horizons that *do* represent time-planes and which can, therefore, form a basis for dating. As with stratigraphic boundaries in the older geological record, these cannot be used in the first instance to date sedimentary successions directly because other methods are required to determine their age. But once dated by radiometric or incremental techniques at one site, they enable those age estimates to be transferred to other sedimentary sequences where that particular marker horizon is also present. In addition, they provide a means for time-stratigraphic correlation between individual depositional records. Four approaches to dating based on this principle of age equivalence are considered here: **oxygen isotope chronostratigraphy**, which rests on the premise that changes in oxygen isotope ratios in deep-ocean fossils reflect globally synchronous shifts in the isotopic signal in ocean water; **tephrochronology**, which uses volcanic deposits (ashes, etc) as a basis for correlation and dating; **palaeomagnetism**, which employs the record of changes in the earth's magnetic field contained in rocks and sediments to develop a chronology; and **palaeosols** (fossil soils), which can form time-planes in sediment profiles and which, in some circumstances, may be used to establish age equivalence between stratigraphic sequences.

7.2 Oxygen Isotope Chronostratigraphy

In Chapter 1 (section 1.5), we saw how the oxygen isotope record in deep-ocean sediments constituted a proxy record of long-term climate change. What is particularly remarkable about this record is that not only is it continuous and spans the entire duration of Quaternary time (and beyond), but that the isotopic signal is largely geographically consistent. Irrespective of whether cores are taken from the Atlantic, Pacific or Indian

Oceans, the isotopic profiles are essentially the same. This means that the MOI trace forms a proxy record of global significance.

7.2.1 Marine Oxygen Isotope Stages

It was also explained in section 1.5 that the isotopic signal from deep-ocean cores could be divided into isotopic stages based on inflections in the oxygen isotope curve. Accordingly, those parts of the curve representing warmer (interglacial) intervals (reflected in lighter $\delta^{18}O$ values) were given odd numbers and the cold (glacial) stages (heavier $\delta^{18}O$ values) even numbers. This notation is applied on a 'count from the top' principle, and hence MOI stage 1 represents the Holocene period, and higher numbers indicate successively older cold and warm stages. In the majority of cases, the numbers refer to episodes of full interglacial or glacial rank, but there are some exceptions. For example, MOI stage 3 is anomalous in that although recognised as a warm stage, it is only considered to reflect a period of interstadial status. Also, the preceding 'interglacials', MOI stages 5 and 7, have been subdivided into separate warmer and colder episodes. MOI stage 5 includes five substages, of which 5a, 5c and 5e are warmer episodes while 5b and 5d are cooler. The last interglacial, *sensu stricto*, is considered to be represented by MOI substage 5e. Similarly, MOI stage 7 is now divided into two warmer phases (substages 7a and 7c) separated by a colder interval (substage 7b).

A distinctive feature of most deep-ocean oxygen isotope profiles is the 'saw-tooth' shape of the curves (Figures 1.4 and 7.1), with the most rapid isotopic changes occurring at the transition from cold to warm episodes. These seemingly abrupt shifts in the earth's climate system from glacial to interglacial modes are referred to as **Terminations** (Broecker, 1984). They too are numbered on a top down basis: Termination I therefore refers to the marked $\delta^{18}O$ shift at the end of the last glacial, but because MOI stage 3 is not considered to be of full interglacial rank, Termination II marks the MOI stage 6/5 transition. Terminations therefore constitute major **events** in the oxygen isotope record and can be used as universal reference points for inter-core correlation.

Because the mixing time in the oceans is relatively rapid, and because of the relatively slow rate of sediment accumulation on the deep-ocean floor, effectively masking any time transgression in the isotopic signal, the isotopically defined stage boundaries, and especially the Terminations, are essentially time-parallel. They therefore form key age-equivalent marker horizons within ocean sediment sequences, and hence provide the basis for a high-resolution correlative chronology that is globally applicable (Jansen, 1989).

7.2.2 Dating the Marine Oxygen Isotope Record

There are two principal ways in which the MOI record can be dated. One involves the use of the palaeomagnetic record in the ocean sediments, and particularly the location in the isotopic sequence of the major geomagnetic boundaries which reflect reversals of the earth's magnetic field. This is explained more fully in section 7.4.5. Using this approach, individual stage boundaries cannot be dated directly, but by interpolation (based on estimated rates of sediment accumulation) from the horizons whose ages can be inferred from the palaeomagnetic timescale.

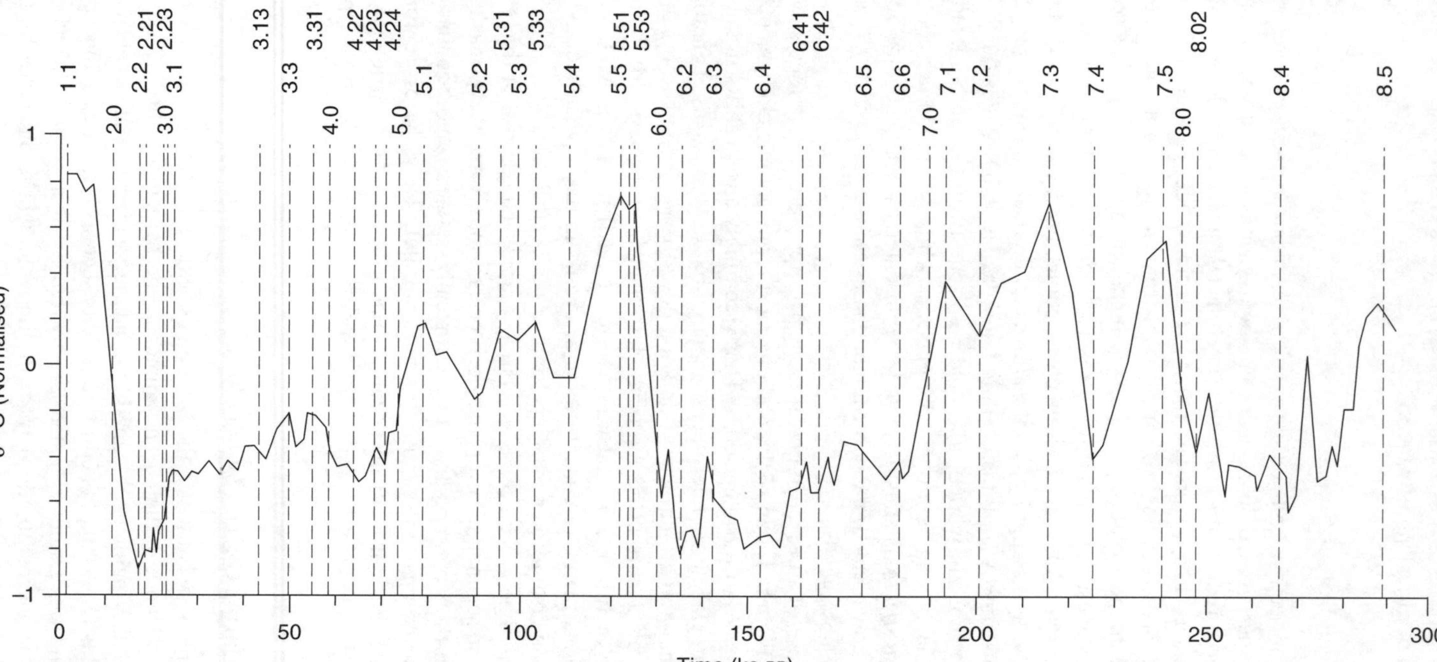

Figure 7.1 *Orbitally tuned chronostratigraphy for a composite ('stacked') deep-ocean isotopic record spanning the last 300 000 years (after* Martinson et al., *1987). The numbered vertical lines indicate distinctive features in the record; the basis for the notation scheme is explained in the text. Reproduced with permission of Elsevier*

An alternative method involves the procedure known as **orbital tuning** and this enables ages to be inferred for each of the isotopically defined stage boundaries. Time-series analysis of oxygen isotope records has revealed periodicities (or regularities) in the isotopic trace that can be related to the astronomical (Milankovitch) variables (Chapter 1: Note 4). Since these orbital parameters are constant and their frequency is known, they provide a basis for timing the cycles reflected in the oxygen isotope records. Thus, the age of each cycle (and hence each stage boundary) can be calculated by extrapolating back from the present day. This approach was first applied by Imbrie *et al.* (1984) who obtained a timescale for the past 800 000 years (known as the **SPECMAP timescale**), by 'tuning' an amalgamation of several isotopic records ('stacked records') to the known frequencies of the astronomical variables. Subsequently, other orbitally tuned timescales have been developed for the Middle and Early Pleistocene periods (Shackleton *et al.*, 1990), while the technique has also been used to develop chronologies for Chinese loess sequences (Ding *et al.*, 1994) and trace gas records in Antarctic ice cores (Ruddiman and Raymo, 2003).

A high-resolution orbitally tuned chronostratigraphy for the last 300 000 years is shown in Figure 7.1. This diagram also includes an alternative approach to the designation of oxygen isotope stages, in that each event within a stage is given a decimal notation so that negative (interglacial) and positive (glacial) excursions are assigned odd and even numbers, respectively (Martinson *et al.*, 1987). Thus the isotopic events within MOI stage 5 which were previously referred to as 5a, 5c and 5e are now designated 5.1, 5.3 and 5.5. The boundary between MOI stages 5 and 6 is now designated 6.0. The rationale behind this revision to the oxygen isotope stage notation is that the precise stratigraphic level of each peak or trough in the isotopic curve can be clearly defined in one curve and correlated with exactly the same level in another. In this way, age equivalence is much more readily established between different isotopic sequences.

7.2.3 Problems with the Marine Oxygen Isotope Record

Although the oxygen isotope record provides a powerful correlative and chronological tool, interpretation of the isotopic evidence is not always straightforward (Patience and Kroon, 1991). Sedimentation rates in the deep oceans are variable, but in many sequences the rate of sedimentation is very low. The result is that a single sample for isotopic analysis may span a time interval of several thousand years. This relatively low stratigraphic resolution may be exacerbated by sediment mixing on the sea floor, either through the movement of bottom currents or as a result of bioturbation caused by benthic (deep-ocean) organisms. In some cases, scouring by ocean bottom currents may lead to removal of sediments leaving a hiatus in the depositional record which will further complicate the isotopic signal. In deep oceans, the carbonate fossils (foraminifers) that carry the isotopic record can dissolve as they settle down through the water column, with the result that a biased assemblage consisting only of the more robust forams will accumulate on the ocean floor. The critical depth is between 3 and 5 km (the **carbonate compensation depth** or **CCD**) and oxygen isotope studies therefore tend to be confined to water that do not exceed the CCD. Other sources of error arise through incorrect correlation of stages or stage boundaries between different isotopic profiles, through the incorrect attribution of warm stages to interglacial as opposed to interstadial episodes (e.g interstadial

MOI stage 3), and because of the complexities of certain isotopic stages (e.g. MOI stage 5 with five substages). All of these mean that consistency of interpretation between different isotopic profiles may not be easy to achieve, especially in those records where stratigraphic resolution is poor.

The foregoing difficulties notwithstanding, there is no doubt whatsoever that the marine oxygen isotope sequence provides a remarkable, indeed unique, record of Quaternary climate change. The stage boundaries form time-stratigraphic markers and offer a basis for establishing age equivalence not only between records in different ocean cores, but between marine sequences, terrestrial records and ice cores. As such, the deep-ocean oxygen isotope signal constitutes the basis for a scheme of Quaternary correlation that is global in terms of its application.

7.3 Tephrochronology

During the course of a volcanic eruption, large quantities of ash and other materials are ejected into the atmosphere and these are subsequently dispersed over a large area downwind of the volcanic source. The tephra will rapidly accumulate as an airfall deposit in lake sediments, on peat surfaces, on river terraces, on the sea bed and in glacier ice. As the deposition of the **tephra** layers is essentially instantaneous on a geological timescale, these horizons provide distinctive and often widespread isochronous (i.e. of the same age) marker horizons that offer a valuable basis for inter-site correlation. Each tephra horizon contains a unique geochemical fingerprint relating it to the source of the eruption. Moreover, it is frequently possible to date the tephra, either directly using argon isotope, ESR or fission track dating, or indirectly by radiocarbon dating of organic material associated with the tephra layer (section 7.3.2). Tephra horizons therefore constitute important chronostratigraphical event markers, and as such provide a basis for dating (**tephrochronology**) and also for correlation (**tephrostratigraphy**).

The dating and correlation of Late Quaternary deposits by means of tephrochronology began in the 1920s and 1930s with work in Japan, New Zealand, South America and, especially, Iceland (Einarsson, 1986). The pioneering work of the Icelandic scientist Sigurdur Thorarinsson was especially significant, as this formed the basis for the first detailed tephrochronology from any region of the world. Indeed, it was Thorarinsson who introduced the word **tephra** (taken from the Greek τεφρα meaning 'ash') into the literature. Tephra is now used as a collective term for all airborne pyroclasts,[1] including both airfall and pyroclastic flow material, and may include a wide range of different-sized fragments, including ash (<2 mm), lapilli (2–64 mm) and blocks and bombs (>64 mm). Tephrochronologists who work with materials deposited a considerable distance away from the volcanic source are primarily concerned with tephra in ash-size fragments (Haflidason *et al.*, 2000). Such ash-fall deposits are usually referred to as **distal tephras** to distinguish them from **proximal tephras** which are found close to the point of eruption.

7.3.1 Tephras in Quaternary Sediments

In areas close to volcanic eruptions, tephras are visible as light-coloured layers in stratigraphic sections of soils, peats or lake sediments. Distal tephras, by contrast, are seldom visible to the naked eye, although they do occur occasionally. A good example is the widely dispersed Vedde Ash (dated to ca. 12 000 cal. years BP) which can be seen in lake sediment cores from Loch Ashik on the Isle of Skye, Scotland (Davies *et al.*, 2001), more than 1000 km from its source in the Katla volcanic complex of southern Iceland (Figure 7.2). In the majority of cases, however, distal tephras are represented in peats, in lake sediments or in deep-ocean sediments at the microscopic level. Typically these consist of tiny glass shards, and the isolation and identification of these **microtephra** (or **cryptotephra**),

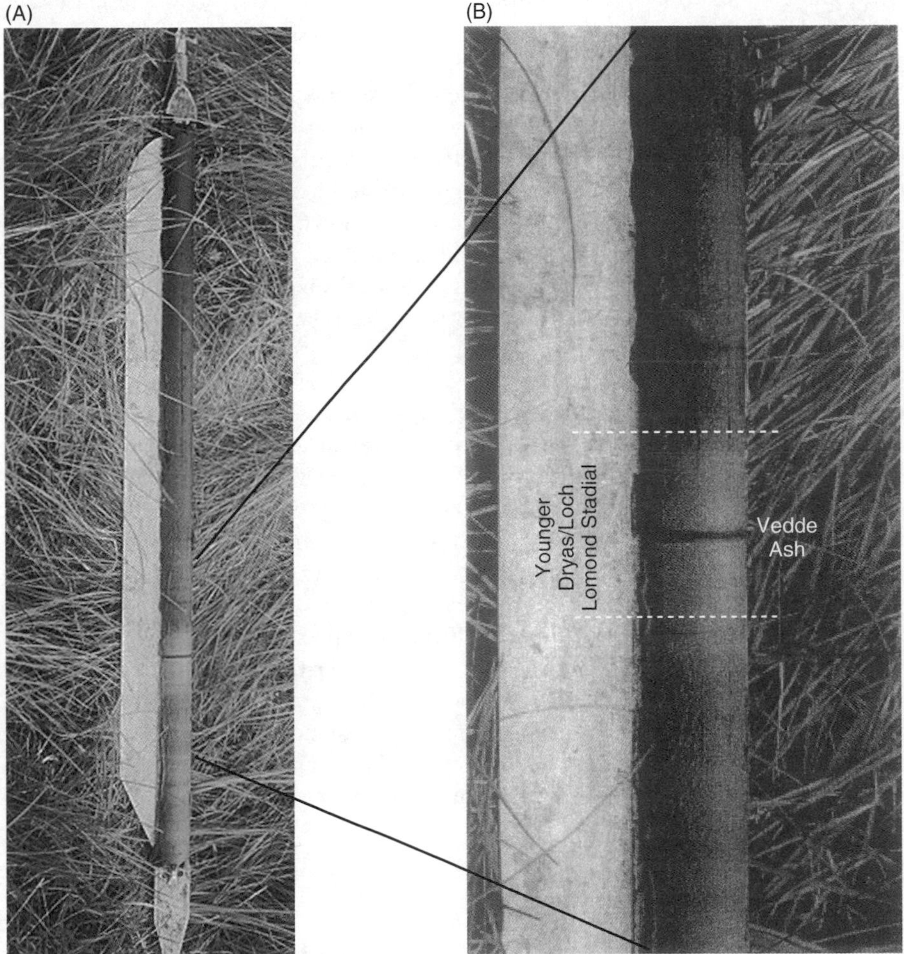

Figure 7.2 *(A) The Vedde Ash (dark band) in a 5 cm diameter by 1 m long core from Loch Ashik, Isle of Ske, Scotland. The lowermost deposits in the core (left) are organic lake muds of Lateglacial Interstadial age (Figure 1.5). These are overlain by light-coloured silts and clays of the Loch Lomond/Younger Dryas Stadial and by organic lake muds and peats of Holocene Age (B). The Vedde Ash occurs towards the top of the Stadial sediments (photo: John Lowe)*

and their subsequent extraction from both organic and minerogenic sediment, entails meticulous laboratory preparation techniques. These include ashing of organic matter from peats (Pilcher and Hall, 1992), the removal of biogenic silica[2] especially from lake sediments (Rose *et al.*, 1996), X-ray analysis (Dugmore and Newton, 1992), density separation (Turney, 1998) and magnetic separation (Mackie *et al.*, 2002). The last two methods have proved especially valuable for the detection of microtephras in minerogenic lake sediments where volcanic glass shards are often present in very low concentrations (Turney and Lowe, 2001).

Tephras have long been identified on the basis of their physical properties, including colour, grain-size distribution, thickness, lithic content, refractive indices of glass shards and mineral inclusions. These characteristics may be valuable diagnostic criteria in regions near the source vents, especially those deposited during historical times when there may also be documentary records of particular volcanic events (Haflidason *et al.*, 2000). For more distal ashes, however, the chemical composition of volcanic glasses has proved to be a more appropriate method for identifying particular tephras. This **geochemical fingerprinting** not only enables a unique 'signature' to be obtained for each tephra, but it also provides a basis for linking tephras directly to known centres of eruption. In many volcanic centres, such as Iceland, the individual volcanic systems are characterised by distinctive rock suites which are petrologically and geochemically distinguishable from each other (Jacobsson, 1979). Distal tephras can be related to these volcanic centres because the glass shards that comprise the tephras are formed during the rapid cooling of volcanic magma and therefore have a composition that is representative of the bulk geochemistry of a particular magma.

The technique most widely applied to determine the geochemical composition of a tephra is **electron probe micro-analysis (EPMA)**. The glass surface of each tephra shard is bombarded with an electron beam, the X-ray energy produced being unique to each chemical element, while the intensity of the signal is proportional to the amount in the glass shard (Hunt and Hill, 1993). Usually about ten major elements (expressed as oxides) are selected for geochemical analysis. Measurement is either by **energy dispersive spectrometry (EDS)** or by **wave-length dispersive spectrometry**, the latter normally being preferred because although requiring a higher electron beam current and longer counting time, it offers a greater degree of analytical precision (Davies *et al.*, 2002). Considerable efforts have been made in recent years to standardise the analytical methods used by different laboratories, the reporting procedures, and the ways in which tephrochronological data are presented (Hunt and Hill, 1993; Hunt *et al.*, 1998).

7.3.2 Dating of Tephra Horizons

Tephra horizons may be dated directly by a number of methods. Where a tephra is associated with organic materials, such as wood, peat, or lake sediments, the age of the ash bed can be determined by radiocarbon dating. For example, in the sediments of Kråkenes Lake in western Norway, the mean of four AMS radiocarbon dates on samples of *Salix herbacea* leaves adjacent to the Icelandic Vedde Ash (Figure 7.2) date the tephra to $10\,310 \pm 50$ ^{14}C year BP (Birks *et al.*, 1996). In Ireland, radiocarbon dates on 20 separate tephra horizons in peat and lake sequences underpin a tephrochronology extending back to the eighth millennium BP (Hall and Pilcher, 2002). In some cases, tephras have been dated directly by radiometric determinations on the primary mineral constituents. Examples include $^{40}Ar/^{39}Ar$ dating of the widely dispersed Laacher See Tephra (12 900 cal. years: Figure 7.5) from the Eifel region of Germany (Van den Bogaard, 1995); ESR dating of the Youngest Tuff (81 000 ± 17 000 years) from the Toba Volcano in Sumatra (Wild *et al.*, 1999); and fission-track dating of the Rockland Tephra (400 000 years), an ash layer that is widespread in northern California and western Nevada (Meyer *et al.*, 1991).

Calendar ages for tephras can be obtained from varved sediment sequences (Zillen *et al.*, 2002); for instance, the Laacher See Tephra referred to above has also been dated by varve chronology and an age (12 880 years) similar to the $^{40}Ar/^{39}Ar$ date obtained (Brauer *et al.*, 1999). Ice-core records provide a further means of assigning precise ages to tephras; a good example is again the Vedde Ash, which has been dated to 11 980 ± 80 years in the Greenland GRIP Core (Grönvold *et al.*, 1995), an age which also equates well with the calendar age estimate 12 045–11 975 cal. years BP based on radiocarbon dating (Wastegård *et al.*, 1998). Further means whereby tephra can be dated include stratigraphical position in relation to other tephra layers, palaeomagnetic correlations and relationships to oxygen isotope stage boundaries in deep-ocean sediments (Wastegård and Rasmussen, 2001).

7.3.3 Problems with Tephrochronology

While tephrochronology is undoubtedly a technique of considerable potential, it is not without its problems. These relate partly to the fact that it is, at best, a dating method that is regionally or areally specific, for the dissemination of tephra from the volcanic source, and hence the distribution of individual ash layers, will depend on a range of factors, including the magnitude and type of volcanic eruption, the strength of the prevailing wind, the direction of the wind at the time of the eruption, etc. Many volcanic plumes appear to have been relatively narrow, and hence the spatial distribution of tephra fall-out from the atmosphere will be restricted. This means, of course, that tephrochronology can only be applied in those areas that are immediately downwind of a major volcanic source. Post-depositional processes may also influence tephra presence in the sedimentary record. In peatland areas, for example, where there has been protracted snow cover, redeposition by wind and meltwater may result in patchy distributions of tephras across the bog surface, and hence only fragmentary tephra records may be preserved in peat sequences (Bergman *et al.*, 2004). A further difficulty arises with tephras in marine sediments. Tephras falling over the high-latitude northern and southern oceans may not reach the sea bed, and hence the marine sediment archive, because of sea-ice cover at the ocean surface. Moreover, tephras incorporated into glacier ice may subsequently be ice-rafted into the oceans and deposited in marine sediments. If so, then older tephras would be incorporated into younger stratigraphic horizons (Bond *et al.*, 2001).

A number of analytical problems also remain to be resolved. Although considerable progress has been made in the standardisation of laboratory procedures (section 7.3.1), there is still a measure of variability in laboratory and operator practice (Hunt and Hill, 1996). Moreover, while individual tephra layers may be shown by EPMA analysis to be geochemically distinct, when the geochemical envelopes representing the major element chemistry of different volcanic provinces are plotted, there are often overlaps between them (Figure 7.3). Hence, it may be necessary to employ other diagnostic criteria, such as

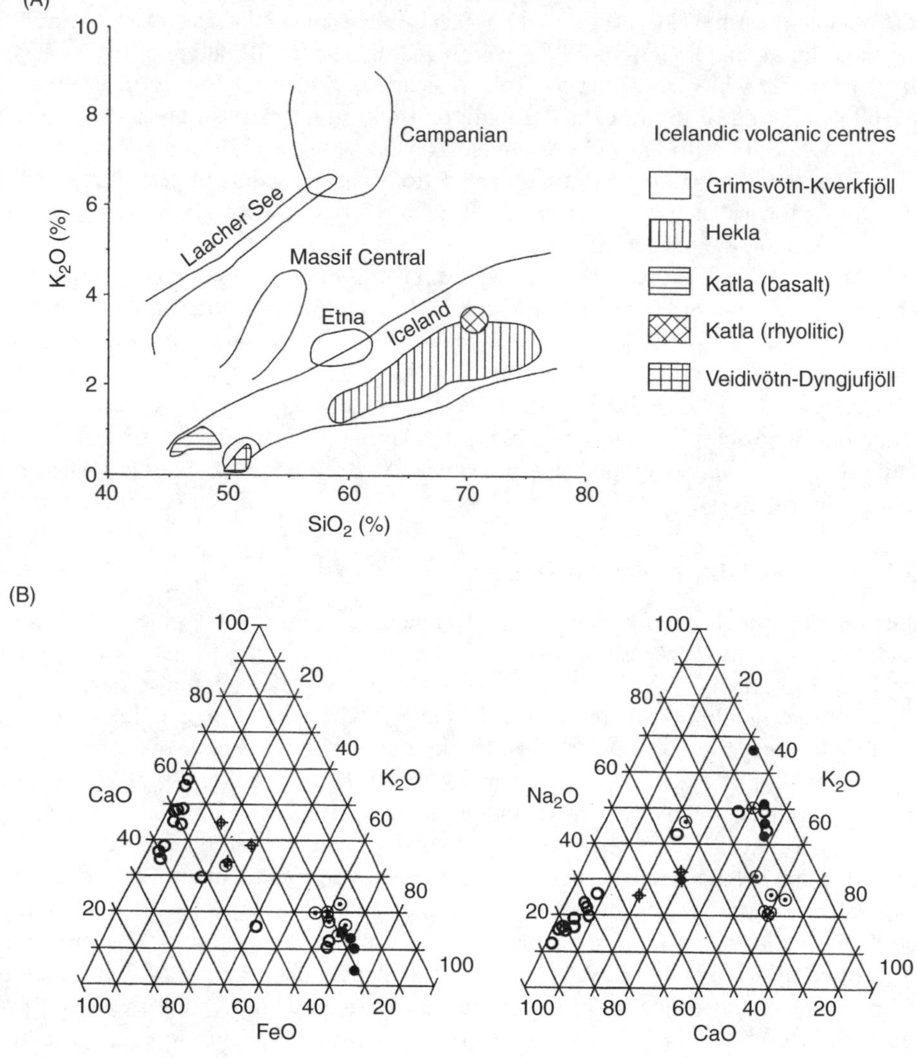

Figure 7.3 (A) Bi-plot of SiO_2 and K_2O concentrations in tephras from the main European ash provinces. Note the overlaps between the Laacher See and Campanian (northern Italy) envelopes. Reproduced by The Royal Society. (B) Ternary plots showing variations in the proportions of FeO, CaO and K_2O (left) and CaO, Na_2O and K_2O (right). Again, note the overlaps between the tephras from different volcanic sources (after Davies et al., 2002). Reproduced by The Royal Society

the physical properties of the glass shards and the stratigraphic position of the tephra layer in order to distinguish between individual tephras and to determine tephra provenance (Davies *et al.*, 2002). Even then, a correct sourcing of the tephra to a particular eruptive event may not be possible, simply because of a lack of data on the eruptive histories and diagnostic geochemistries of different volcanic events. Finally, theoretical stability modelling suggests that some tephras (particularly basaltic tephras) may have relatively poor chemical stability in the post-depositional environment, leading to major changes in the tephra geochemistry (Pollard *et al.*, 2003). If so, then this could provide a further difficulty for the correct geochemical typing and provenancing of distal tephras.

7.3.4 Applications of Tephrochronology

Significant advances have been made in tephrochronology over the course of the last 20 years, and it is now clear that tephrochronology as a Quaternary dating and correlation tool has come of age. While the technique has continued to be applied in the western Cordillera of North America, in Japan, in New Zealand and in Iceland, areas where tephrochronology has had a long and distinguished record (e.g. Machida, 1981; Thorarinsson, 1981; Lowe, 1988), it has been increasingly employed around the North Atlantic province and in Europe, where a greatly expanded Icelandic tephra database (Haflidason *et al.*, 2000) can now be combined with tephra data from the major European volcanic provinces of Germany, France and the northern fringes of the Mediterranean. Some archaeological and environmental applications of tephrochronology are considered in the following sections.

7.3.4.1 Dating the first human impact in New Zealand using tephrochronology. New Zealand, an isolated archipelago in the southwest Pacific, was the last substantial landmass to be settled in by humans. The orthodox view has been that the colonisers migrated southwestwards from eastern Polynesia, and that the first settlement occurred between ca. AD 750 and 950. A Polynesian origin for the New Zealand Maori people has never been questioned, but in recent years two schools of thought have emerged relating to the timing of settlement, one favouring an early settlement between AD 0 and 500 while the opposing view arguing for a later colonisation between AD 1150 and 1300. Discussion has centred on both the evidence for human impact (vegetational disturbance reflected in pollen records, for example), and the reliability of radiocarbon dates associated with human-related evidence, particularly as the relatively short-time intervals involved limit the effectiveness of radiocarbon dating for defining the order of archaeological events.

An alternative approach to resolving this problem is to use tephrochronology (Lowe *et al.*, 1998). A number of tephras from eruptive centres in North Island are found in stratified contexts and several have archaeological significance. The most important of these is the regionally extensive Kaharoa Tephra, which erupted from Mt Tarawera in the Okataina Volcanic Centre (Figure 7.4A), and which has been dated to 665 ± 15 ^{14}C years BP on the basis of more than 20 radiocarbon measurements on a range of materials, including wood, peat and charcoal. This age determination is equivalent to calibrated dates ranging from AD 1300 to 1390 (1σ), or AD 1290 to 1400 at the 2σ level (Lowe *et al.*, 1998). Wiggle-match radiocarbon dating (section 2.6.5) using wood samples gives a more precise estimate of age, AD 1314 ± 12, again at the 2σ level (Figure 7.4B). The Kaharoa Tephra is a critical marker horizon for two reasons. First, no cultural remains are known to occur beneath it and so it provides a maximum age for settlement in North Island of ca. AD 1300. Second, palynological and charcoal analyses from a substantial number of sites, all of which contain the Kaharoa Tephra datum (Figure 7.4A), indicate sustained deforestation very close to the time of its deposition (Lowe *et al.*, 1998; Horrocks *et al.*, 2001). This implies that the earliest inferred human impacts in eastern North Island

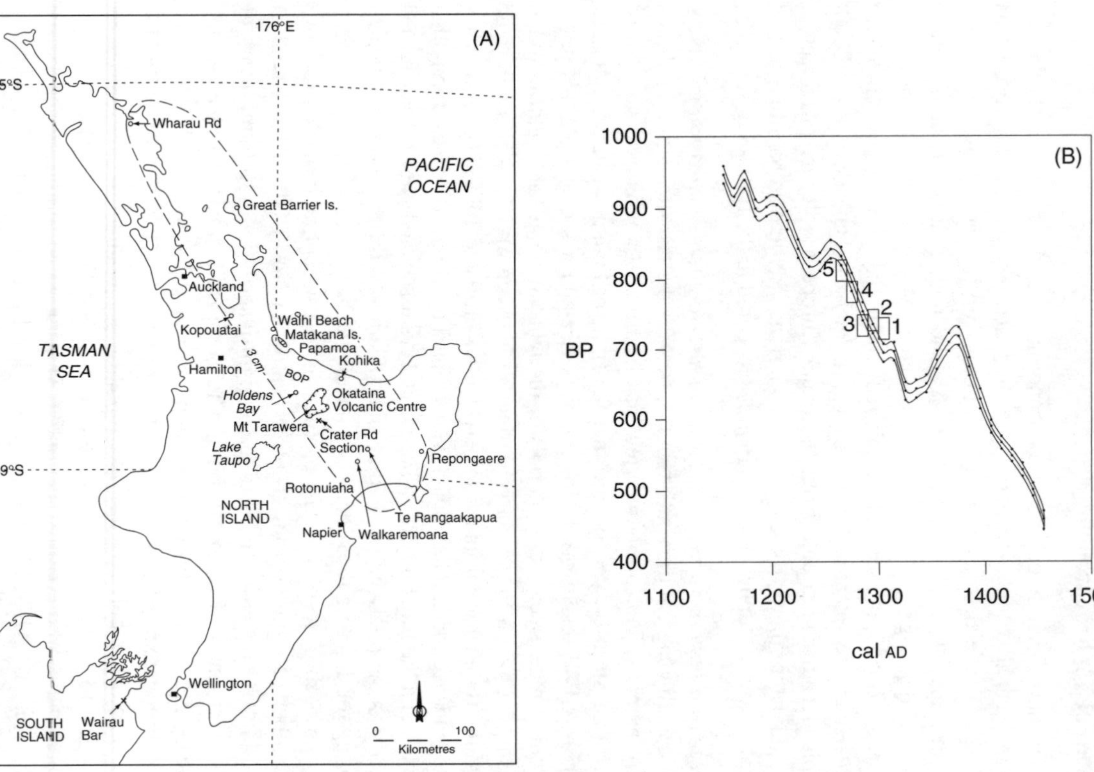

Figure 7.4 *(A) North Island, New Zealand, showing the distribution of the Kaharoa Tephra (dashed line), Mt Tarawera in the Okataina Volcanic centre which is the source of the tephra, and the location of pollen profiles (open circles) where the tephra has been recorded. (B) Five radiocarbon dates on wood samples contained within the tephra at the Crater Road section (Figure 7.5A) wiggle-matched to the southern hemisphere terrestrial radiocarbon calibration curve. The individual data points are shown as rectangles, with the width of each rectangle indicating the number of rings spanned by the radiocarbon-dated sample (10 years) and the height indicating the sample standard error (±1σ). The wiggle-match provides an age for the tephra of* AD *1314± 12 years (after Hogg* et al.*, 2003). Reprinted by permission of A.G. Hogg and Antiquity Publications Ltd*

as reflected in pollen and other records occurred no earlier than ca. AD 1290. In turn, this lends support to the 'late arrival' hypothesis for the first settlements in New Zealand. It also suggests that the earliest Maori people in New Zealand may have witnessed the Kaharoa eruption very soon after their arrival (Lowe *et al.*, 1998)!

7.3.4.2 Dating and correlating events in the North Atlantic region during the Last Glacial–Interglacial transition using tephrochronology. The transition from the Last Glacial to the Holocene, often referred to as 'Termination 1' in the MOI sequence (section 7.2.1) or the Lateglacial in the terrestrial record (Figure 1.5), is one of the most intensively studied episodes in Quaternary science. During the period from approximately 15 000 to 10 000 years ago, climatic changes of interstadial/stadial amplitude occurred and are reflected in a wide range of environmental data (Lowe and Walker, 1997). These often dramatic climatic oscillations are the most recent manifestations in the geological record of the unstable nature of the global climate system, and they therefore provide atmospheric scientists with natural analogues for models of climate change. The Lateglacial is also of great interest to earth scientists who are seeking to understand the spatial and temporal interactions between the atmosphere, cryosphere, ocean and terrestrial biosphere on millennial timescales. Establishing a chronology for the sequence of climatic events is of paramount importance, but the most widely used method, namely radiocarbon dating, is known to be particularly problematic for this time period (Lowe *et al.*, 2001). Hence, alternative bases for dating and correlation are being sought and one of these is tephrochronology.

A key area for research on environmental change during the Lateglacial and Early Holocene is the North Atlantic region where the potential for tephrochronology is considerable (Lowe, 2001; Turney *et al.*, 2004). This is because there are a number of discrete volcanic centres in the region, of which those in Iceland, the Eifel region of Germany, the Massif Central of France, and the Italian provinces of Campania (around Naples), Etna and the Aeolian Islands (Figure 7.5A) are known to have erupted during this time period. In all, 34 distal tephras from these volcanic complexes have been discovered in sequences spanning the Last Termination (Davies *et al.*, 2002). Key tephras include the widely dispersed Icelandic tephras, including the Saksunarvatn Tephra (ca. 10 200 cal. years BP), the Vedde Ash (ca. 11 980 cal. years BP) and the Borrobol Tephra (ca. 14 400 cal. years BP[3]); the Laacher See Tephra (ca. 12 900 cal. years BP) from Germany; and the Neapolitan Yellow Tuff (ca. 14 400 cal. years BP) from Italy (Figure 7.5B). Not only do these tephras constitute key dated marker horizons in terrestrial sequences, particularly in lake sediments, but the fact that the tephras can also be identified and provenanced in cores from the North Atlantic ocean and from the Greenland ice sheet means that they form a basis for time-stratigraphic correlation between terrestrial, marine and ice-core records. As such, they provide a broad chronological framework for climatic and environmental change throughout the entire North Atlantic province during the transition from the Last Cold Stage to the present interglacial.

7.3.4.3 Dating Middle Pleistocene artefacts and cultural traditions in East Africa using tephrostratigraphy. During the Middle Pleistocene, African hominids abandoned the manufacture of handaxes and cleavers and began making points, a cultural transition which archaeologists interpret as marking the end of the Acheulian and the beginning of the Middle Stone Age. This is an important event in the record of human technological

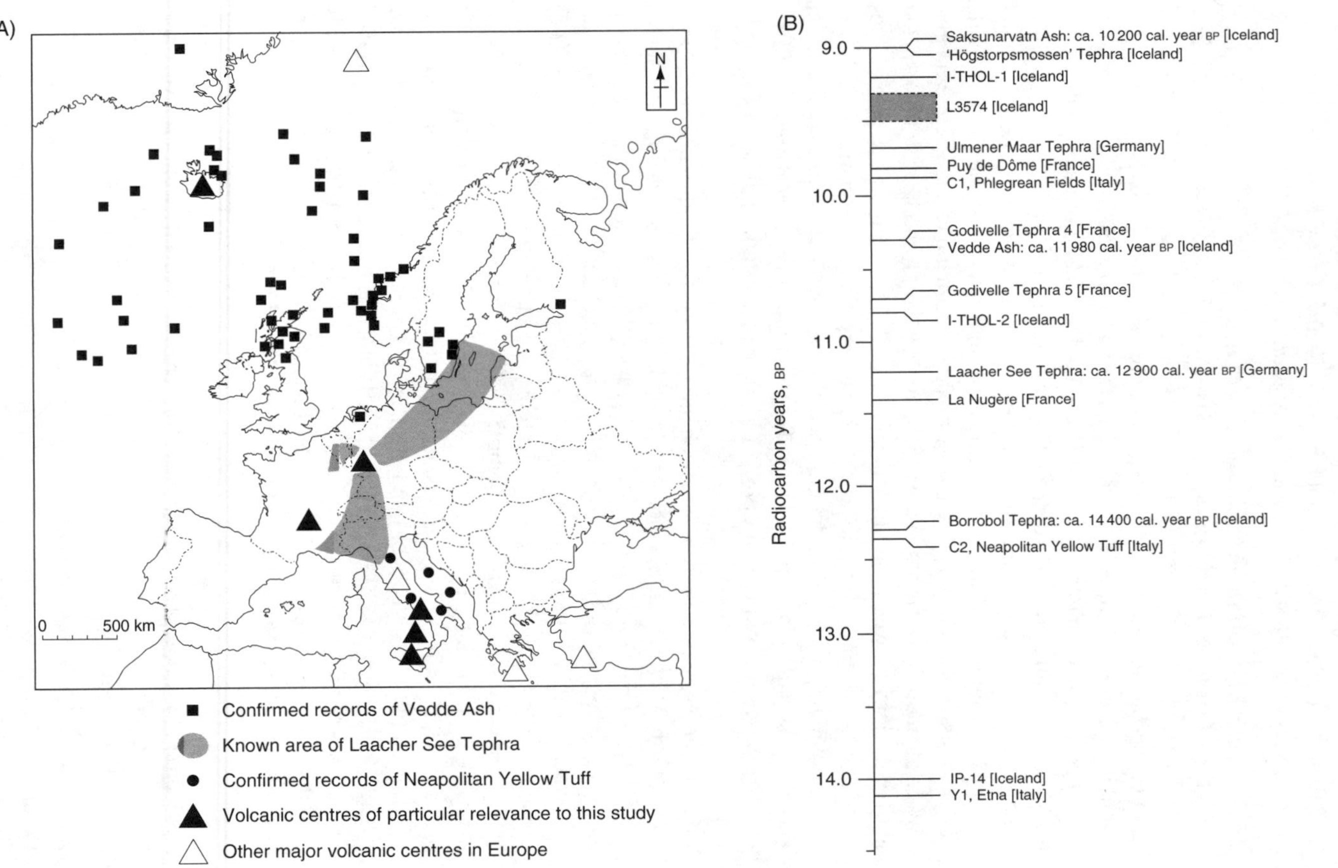

Figure 7.5 *(A) Location of the principal volcanic centres that were active during Termination 1 (after Davies* et al.*, 2002). Reproduced by The Royal Society. (B) Chronology of tephra layers dated to between 14 500 and 9500 ^{14}C years BP from sites in Europe and the Mediterranean Sea (after Lowe, 2001). Reproduced with permission of Elsevier*

development, and may possibly reflect a major behavioural shift or even a speciation event reflected, perhaps, in the first appearance of *Homo sapiens* (McBrearty and Brooks, 2000). Few locations, however, preserve a continuous well-dated sedimentary or occupational record across the Acheulian–Middle Stone Age transition.

Sites containing both Acheulian and Middle Stone Age artefacts occur in several stratified sequences in the Tugen Hills of western Kenya (Figure 7.6A). The artefacts are found within and below the Bedded Tuff, a widespread volcaniclastic unit consisting of up to 12 tephra horizons interbedded with sediments and palaeosols (Tryon and McBrearty, 2002). Two tephra units high in the sequence have been dated by $^{40}Ar/^{39}Ar$ to $235\,000 \pm 2000$ years and $284\,000 \pm 12\,000$ years ago, the latter date providing a minimum age for the archaeological materials, all of which underlie this particular tephra horizon (Figure 7.6B). The Bedded Tuff outcrops within the study area (Figure 7.6A) have been correlated on the basis of field stratigraphic relationships, and on the geochemical signatures of the individual tephra horizons. Distinctive geochemical changes within Bedded Tuff units permit a chronological ordering for the archaeological sites. The results show that: (a) the Acheulian–Middle Stone Age transition in this part of Africa began well before 285 000 years ago; (b) the projectile points found in three of the sites (GnJh-17, GnJh-63 and GnJi-28; Figure 7.6B) are the oldest yet known from Africa; and (c) perhaps most interestingly, there is no unidirectional succession at all of the sites from Acheulian to Middle Stone Age industries for, in some cases, Acheulian artefacts (handaxes) overlie strata containing artefacts of Middle Stone Age affinity (points). Whether this technological diversity reflects more complex human adaptations, diverse palaeoenvironments within the area, or perhaps even the presence of multiple human species remains unclear. Whatever the case, it is clear that the Acheulian–Middle Stone Age transition in this part of Africa was a more complex cultural process than may hitherto have been considered.

7.3.4.4 Dating Early and Middle Pleistocene glaciations in Yukon by tephrochronology. In the mountains of western North America tephras, mainly in the form of ancient tuffs, have long been recognised as important time markers in Late Pliocene and Early and Middle Pleistocene deposits. These include the Bishop Ash (K–argon age 738 000 years), the Pearlette 'O' Ash (610 000 years), the Pearlette 'S' Ash (1.27 million years) and the Pearlette 'B' Ash (2.01 million years), and all are found throughout the High Plains, Rocky Mountains, Colorado Plateau and the Great Basin, where they are frequently inter-stratified with glacial deposits. They have provided a basis for correlating these glacial sediments, while dating of the tuffs has enabled an outline chronology of Early and Middle Pleistocene glacial events to be developed (Richmond and Fullerton, 1986).

Further north, in Alaska and Yukon, tephras also occur inter-stratified with glacial sediments. Central Yukon lies within the sphere of influence of volcanoes in the Wrangell Mountains and the more distant Aleutian arc-Alaskan Peninsula (Figure 7.7), so that the Late Cenozoic sediments contain a large number of distal tephra beds (Preece *et al.*, 2000). Westgate *et al.* (2001) have shown how these tephras can be used to determine the age of extensive cordilleran and alpine glaciations in central Yukon. At three sites, between Dawson City and Whitehorse (Figure 7.7), sequences of sediment containing both tephras and basalt lava flows were discovered. At two of these sites, Fort Selkirk and Midnight Dome Terrace, the tephra horizons overlie glacial and/or glacial outwash deposits. Fission-track dating of the tephras gave ages of 1.48 ± 0.11 million years for the Fort Selkirk

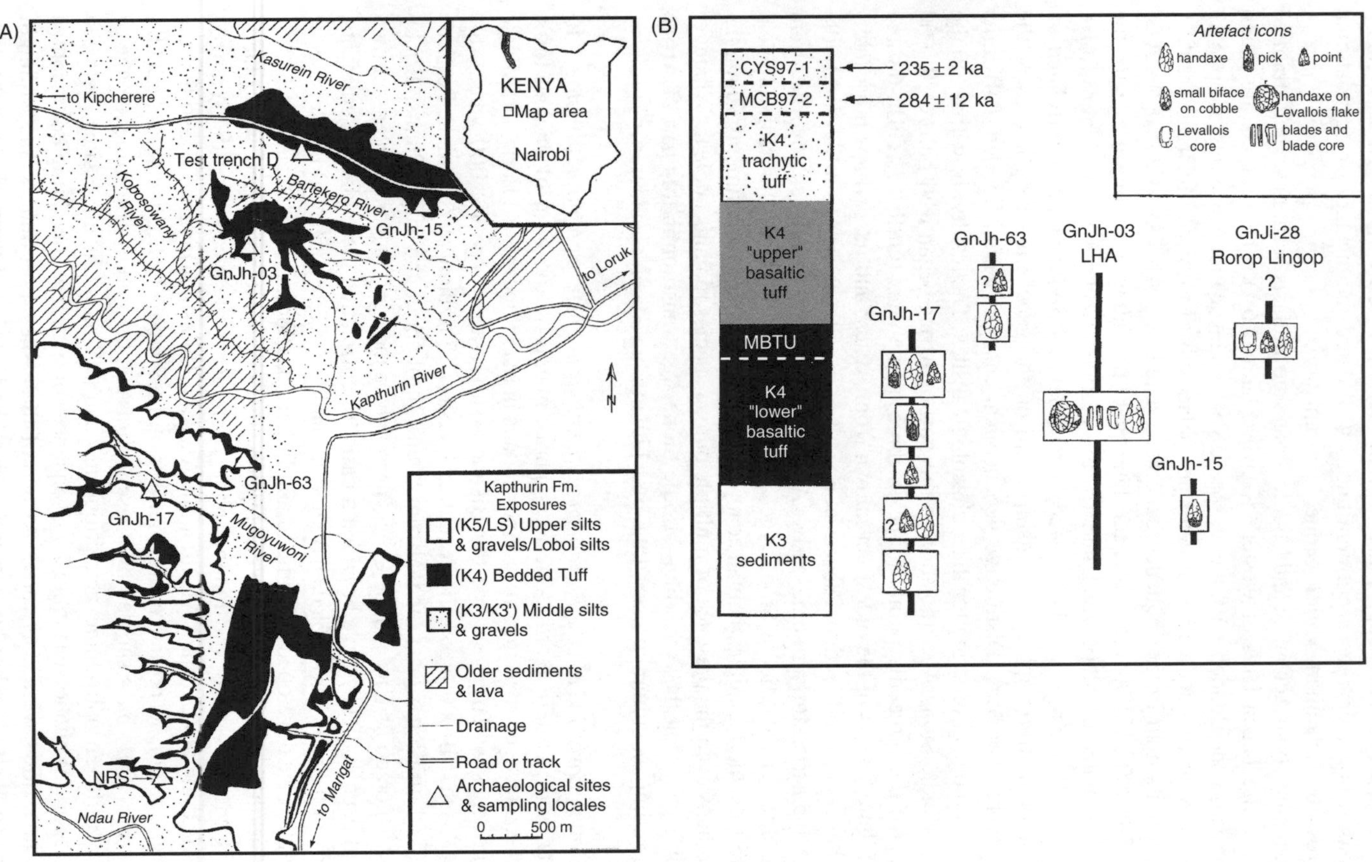

Figure 7.6 *(A) The Kapthurin Formation in part of the Tugen Hills, western Kenya, showing the location of the sampling sites. (B) A summary of the stratigraphic relationships between units of the Bedded Tuff, radiometric dates and archaeological assemblages (after Tryon and McBrearty, 2002). Reproduced with permission of Elsevier*

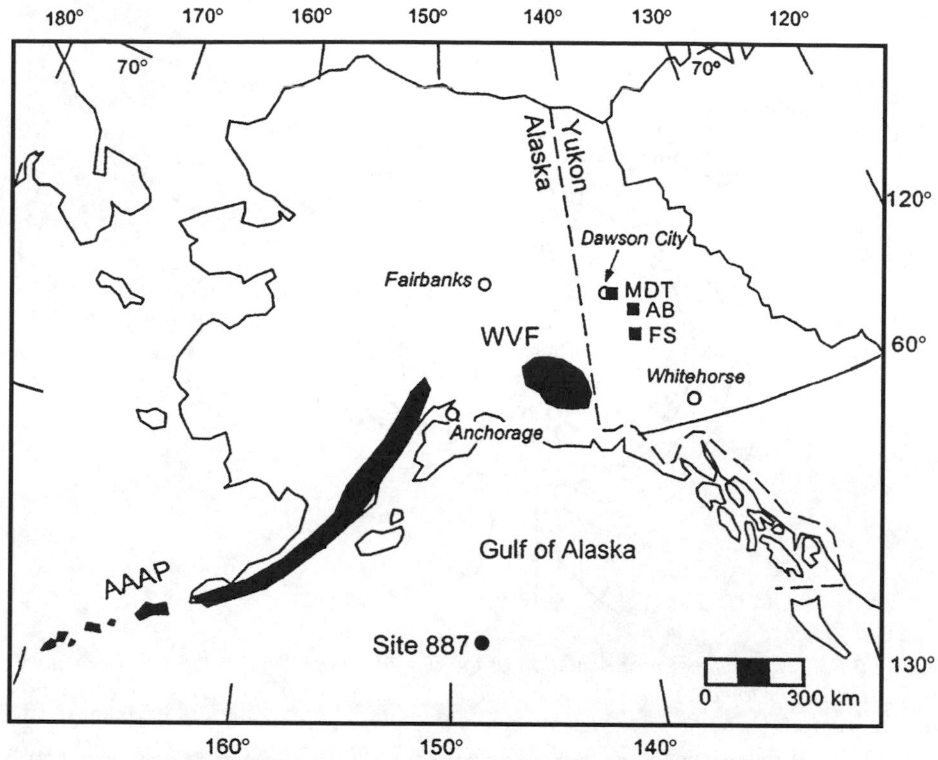

Figure 7.7 The two major volcanic zones in Alaska responsible for the deposition of silicic tephras across Alaska and the Yukon territory: AAAP – Aleutian arc Alaskan Peninsula; WVF – Wrangell Volcanic Field. The three study sites referred to in the text are indicated by the filled squares: MDT – Midnight Dome Terrace; AB – Ash Bend Terrace; FS – Fort Selkirk (after Westgate et al., 2001). Reproduced with permission of Elsevier

Tephra and 1.45±0.14 million years for the Midnight Dome Tephra. When set alongside an age of 1.60±0.08 million years on an basalt flow from beneath glacial deposits, this constrained the age of an extensive Early/Middle Pleistocene glaciation in the area to ca. 1.5 million years BP. At the Ash Bend site (Figure 7.7), a younger tephra overlying a thick suite of glacial deposits (Reid Glaciation) could be linked geochemically with the Sheep Creek Tephra of central Alaska (TL date: 190 000 years). This showed the Reid Glaciation to be older than MOI stage 6, and an MOI stage 8 is suggested, thereby placing the glacial event around 250 000 years ago.

7.4 Palaeomagnetism

Magnetism is something that is known to most of us. We are familiar with the idea of a magnet consisting of two poles (a **dipole**) which we often refer to as north and south. Such magnets attract ferrous objects that are themselves not magnets. We can also visualise,

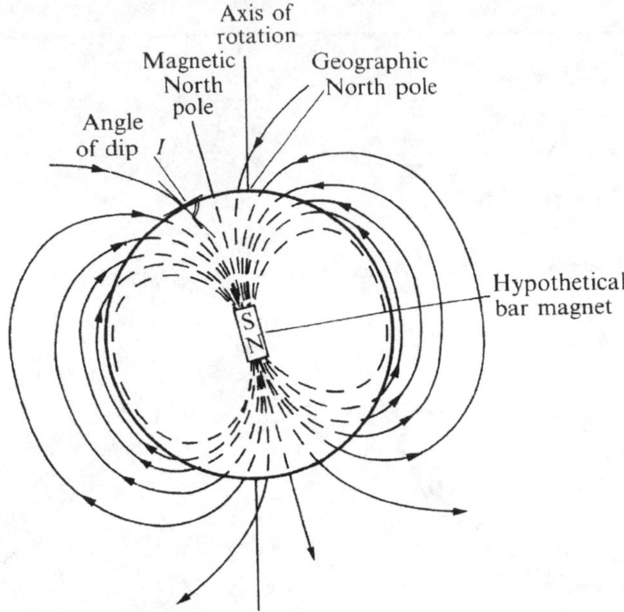

Figure 7.8 *A bar-magnet representation of the dipole component of the earth's magnetic field. The lines of force represent, at any position, the direction in which a small magnetised needle would try to point. The concentration of these lines is a measure of the* field strength *or* field intensity *(after Aitken, 1990). Reproduced by permission of Pearson Education Ltd*

at least in very general terms, a magnetic field associated with such a magnet where lines of force emanating from one pole of the magnet pass through free space in a series of loops to reach the other pole (Smith, 1999). Many will also be aware that the earth itself behaves like a giant magnet and creates a magnetic field that is analogous to the magnetic field generated by a small hand-held magnet (Figure 7.8). What may be less well known, however, is that the earth's magnetic field is not constant; rather it changes over time, and a record of these changes will be contained in rocks and sediments. As was the case with tephras, these **palaeomagnetic changes** form time-planes in stratigraphic records and constitute a basis for inter-site correlation. If they can be assigned an age by radiometric dating (argon isotope, fission track) or other means (varve chronology, for example), then they also provide a basis for dating.

7.4.1 The Earth's Magnetic Field

The source of the earth's magnetic field lies deep in the fluid part of the earth's core. There, a dynamo mechanism and the associated electrical currents generate the magnetic field that we experience at the earth's surface (Figure 7.8). About 85% of the spatial variation of the geomagnetic field at the earth's surface can be attributed to the pattern expected for a **dipolar field** (Sternberg, 2001). In addition, however, there is a **non-dipolar** component in the geomagnetic field arising, for example, from irregularities in current flow near the earth's core-mantle boundary. The effects of these are localised in that they affect restricted regions of the earth's surface (Aitken, 1990). Over time, the earth's magnetic field varies in both field strength and direction. There are two elements involved in these changes: major **dipole changes**, where the dynamo currents that generate the main dipole field become reversed (**polarity changes**[4]) so that effectively the north pole becomes the south pole and vice versa, and **secular variations**, which operate largely (although not entirely, because of the influence of non-dipole effects) through a process known as 'dipole wobble' whereby the axis of the dipole field (magnetic north) precesses[5] around the earth's axis of rotation (true north: Figure 7.8). Secular variations occur over centuries or less during which field direction changes at a rate of 1° every few decades, while field strength changes several per cent per century (Sternberg, 2001). The major polarity changes, by contrast, are measurable over timescales of thousands or millions of years. It is a combination of secular changes and polarity changes that forms the basis for palaeomagnetic dating. Good reviews are provided by Tarling (1983), Aitken (1990), Eighmy and Sternberg (1990) and Sternberg (1997).

There are three components in the earth's magnetic field. **Declination** is the angle between magnetic north and geographic (true) north. Currently this is 11.5°, but will vary with the dipole wobble described above. **Inclination** refers to the angle of dip of the prevailing magnetic field, in other words the angle between the prevailing lines of force in the magnetic field and the horizontal. If a needle is suspended in space exactly at its centre of gravity, its north-seeking end will dip below the horizontal in line with the prevailing magnetic field. Inclination varies from 0° at the magnetic equator to 90° at the magnetic poles. Third, there is the **intensity** or strength of the earth's magnetic field. At present the field strength at the geomagnetic poles is twice that at the geomagnetic equator.

7.4.2 The Palaeomagnetic Record in Rocks and Sediments

Rocks and sediments that contain magnetic minerals, either in crystal form in the case of igneous rocks or as particles in sediments, can acquire a **permanent** or **remanent** magnetism. This **natural remanent magnetism (NRM)**, typically in the ferromagnetic minerals magnetite and hematite, will reflect the earth's geomagnetic field at the time of rock or sediment formation. Volcanic rocks will acquire a remanence through heating, which is referred to as **thermoremanent magnetism (TRM)**. Archaeological materials that have been heated, such as pottery and hearths, will also contain a TRM signal, and again this will reveal the direction of the earth's magnetic field at the time of firing. Sedimentary rocks and unconsolidated sediments that have accumulated on the sea floor or in lakes, for example, contain a different form of magnetic remanence – a depositional or **detrital remanent magnetism (DRM)**. As ferromagnetic sediment particles settle through the water column or in water-saturated sediments, the grains become aligned with the prevailing geomagnetic field. This DRM signal, although sometimes weak, can easily be measured in cores of sediment from lakes and from the deep ocean floor, and again enables the changes in the direction of the earth's magnetic field to be measured over time. Further details of the magnetic properties of minerals, rocks and sediments are provided by Thompson and Oldfield (1986), while various aspects of environmental magnetism are described by Walden *et al.* (1999) and Evans and Heller (2003).

7.4.3 Magnetostratigraphy

The study of variations in magnetic properties through a sequence of rocks or sediments is termed **magnetostratigraphy**. Even very small amounts of ferromagnetic minerals in a body of rock or sediment can provide a record of past geomagnetic field variations, and the identification of major polarity changes in volcanic rocks or in deep-sea sediment records, or of variations in inclination and declination (secular changes) in different depositional sequences, provides a basis for relative dating and for correlation. In addition, **mineral magnetic potential**, in other words the concentration and magnetic susceptibility of magnetic minerals, can also provide a basis for correlating between sedimentary sequences.

7.4.3.1 Polarity changes and the palaeomagnetic timescale. During the course of geological time there have been numerous changes in the earth's magnetic polarity, with the magnetic poles changing relative positions through 180°. These polarity or **field reversals** are of considerable importance in magnetostratigraphy, for they can be detected in the geological record where they form time-parallel marker horizons in stratigraphic sequences. At present, the earth's magnetic field is regarded as possessing **normal polarity**, but when the field is inverted from the present it is referred to as **reversed polarity**. Periods of long-term fixed polarity, whether normal or reversed, are termed **polarity epochs**, and have a time range from 100 000 up to 10 million years. Within these long-term polarity episodes are polarity changes of much shorter duration. **Polarity events** typically last from 10 000 to 100 000 years, while **polarity excursions**, in which the geomagnetic pole changes direction through 45° or more for a short-time interval only, are of much shorter duration (1000–100 000 years).

Polarity epochs and polarity events are global phenomena and hence they form a basis for world-wide correlation. Many of these have been detected in volcanic rocks where they have been dated by K–argon (Tarling, 1983). The palaeomagnetic timescale can also be dated using the deep-ocean sediment record. In ocean water, ferromagnetic minerals settling through the water column adopt the direction of the earth's magnetic field, and hence a core through ocean-floor sediments will contain a long-term record of geomagnetic changes. As we saw in section 7.2.2, ocean cores can be dated by 'orbital tuning' and this enables an age to be assigned to those parts of the core where geomagnetic changes are recorded (Funnell, 1995). These dated horizons form the basis for a **palaeomagnetic timescale** for Quaternary and Late Tertiary sequences (Figure 7.9). Over the course of the past 3 million years, three polarity epochs have occurred, the Brunhes Normal (which extends to the present day), the Matuyama Reversed and the Gauss Normal. Dating the epoch boundaries places the Brunhes–Matuyama at 0.78 million years and the Matuyama–Gauss at 2.58 million years. Important polarity events include the Jaramillo Normal (0.99–1.07 million years) and the Olduvai Normal (1.77–1.95 million years), both of which occur within the Matuyama Reversed polarity epoch.

7.4.3.2 Secular variations. Secular variations, which are due to both dipole and non-dipole field behaviour and are variable over time and space, are reflected in changes in declination, inclination and intensity. Past variations in these parameters are referred to as **palaeosecular variations (PSV)**. In Britain, an observational record of the geomagnetic field extends back to AD 1576, and a plot of inclination against declination for London shows geomagnetic field changes over the course of the last 400 years (Figure 7.10A).

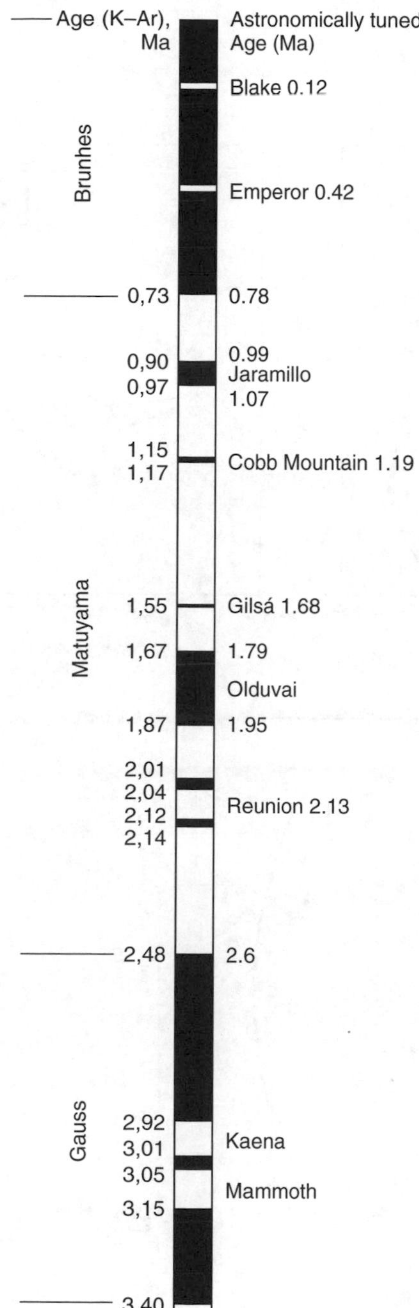

Figure 7.9 *The palaeomagnetic timescale of the last 3.5 million years. Black areas indicate periods of normal polarity; white areas show episodes of reversed polarity. K–argon ages are shown on the left-hand side; astronomically tuned ages from deep-ocean cores are on the right (after Mankinen and Dalrymple, 1979; Cande and Kent, 1995; Funnell, 1995)*

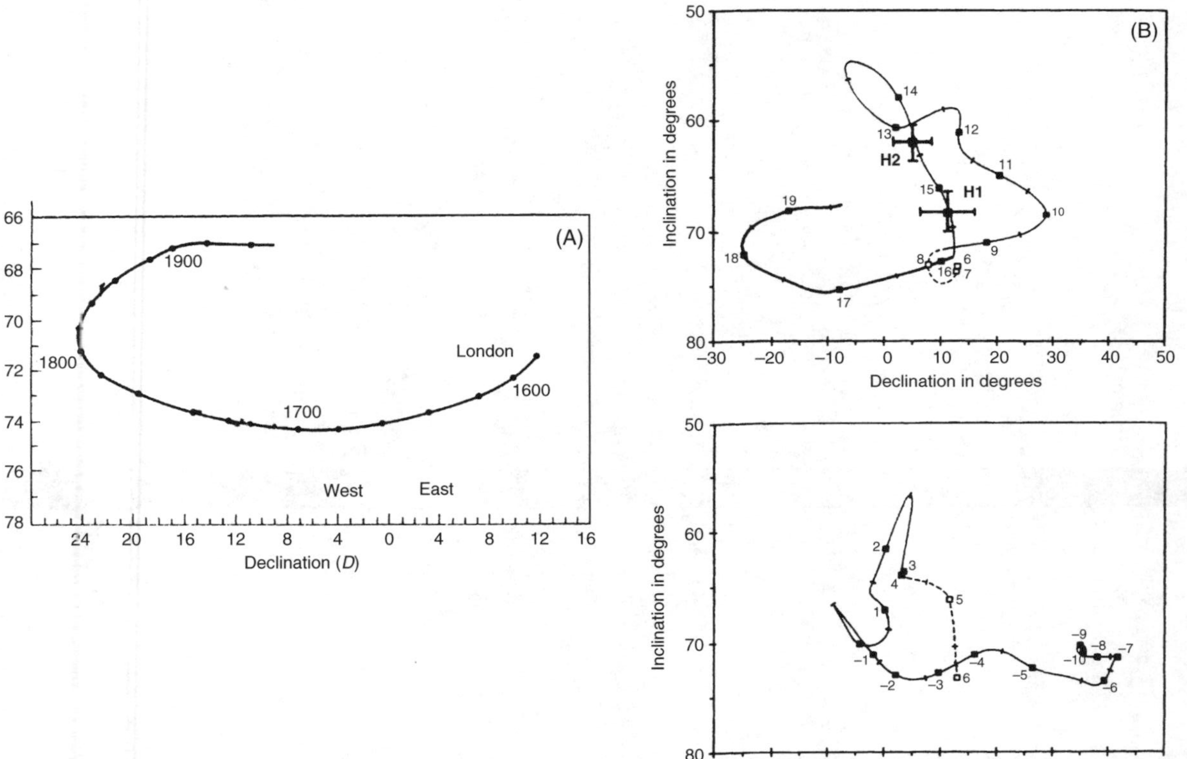

Figure 7.10 *(A) Secular changes in declination and inclination as observed in London from the late sixteenth century onwards (from Aitken, 1990). Reproduced by permission of Pearson Education Ltd. (B) The current British archaeomagnetic calibration curve. Upper:* AD *600–1975; lower: 1000* BC*–*AD *600. Also shown are the remanence measurements for samples from two medieval tiled hearths from York. The plots give archaeomagnetic ages of* AD *1625±75 and* AD *1475±75 (after Batt, 1997). Reproduced by permission of Blackwell Publishers*

During this period, declination has varied from 11°E in AD 1570 to 24°W in AD 1820, has subsequently decreased to its present value of around 5°W, and continues to decrease at a rate of 9 min each year. Inclination has varied from more than 74° in AD 1700 and is now near to 66° (Thompson and Oldfield, 1986). Such curves of secular geomagnetic variations based on observational data can be extended into the historic period by using **archaeomagnetic measurements** from features dated by documentary sources, dendrochronology, pottery typology, coins, radiocarbon or other methods (Batt, 1997).

These curves can be used as a basis for dating, for artefacts or other objects of previously unknown age, but whose remanent magnetic properties (inclination and declination) have been determined, can be matched to the curve and a date obtained. In order to do this, however, the magnetic measurements must be fitted to a **calibration curve** (or **master curve**) which shows the changes through time in declination and inclination of the earth's magnetic field for a specific geographical area (Figure 7.10B). As the geomagnetic field reflects spatial as well as temporal variations, however, all archaeomagnetic measurements must be corrected to a central location, and only data from an area approximately 1000 km across can be used to construct a calibration curve without introducing significant errors (Nöel and Batt, 1990). Geomagnetic master curves have been developed from many parts of the world, in some cases extending back over two or three millennia. Examples include those from Britain (Clark *et al.*, 1988), Meso-America (Wolfman, 1990) and the American Southwest (Lengyel and Eighmy, 2002).

Longer-term records of secular variations can be obtained from geomagnetic changes recorded in lake sediment sequences (King and Peck, 2001). These can be dated by, for example, radiocarbon or varve chronology, and will then constitute regionally specific type profiles against which other secular magnetic records can be matched. These 'master curves', some of which extend back into the Early Holocene, are often at a very high degree of stratigraphic resolution, particularly those dated by varve chronology (section 7.4.5.1). The master curves may be applicable to sediments found up to 2000 km from the type site. Correlation between individual cores is based on distinctive inflexions (**turning points**) in the magnetic profiles. Long lake sequences showing geomagnetic secular variations are now available from many areas including Britain (Turner and Thompson, 1981), North America (Lund, 1996) and Sweden (Snowball and Sandgren, 2002). In addition to inclination and declination data, some lake sediment sequences also contain a record of relative geomagnetic palaeointensity (Ojala and Saarinen, 2002).

7.4.3.3 Mineral magnetic potential. Two other magnetic characteristics have sometimes been used in magnetostratigraphy, namely **mineral magnetic susceptibility (MS)** and **isothermal remanent magnetism (IRM)**. Neither of these depends on changes in the earth's magnetic field, but rather they reflect variations in the nature and origins of magnetic minerals in sediments (Dearing, 1999; Walden, 1999). These properties cannot in themselves form a basis for a chronology, but they offer a potential basis for correlation. This is because changes in MS and IRM are often a reflection of environmental change, and as these will be broadly synchronous, the magnetic signal in sediments may provide a basis for correlation between depositional contexts and across a range of timescales (Lowe and Walker, 1997).

Mineral magnetic susceptibility measures the degree to which a material can be magnetised; in other words it is an index of its 'magnetisability'. MS has a range of applications in the archaeological and environmental sciences (as an indicator of sediment flux and erosion in lake catchments, for example), but it has also been used as a basis for correlation between cores from Holocene lake sediment sequences and other palaeoenvironmental records (Snowball *et al.*, 1999). Isothermal remanent magnetism is the magnetism generated within a sample of material by being placed in a strong steady magnetic field. With a gradual increase in field strength, IRM will increase until **saturation isothermal remanent magnetisation (SIRM)** is reached. Beyond this point, any further increase in the magnetic field will not be accompanied by an increase in IRM. SIRM measurements reflect variations in the size and types of magnetic material in a sample, and again these properties have been used in the correlation of Holocene lake sediment records (Snowball and Thompson, 1992).

7.4.4 Some Problems with Palaeomagnetic Dating

As with all dating techniques, there are a number of potential error sources in palaeomagnetism (Aitken, 1990). One area of concern relates to the acquisition and retention of the magnetic signal. In some sediments, for example, natural remanent magnetism can be acquired by chemical action, where the crystallisation of ferromagnetic oxides results in a **chemical remanent magnetism (CRM)**. This process may occur later than and under a different magnetic field from that of DRM, a record of which is contained in the same sedimentary unit. In this way a secondary magnetisation is introduced into both volcanic and sedimentary materials which, of course, serves to mask the true record of palaeomagnetic variations. In some volcanic rocks, acquisition of TRM in the *opposite direction* to the prevailing magnetic field has been demonstrated for certain special types of magnetic material. These **self-reversals** will serve to complicate the true palaeomagnetic record. In some sediments, a form of DRM may be acquired subsequent to deposition. This **post-depositional remanent magnetisation (PDRM)** may develop because although sediment will consolidate relatively quickly (within days), in other situations consolidation is a much longer process and might span thousands of years. The consequence is that there may be a substantial lag of the magnetic record behind the date of deposition. Indeed, the 'smoothing-out' of directional changes that do not last very long could mean that short-lived polarity events could be missing altogether from a record. Also, in lake sediment sequences, bioturbation (disturbance of the surface sediments by biological activity) might result in any immediately formed magnetisation being destroyed.

Problems also arise in the construction of secular calibration curves and in dating the palaeomagnetic timescale. With regard to the former, measurements of archaeomagnetic materials used to construct the curves are not always straightforward, and a range of values may well be obtained. In a weakly magnetised structure, such as a hearth or kiln, it is possible for the local magnetic field recorded in the cooling sample to be distorted by the presence of nearby iron objects or by iron slag. In the dating of burnt structures, there may also be a distortion in the local field due to the magnetism of the baked clay itself (magnetic refraction). Hence a range of measurements will be needed in order to obtain meaningful values for declination and inclination. When these data are used to plot the calibration curve, as with radiocarbon dates, they are expressed in terms of probabilities.

On the calibration plots, therefore, there is **circle of confidence** (reflecting 95% probability) about each individual data point that has been used to construct the curve. With reference to the palaeomagnetic timescale, there are discrepancies between K–argon dates and those based on the tuning of the deep-ocean isotope record (Figure 7.9). While it is likely that, at least to some extent, these age differences reflect often poor stratigraphic (and hence temporal) resolution in deep-ocean sequences, it is also the case that the K–argon method is not sufficiently precise to date some of the relatively short-lived polarity events. In this case, their positions on the palaeomagnetic timescale have been established by extrapolation based on the ages of epoch boundaries. The consequence is that the dating of some polarity events tends to be less secure.

7.4.5 Applications of Palaeomagnetic Dating

Palaeomagnetism has long been used as a basis for dating and correlation in the geological sciences and a sequence of magnetic reversals has been established back beyond 100 million years. In the Quaternary, the palaeomagnetic timescale has proved especially valuable for the dating of Early and Middle Quaternary stratigraphic sequences, including those that contain human remains and artefacts, while archaeomagnetic dating based on secular variations has proved to be a useful chronometer for the historic and late prehistoric periods. It has also been used to date key levels in the MOI record using, in particular, the Brunhes-Matuyama boundary which is located in MOI stage 20, the onset of the Olduvai event at the base of MOI stage 63 and the Matuyama–Gauss boundary in MOI stage 104 (Shackleton *et al.*, 1990). Four further examples of these applications are considered here.

7.4.5.1 Dating lake sediments using palaeosecular variations. In some lake sequences where organic carbon content is low or where there are likely to be significant amounts of older carbon residues, reliable radiocarbon ages cannot be obtained from the lake sediments. It may also be difficult to find suitable terrestrial macrofossils for AMS radiocarbon dating, particularly in deeper lake basins. One way around this problem is to use the palaeomagnetic record contained within the lake deposits. Providing that this can be related to a regionally specific master curve which itself has been calibrated by an independent dating method, the record of palaeosecular variations can be used to develop a timescale for a lake sediment sequence.

An example of this approach is provided by Saarinen (1999) in a study of Late Holocene lake sediments in northern Europe. A high-resolution record of palaeosecular variations dated by varve counting had previously been obtained from Lake Pohjajärvi in southern Finland (Saarinen, 1998), and this calibrated master curve was used to date three other lakes in Finland and one in northwest Russia (Figure 7.11). A number of distinctive fine-scale declination and inclination fluctuations are apparent in the PSV records for the last 3200 years and these 'turning points' form a basis for correlation between the individual lake cores. They also enable each lake core to be linked to the master curve from Lake Pohjajärvi. Figure 7.11 shows how an absolute timescale for Lake Päijänne can be obtained in this way. The dating error for the last 3200 years is estimated to be less than ±50 years. This study shows that palaeomagnetic dating can be an accurate and relatively rapid dating tool for lake sediments that cannot be dated using more conventional techniques.

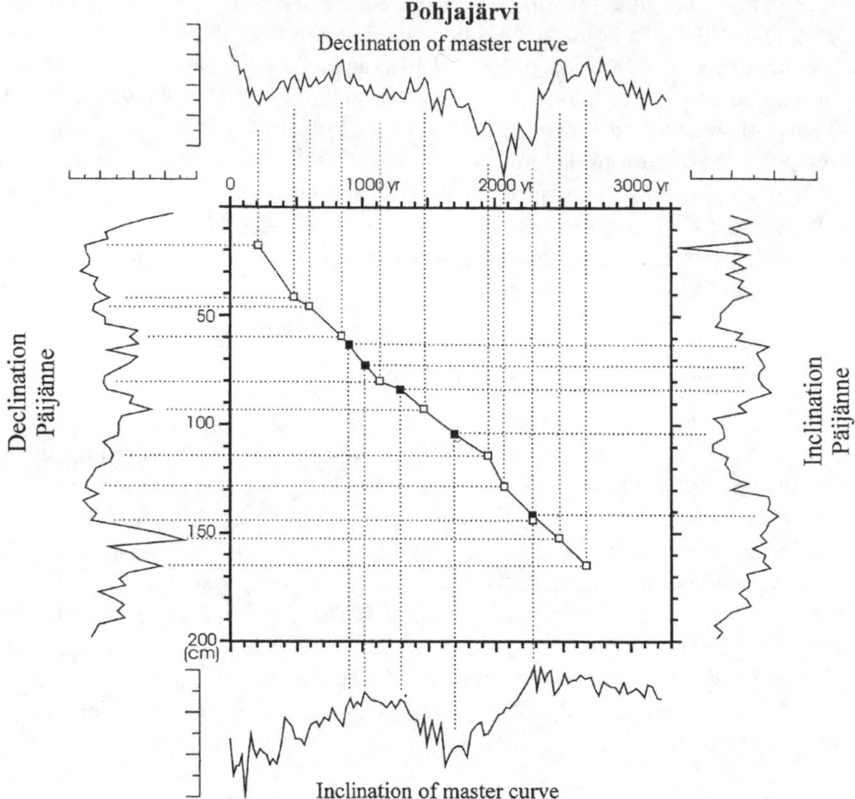

Figure 7.11 Palaeomagnetic dating of sediments from Lake Päijänne, Finland, based on a correlation of the declination (left) and inclination (right) records with the dated PSV master curves from Late Pohjajärvi. Open boxes indicate matching declination features and solid boxes show matching inclination features. Ages are in varve years before present (after Saarinen, 1999). Reproduced with permission of Elsevier

7.4.5.2 Palaeomagnetic correlations between Scandinavian Ice Sheet fluctuations and Greenland ice-core records. As noted in Chapter 5, during the Last Cold Stage a number of rapid and large amplitude climatic events [Dansgaard–Oeschger (D–O) cycles] are recorded in the oxygen isotope records from the Greenland Ice Sheet (section 5.5.3.1). These D–O events are also recorded in palaeoceanographical changes in the North Atlantic and in vegetational changes in the Iberian Peninsula (Peterson *et al.*, 2001; Roucoux *et al.*, 2001). There are, however, few data on how the Scandinavian Ice Sheet reacted to these often abrupt climatic changes, partly because much of the evidence has been removed by erosion as the ice sheet advanced to its maximum around 20 000 years ago, but also because the evidence that is preserved in pre-Last Glacial maximum sediments is difficult to date with sufficient accuracy to allow the terrestrial record to be correlated with the relatively short-lived D–O events.

One way in which correlations (and, by implication, dating) can be achieved between the Scandinavian and Greenland ice-core records is by means of palaeomagnetism. Mangerud *et al.* (2003) describe evidence from three large caves in western Norway, which were blocked by advancing glacier ice, and which contain thick sequences of laminated sediment. Two clearly defined palaeomagnetic excursions were identified in stratigraphic superposition within these deposits: the Skjong which is correlated with the well-documented Laschamp magnetic reversal and the Valderhaug correlated with the Mono Lake excursion. These events have been dated radiometrically elsewhere to ca. 44 000–47 000 years BP and ca. 31 000–33 000 years BP, respectively (Benson *et al.*, 2003). The excursions are also reflected in the Greenland ice cores, for during a magnetic reversal or excursion, the earth's geomagnetic field is weakened and this will be marked by increased concentrations and fluxes of cosmogenic nuclides such as ^{10}Be and ^{36}Cl (section 3.4). Such increases occur in the ice cores ca. 41 000 and ca. 34 000 years ago, dates that are closely comparable with those described above. Assuming that the geomagnetic excursions recorded in the Norwegian caves are the Laschamp and Mono Lake events, this allows the Norwegian terrestrial record to be correlated with that from Greenland (Figure 7.12) and enables advances and retreats of the Scandinavian Ice Sheet to be linked with the Greenland D–O cycles. For example, the major readvance in western Norway during the Valderhaug/Mono Lake excursion took place just at the transition from Greenland Interstadial (GIS) 7 to Greenland Stadial (GS) 7 in the Greenland cores. In other words, the Scandinavian Ice Sheet was responding rapidly to the climatic downturn reflected in the Greenland isotope record. This cause and effect relationship can only be demonstrated because of the palaeomagnetic-based correlation between western Scandinavia and Greenland.

7.4.5.3 Palaeomagnetic dating of the earliest humans in Europe. There has been a lively debate over the timing of the first human migrations into Europe. While there are signs of early humans in Asia 1.7 million years ago, and in the Middle East 1.5 million years ago, conclusive evidence for humans of comparable antiquity in Europe has, until relatively recently, proved to be frustratingly elusive. This has led some to advocate a relatively 'short chronology' of human occupation, extending over no more than 500 000 years (Roebroecks and von Kolfschoten, 1994). Others, by contrast, have argued for a more protracted period of human colonisation ('long chronology') extending, perhaps, to 1 million years or even earlier (Arribas and Palmqvist, 1999).

A key area in this debate has been the Iberian Peninsula, as its proximity to North Africa means that it could have been one of the first areas of Europe to be colonised by humans moving northward out of Africa, and a key site has proved to be the cave of Gran Dolina in the Sierra de Atapuerca, near Burgos in northern Spain. The cave contains a number of distinct fills, and one of the strata, the Aurora Bed, has yielded 85 human remains, almost 200 stone artefacts and more than a thousand fragments of vertebrate bone (Aguirre and Carbonell, 2000). Natural remanent magnetisation (NRM) measurements show that ca. 1 m above the Aurora Bed, there is a marked change in polarity from normal to reversed. This has been interpreted as representing the Brunhes–Matuyama boundary, while the Jaramillo Event (Figure 7.9) may be recorded some 8 m further down the sequence (Parés and Pérez-Gonzalez, 1999). If so, the hominid-bearing strata have an age between 780 000 and 980 000 years. The Grand Dolina evidence, therefore, lends support

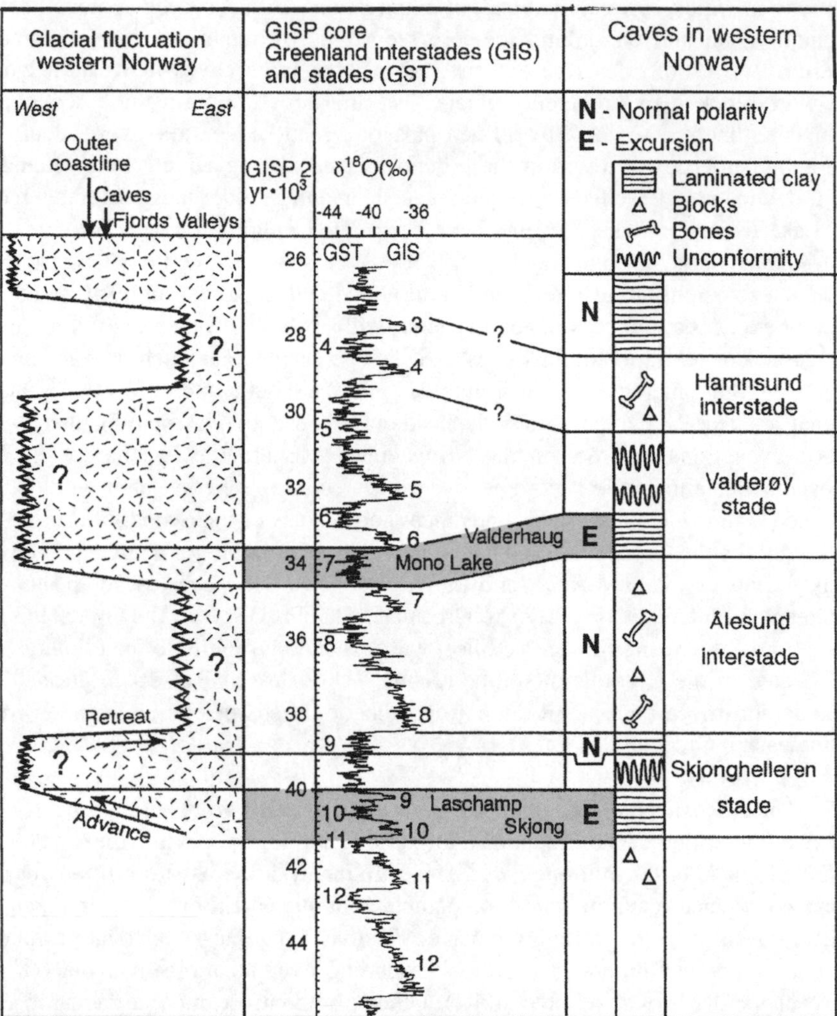

Figure 7.12 Correlation between glacier fluctuations in western Norway (left-hand column), Greenland ice-core data (centre) and cave sequences in western Norway based on two palaeomagnetic excursions. The ice-core data is in the form of the oxygen isotope ($\delta^{18}O$) signal, with the Greenland Stadials (GS) reflecting colder periods on the left and the Greenland Interstadials (GIS) reflecting warmer periods on the right. The timescale is in ice-core years BP. The cave records show the generalised stratigraphy of the cave sequences and magnetic polarity (after Mangerud et al., 2003). Reproduced with permission of Elsevier

to the 'long chronology' for human occupation of Europe, and adds to a growing body of evidence from other sites around the northern Mediterranean for early human presence in Europe between 700 000 and 1 million years ago (Balter, 2001).

7.4.5.4 Palaeomagnetic dating of the Sterkfontein hominid, South Africa. Sterkfontein Cave near Johannesburg in the Gauteng Province of South Africa is one of the most important early hominid sites on the African continent. Since the discovery in 1936 of the first adult *Australopithecus*, the older cave breccias[6] have yielded large concentrations of *A. africanus* fossils, while successively younger breccias have been found to contain not only hominid fossils, but also stone tools of Oldowan (ca. 2.0–1.7 million years) and Late Acheulian (ca. 1.7–1.4 million years) Age (Kuman and Clark, 2000).

Perhaps the most striking of the recent finds at Sterkfontein was the discovery in 1997 of parts of a skeleton, including the skull, of *Australopithecus* (Figure 7.13A). This is the first discovery of such a complete *Australopithecus* skull and skeleton (Clarke, 1998), and the stratigraphic position of the fossil remains in the cave, and possible affinities with early hominids from other parts of Africa, suggested that the skeleton may be between 2.7 and 4.0 million years in age. Precise dating of the cave deposits proved to be difficult, however, as radiometric techniques could not be applied to the Sterkfontein cave fills. Palaeomagnetism had been attempted, but the cemented breccias in which the bones are found do not preserve a stable palaeomagnetic signal. Partridge *et al.* (1999), however, tried an alternative approach, which involved the extraction of a palaeomagnetic record from flowstone deposits that are interbedded with the cave breccias. Nine samples were measured and five changes in polarity identified. The determined polarity record was compared with the palaeomagnetic timescale and this places the skeleton in the time range of 3.22–3.58 million years (Figure 7.13B). Interpolation of sedimentation rates over the small intervals between the magnetic reversals allows this age estimate to be reduced to 3.30–3.33 million years. This dating is highly significant in studies of human evolution, for it means that this is the oldest hominid skeleton yet discovered. Moreover, the bones may belong to a species of *Australopithecus* other than *africanus*, and could possibly be a contemporary of *A. afarensis* from East Africa.

7.5 Palaeosols

Soils develop through the combination of physical, chemical and biological processes operating on the earth's surface. Once a soil becomes buried by younger sediments, however, they are no longer affected by contemporary soil-forming processes, and they become relict or fossil features. These fossil soils, or **palaeosols**, are frequently encountered in Quaternary stratigraphic sequences (Figure 7.14). Two of the principal soil-forming factors are climate and vegetation, and as these will be characteristic of often broad geographical regions where topography and parent material are also uniform, soils that develop under a particular climatic regime or vegetation cover (**zonal soils**) will, at least in theory, possess a common set of pedological characteristics. Hence, in any one climatic region, palaeosols that formed in different localities during the same climatic episode will not only be broadly coeval, but will be comparable in terms of soil properties. These now buried palaeosols therefore constitute time-planes or marker horizons in stratigraphic sequences and, as such, can form a basis for establishing age equivalence. Where the palaeosols can be dated by other means, for example by radiocarbon dating of contained organic materials, or by luminescence dating of overlying and underlying sediments (assuming, of course, that no erosion has occurred), the palaeosols can form chronostratigraphic marker horizons.

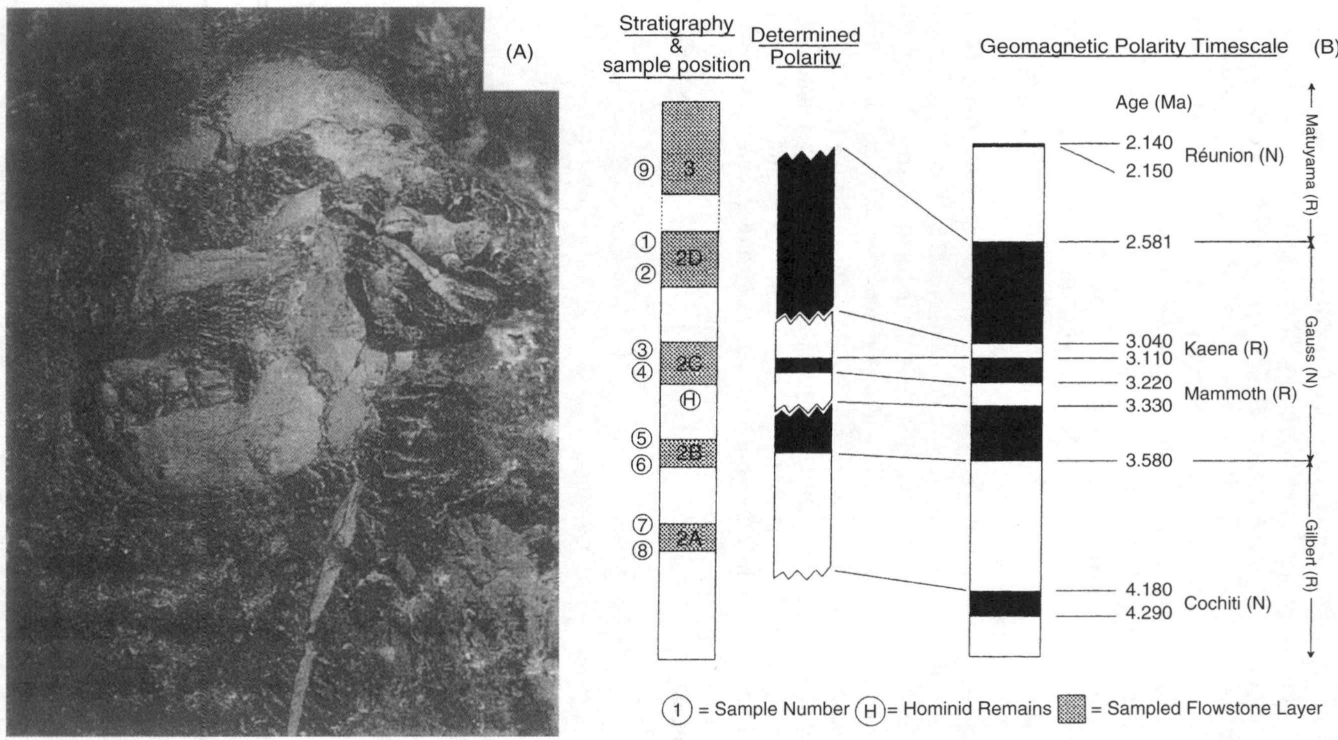

Figure 7.13 *(A) The Sterkfontein skull (StW 573) and left humerus. (B) Correlation of the determined Sterkfontein magnetostratigraphy to the geomagnetic polarity timescale (after Partridge* et al.*, 1999). Reproduced by permission of John Wiley & Sons Ltd*

Figure 7.14 *The 'Eemian' palaeosol (dark layer) in a sedimentary sequence at Straubing, Bavaria, Germany. Although the palaeosol may have formed initially during the Last Interglacial (MOI substage 5e), it probably reflects the operation of pedosedimentary processes during the whole of MOI stage 5 (photo: Rob Kemp)*

7.5.1 The Nature of Palaeosols

Buried palaeosols may not always be easy to identify in Quaternary stratigraphic sequences, for a wide range of weathered materials are interstratified in Quaternary sediments, not all of which may be soils *sensu stricto* (Lowe and Walker, 1997). There are, however, a number of properties of buried soils which enable them to be distinguished from other deposits. When a soil forms on a land surface, a distinctive, vertically differentiated series of layers or horizons evolve in response to variations in weathering processes and to the movement of weathering products up and down the profile (Catt, 1986). In younger palaeosols, such as those that are found on later prehistoric and historic archaeological sites, both the A and the B horizons[7] may be present. Sometimes these are buried by younger sediments, such as blown sand, colluvial deposits or peat growth, whereas in other situations they may be found beneath structures such as ancient field boundaries (Molloy and O'Connell, 1993). In older palaeosols, however, the organic A horizon is no longer preserved having been lost through prolonged decomposition post-burial or erosion. In some situations, however, the mineral part of the A horizon may remain, and may be recognised on the basis of its distinctive clay content which differs from that of the B horizon below (Catt, 1988). In most buried palaeosols, it is the B horizon which enables **pedogenic (soil forming) processes** to be identified (Catt, 1990). Important diagnostic features of the B horizon include colour, textural variations, weathered minerals and enrichment or depletion in carbonate content. For example, soils in semi-arid areas will tend to be enriched in carbonate, while those in wetter environments (acid soils) will show evidence of carbonate depletion, a process known as **decalcification**.

There are two other diagnostic characteristics of buried soils, **soil micromorphology** and **mineral magnetic properties**. The former refers to the distinctive arrangement of particles and voids which together make up what is referred to as the **soil fabric**, and which is determined by examining thin sections of soils under a microscope. Soil micromorphology is now regarded as one of the most important methods for detecting evidence of pedogenesis, for elucidating processes that contributed to the formation of a particular soil, and for inferring former environmental conditions (Kemp, 1998; Davidson and Simpson, 2001). The magnetic properties of the mineral constituents of soils (especially magnetic susceptibility – section 7.4.3.3) are also a valuable diagnostic tool in the detection and characterisation of buried palaeosols, for magnetic minerals produced or enriched through pedogenic processes often generate distinctive mineral magnetic signals (Thompson and Oldfield, 1986).

7.5.2 Palaeosols as Soil-Stratigraphic Units

Well-developed palaeosols that developed during the course of a specific soil-forming interval, and which are sufficiently distinctive in terms of their pedological characteristics that they can be traced over a relatively wide area, can be considered as **soil-stratigraphic units** (Morrison, 1978). A soil-stratigraphic unit was defined initially by the American Commission on Stratigraphic Nomenclature (1961) as 'a soil with physical features and stratigraphic relationships that permit its consistent recognition and mapping'. Subsequently, the North American Commission on Stratigraphic Nomenclature (1983) introduced the term '**pedostratigraphic unit**' to refer to 'a buried, three-dimensional body of rock that consists of one or more differentiated pedological horizons'. Such well-defined palaeosols (or **geosols**) constitute marker horizons and form the basis for inter-site correlation and for establishing age equivalence.

Some distinctive palaeosols will develop over a relatively short weathering interval. For example, radiocarbon dates on the Lateglacial Interstadial 'Pitstone Soil' (or 'Allerød Soil') which is found throughout southeast England (Rose *et al.*, 1985) suggest a period of formation of less than 1000 years (Preece *et al.*, 1995), while the Farmdale Geosol (25 000–28 000 ^{14}C years BP) which is widespread in Illinois and adjacent areas may have a development duration of around 1000 years (Grimley *et al.*, 2003). Other palaeosols have formed over a more protracted time period and may be **polygenetic**, in other words they have evolved under a range of different climatic/vegetational conditions. For instance, the Sangamon Soil of the American Midwest appears to have formed over a period of 100 000 years (ca. MOI stages 3–5) during which the climate shifted from cold to warm to cold again (Hall and Anderson, 2000), while the older Yarmouth Soil evolved over the course of 180 000 years of interglacial weathering, including MOI stages 7, 9 and 11 (Grimley *et al.*, 2003).

In situations where landscapes have been responding to cyclical changes in climate, pedogenesis may have been interrupted during phases of instability, only to resume once more during a succeeding stable phase. The result is series of palaeosols preserved in a stratigraphic sequence. This is a characteristic phenomenon of the loess regions of the world, where numerous soil-stratigraphic units (reflecting landscape/vegetational stability) are interbedded with suites of aeolian (wind-blown) sediment (Derbyshire, 2001; 2003). These are found, for example, in central Europe, in the American Midwest and in the Pampas of Argentina, but the most spectacular record of loess and interbedded palaeosols occurs on the Chinese Loess Plateau, where the loess–palaeosol sequence extends back to around 2.5 million years. The soil-stratigraphic units within these more or less continuous sequences form the basis for correlation between the different loess provinces, and also serve as marker horizons for linking these terrestrial records to that from the deep oceans (section 7.5.4.3).

7.5.3 Some Problems with Using Palaeosols to Establish Age Equivalence

Once a land surface has been exposed to the atmosphere, soil-forming processes begin, and the degree of pedogenesis will reflect a range of site and climatic factors (parent material, vegetation cover, exposure, etc.) as well as the length of time over which pedogenesis has been operating. Time has often been regarded as a major element in this process. Thus, in North America, for example, where palaeosols have been widely employed in the establishment of age equivalence between stratigraphic sequences, the assumption has been that episodes of rapid soil formation will occur during interglacial and, to a lesser extent, interstadial episodes, while negligible soil development will occur during intervening cold stages. Thus, well-developed buried soils (such as the Sangamon Soil) have been considered as indicative of interglacial episodes and have therefore been employed as key marker horizons for time-stratigraphic correlation. It is now accepted that soil development cannot simply be regarded as a function of time, however, but that other soil-forming factors will affect the degree of soil development at both the local and regional scale (Boardman, 1985). Moreover, as we saw above, many older buried palaeosols (including the Sangamon Soil) are polygenetic, and reflect more than one phase of pedogenesis. Hence assigning an older buried soil to a particular interglacial or interstadial episode may not be possible in the absence of additional evidence.

Other difficulties arise in terms of the recognition of buried soils. For example, buried soils may be affected by post-depositional diagenesis due, for example, to changing groundwater conditions, or to the effects of differential compaction arising from the weight of the overburden. These can cause complex physical and chemical changes in the buried soils (e.g. in microfabric structures) and can lead to difficulties in distinguishing true palaeosols from 'pseudosoils', which are distinctive coloured horizons in sedimentary sequences caused by the mobilisation of iron, manganese and other elements during diagenesis (Robinson and Williams, 1994). Also, a buried soil may be modified both physically and chemically by soil-forming processes acting on the new ground surface, and a **welded soil** may develop if the younger soil profile is superimposed upon, or merges with, the buried one. In these cases, it may be difficult to distinguish between separate phases of soil formation (Dahms, 1994), and this clearly poses a problem when correlating soil-stratigraphic units in one area with those in another. The Sangamon Soil of the midwestern United States (see above) is a case in point, for it is evident that in many areas the interglacial Sangamon Soil forms a pedologic continuum with the overlying cold-climate Early Wisconsian loess, and hence the palaeosol, rather than being a time-parallel horizon, is time-trangressive (Follmer, 1983). Furthermore, in some palaeosols, short-range variations have been detected in key diagnostic properties such as micromorphology (Kemp and Faulkner, 1998), and this again creates difficulties when comparing and correlating polygenetic palaeosols over longer distances. Dating of palaeosols is also problematic. Soils are some of the most difficult materials to date by radiocarbon (section 2.5.4), and few other methods are applicable to the dating of ancient soil material, although cosmogenic nuclides such as ^{10}Be have been employed with apparent success in some areas (Markewich *et al.*, 1998; Muhs *et al.*, 2003). In general, therefore, palaeosols have to be dated using other approaches, such as optical dating of overlying or underlying sediments, a technique that has been widely used in loess–palaeosol sequences, or by means of proxy data including fossil fauna or artefact evidence (Chlachula, 2003).

7.5.4 Applications of Palaeosols in the Establishment of Age Equivalence

In spite of these problems, palaeosols have been used successfully to establish time-stratigraphic correlations in a number of different circumstances. On the local scale, they have proved valuable as marker horizons and have formed an important component in the development of site histories (see below). On a wider canvas, in Europe, North America and parts of Asia they have been employed as basis for inter-site correlation at the local and regional scale (Birkeland, 1999). Some examples of these applications are considered here.

7.5.4.1 Buried palaeosols on the Avonmouth Level, southwest England: stratigraphic markers in Holocene intertidal sediments. Around the Severn Estuary in southwest England is a range of wetland landscapes known as the Severn Levels. These include the Gwent Levels in South Wales and the Avonmouth Levels to the north of Bristol (Figure 7.15A). In all, these wetland areas cover some 840 km^2 and have yielded a series of spectacular archaeological discoveries, including Mesolithic human footprints, the remains of Bronze Age, Romano-British and medieval boats, Iron Age buildings, and historic structures such as ancient piers and fishtraps (Rippon, 1996). Dendrochronological work at one of these sites (Goldcliff near Cardiff) was described in section 5.2.4.4. Underlying the present surface of the Levels is a sequence of estuarine muds and/or alluvial silts with interbedded organic horizons known as the Wentlooge Formation (Allen and Rae, 1987). These organic units, which reflect stabilisation of the land surface during periods of episodic flooding, sometimes comprise peats, but in a number of cases are the remains of soils that have been buried beneath the silts and clays during a subsequent marine inundation. When exposed in section, some are indicated by the presence of gleyed[8] horizons, occasionally overlain by a thin layer of organic material. Others comprise a thicker layer of clay with a widespread but less dense organic component which probably reflects the A horizon of a conventional dry-land soil (Locock, 1999). Some of these palaeosols have provided sufficient material for radiocarbon dating, and where they can be identified in the buried stratigraphy, they form key time-stratigraphic marker horizons. As such they are potentially extremely valuable in archaeological excavation, and in the reconstruction of site histories.

On the Avonmouth Levels, several prehistoric and historic settlement sites have been investigated, and excavations have revealed the presence of a number of buried palaeosols

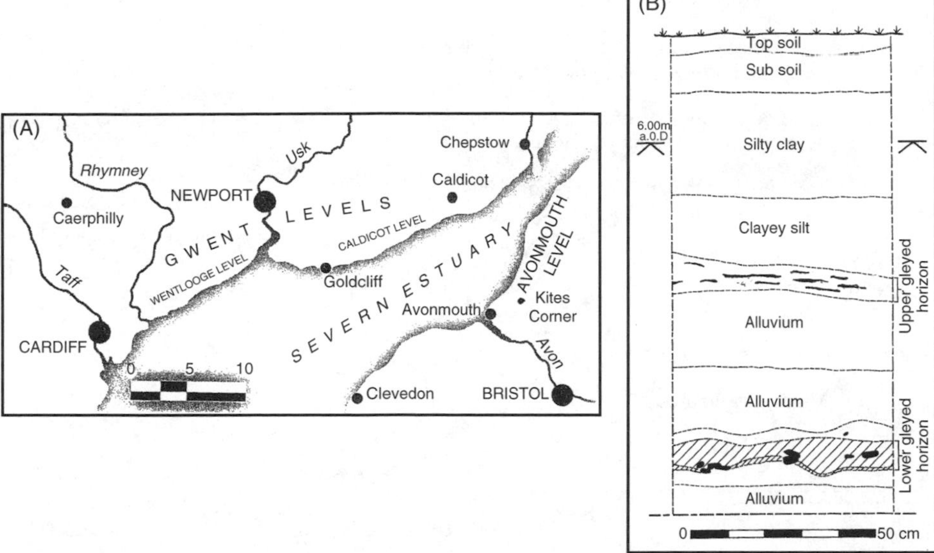

Figure 7.15 *(A) The Severn Estuary Levels in southwest England. (B) The stratigraphic sequence at Kites Corner on the Avonmouth Levels, showing two gleyed horizons interbedded with estuarine silts and clays. The lower palaeosol is dated to 3900–4200* ^{14}*C years* BP, *and the upper palaeosol to 3000–3300* ^{14}*C years* BP *(after Locock* et al., *1998). Reproduced with permission of English Heritage*

(Locock *et al.*, 1998; Allen *et al.*, 2002). At the Kites Corner site to the east of Avonmouth (Figure 7.15A), for example, two palaeosols have been found in stratigraphic sequence: a lower buried soil radiocarbon dated to ca. 3900–4200 ^{14}C years BP and which micromorphological analysis shows to be a ripening or developing soil, and an upper palaeosol dated to ca. 3000–3300 ^{14}C years BP which, on the basis of micromorphological evidence, appears to be a moderately long-lived soil with seasonal drying of surface horizons and earthworm activity (Figure 7.15B). Both of these buried soils are important in the context of the archaeological investigations on the Avonmouth Levels. The lower palaeosol is remarkably extensive and has been traced in trial pits over more than 500 m. It therefore constitutes a key marker horizon in the general stratigraphic sequence. The upper palaeosol is interesting because the dates are older (by more than 400 ^{14}C years) than those on charcoal from a nearby human activity scatter. This suggests that the activity post-dates renewed flooding (and submergence of the soil), and could therefore have been taking place in the inter-tidal area as opposed to the nearby land area (Locock, 1999).

7.5.4.2 The Valley Farm and Barham Soils: key stratigraphic marker horizons in southeast England. Unlike continental Europe and North America, there are few key soil-stratigraphic units forming marker horizons in the British Quaternary record. Two that have been described are the Barham and the Valley Farm Soils of East Anglia (Figure 7.16), which together represent a depositional hiatus within Early and Middle Pleistocene sequences (Kemp *et al.*, 1993). Not only do they provide valuable data on environmental change

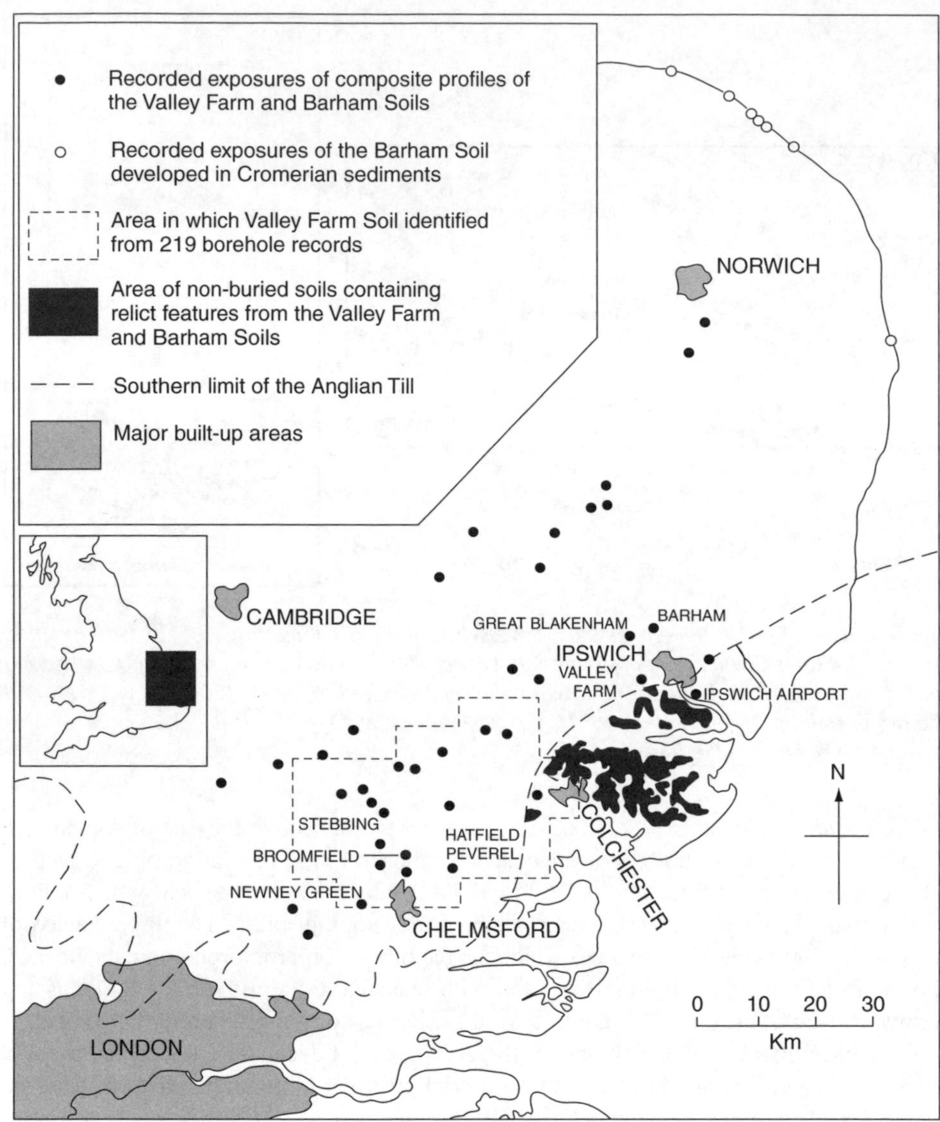

Figure 7.16 The distribution of the Valley Farm and Barham Soils in southeast England (after Kemp et al., 1993, Fig. 1). Reproduced with permission of Elsevier

during time periods not covered by the depositional record, these buried palaeosols, and particularly the Barham Soil, also fulfil an important correlative role as distinctive stratigraphic units.

The Valley Farm Soil is rubified, mottled, and contains large quantities of translocated clay. It is considered to be a complex stratigraphic unit with component soils developed on a series of different age surfaces and perhaps spans a series of cold and temperate intervals over a time period of more than a million years. Its latest stage of development appears to have been during the Cromerian IV Interglacial (Zagwijn, 1996) which equates with MOI stage 13 (ca. 450 000 years ago; Figure 1.4). The Barham Soil, which is buried beneath sediments of the Anglian glaciation (MOI stage 12), formed through pedogenesis in a periglacial environment and is characterised by a variety of cryogenic features. It fulfils the requirement of a soil-stratigraphic unit in that it consists of a series of soil profiles that can be traced regionally and have similar stratigraphic relationships, although the properties may vary from place to place according to local site conditions. Its stratigraphic position, immediately above the Valley Farm Soil and below the overlying Anglian glacial deposits, places it in the Early Anglian cold stage. This closely constrained age, coupled with its widespread distribution throughout East Anglia (Figure 7.16), means that it forms a key time-stratigraphic marker horizon for this part of southeastern England.

7.5.4.3 Correlation between the Chinese loess–palaeosol sequence and the deep-ocean core record for the past 2.5 million years. While the MOI signal (section 7.2) forms an unparalleled record of global climate change during the course of the Quaternary, there are relatively few terrestrial sequences that possess the climatic sensitivity and, above all, the continuity to match this record. One that does is the loess–palaeosol sequence from interior China, the area known as the 'Loess Plateau'. There, great thicknesses of wind-blown sediment (loess) accumulated during cold episodes when a strengthened East Asian winter monsoon transported loess grains from the inland deserts of northern and northwestern China onto the Loess Plateau. During intervening warm periods, however, a reduction in winter monsoonal intensity and a strengthening of the summer monsoon brought moisture from the south. This led to an expansion of the vegetation cover and increased pedogenesis. The result is stacked sequences of loess and interbedded palaeosols which constitute a long-term record of glacial–interglacial climate changes that can be compared directly with the oxygen isotope signal from the deep oceans. Such records are vitally important to climatic modellers and atmospheric scientists seeking to understand the causes of long-term climate change.

Within the area of the Loess Plateau, 37 soil-stratigraphic units have been identified, and these can be correlated between sections using palaeomagnetic reversals and key stratigraphic markers (Kukla and An, 1989; Ding *et al.*, 1993). At one of these sites, Baoji in the extreme south of the Loess Plateau (Figure 7.17A), a continuous record of glacial–interglacial changes has been obtained from grain-size variations down the profile (Figure 7.17B), the grain size in the loess beds being proportionately coarser than that in the palaeosols (Ding *et al.*, 1994). Tuning of this record to the astronomical frequencies (section 7.2.2) shows that major shifts in climatic periodicity occurred around 1.6–1.7 million years ago when the Asian winter monsoon changed from variable periodicities (55 000–400 000 years) to a 41 000-year periodicity, and between 800 000 and 500 000 years ago when an increase in monsoonal intensity was accompanied by the transition to a 100 000-year periodicity (Liu *et al.*, 1999). The more recent change matches that in the oxygen isotope signal from the North Atlantic (Figure 7.17B), but the 1.6/1.7 million year shift is less pronounced in the deep-ocean record. This may reflect different climatic forcing mechanisms between the North Atlantic and eastern Asia in the early and Mid-Quaternary, including the climatic effects of the expanding northern hemisphere ice sheets. The dated loess–palaeosol record from China is therefore proving to be a key source of proxy data in the study of the past global climate system.

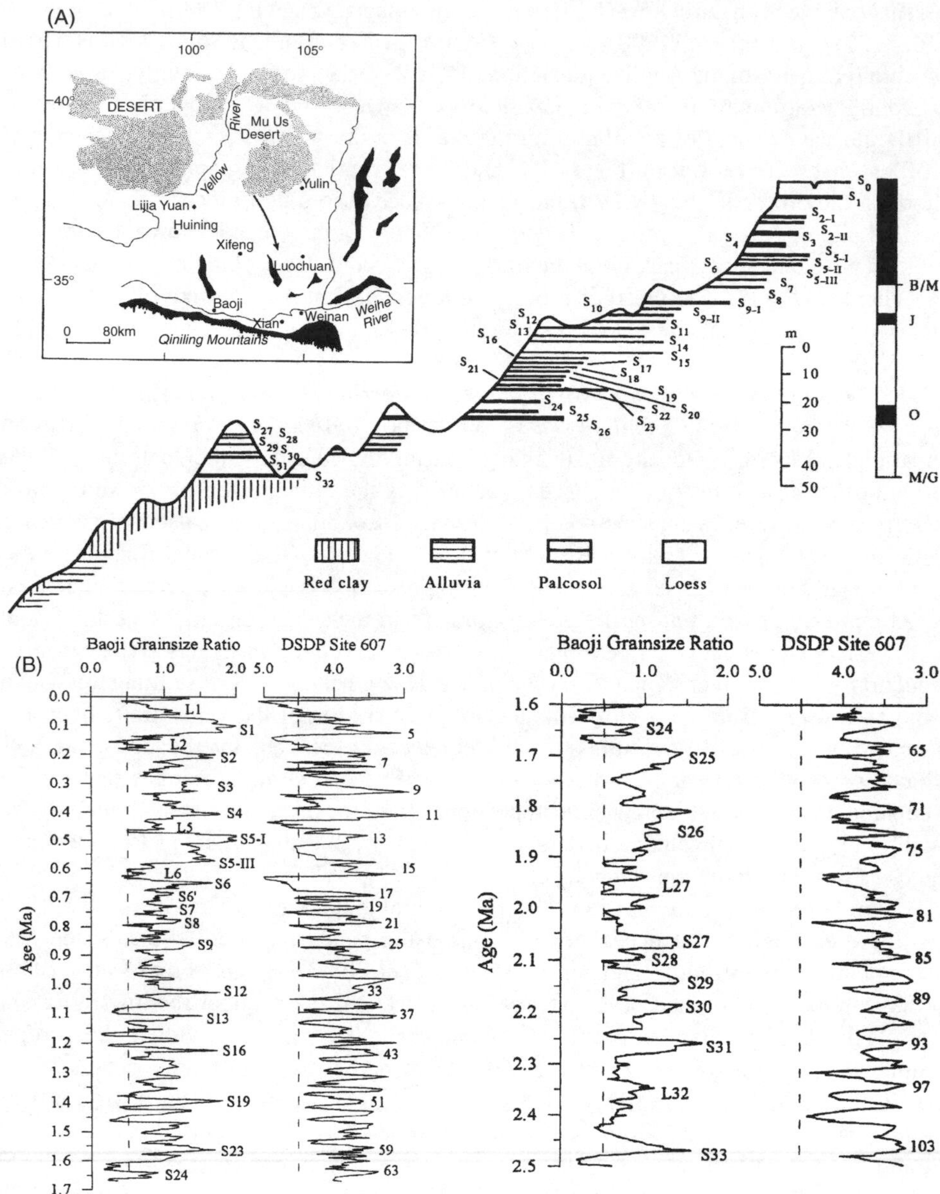

Figure 7.17 (A) The stratigraphy of the Baoji section showing the soils (S) interbedded with loess deposits (after Ding et al.*, 1994). The palaeomagnetic timescale is on the right. The inset map shows the location of Baoji in the lee of the Qinling Mountains in the south of the Loess Plateau. Reproduced with permission of Elsevier. (B) The Baoji grain size record (left) compared with the $\delta^{18}O$ ratio from North Atlantic core DSDP Site 607 (right). The soils are labelled with the prefix 'S', the loess units with the prefix 'L'. The decrease in grain-size ratio to the right reflects finer-grained sediments (palaeosols) that formed during interglacial episodes (after Liu* et al.*, 1999). Reproduced with permission of Elsevier*

Notes

1. The term **pyroclast** refers to all the volcanic material that has been blown into the atmosphere by explosive activity. Pyroclastic rocks are formed entirely of this fragmental volcanic material.
2. Diatom fragments and higher plant silica bodies (including phytoliths) that are found in peats and lake muds may be of a comparable size to volcanic glass, and may have similar optical properties. Removal of this material makes the optical scanning of slides for volcanic glass much more straightforward, especially if the glass is present in very low quantities (Hall and Pilcher, 2002).
3. More recent work at a site in southern Sweden (Davies *et al.*, 2004) suggests an age of 13 900 cal. years BP, for the Borrobol Tephra. It is therefore possible that there may be more than one 'Borrobol Tephra', albeit with similar geochemical characteristics.
4. A **polarity change** is due to the reversal of the dynamo currents producing the main dipole field, and the transition is thought to involve the falling to zero of these currents and the subsequent re-growth in the opposite direction (Aitken, 1990).
5. The term **precession** is used to describe the slow movement of the axis of rotation of a spinning body (such as a gyroscope or spinning top) about a line that makes an angle with it, so as to describe a cone.
6. **Breccia** is a rock that consists of angular fragments of material usually from a limited source. In caves, the rock fragments are cemented together to form a very hard porous rock known as *cave breccia*.
7. The **soil profile** can be considered to consist of three horizons: an **A horizon** which comprises an upper 'topsoil' containing a mixture of organic and mineral material, and a lower 'subsoil' (E horizon) from which clay and sesquioxides of iron and aluminium have been removed; a **B horizon** where organic matter is sparse and where precipitated minerals leached from the overlying A horizon have accumulated; and a basal **C horizon** of parent material comprising partially weathered rock or sediments.
8. **Gleying** in soils occurs under reducing conditions, usually when a soil is saturated. In such circumstances, iron and manganese oxides are reduced and dissolved, thereby imparting a grey colour to the matrix. The reduced iron and manganese may then be redistributed, oxidised and reprecipitated in mottles within the soil body, or removed entirely from the soil profile.

8

Dating the Future

Time present and time past
Are both perhaps present in time future,
And time future contained in time past

T.S. Eliot

8.1 Introduction

The range of dating methods now available to the Quaternary scientists is impressive, and the technical and methodological advances that have been made over the course of the last 50 years have been extraordinary. How the portfolio of dating techniques will look 50 years from now is, of course, impossible to predict, but it is perhaps worth reflecting, albeit briefly, on some areas where development might take place over the coming decades. In this final chapter, therefore, we take a prospective view on what Hedges (2001a) has succinctly described as the 'future of the past'.

8.2 Radiometric Dating

The startling developments that have occurred in radiometric dating in recent years have, in large measure, been a direct result of major advances in measuring instrumentation. The innovation of accelerator mass spectrometry (AMS) in the 1980s, for example, not only revolutionised radiocarbon dating, but paved the way for cosmogenic nuclide dating in the 1990s. In U-series dating, the development of thermal ionisation mass spectrometric (TIMS) techniques in the late 1980s, followed by the use of multi-collector inductively coupled plasma mass spectrometry (MC-ICP-MS), has resulted in significant improvements in sensitivity, speed and analytical precision, and the extension of analyses to materials with lower U and Th concentrations. In due course, this may lead to the analysis of U-series isotopes with shorter half-lives, such as ^{210}Pb (Goldstein and Stirling, 2003). Laser ablation ICP-MS is another innovation that is likely to improve numerous aspects of uranium and thorium analysis (Eggins *et al.*, 2003). Recent developments in mass spectrometry have also enabled measurements to be made of other isotopes, such as ^{40}Ar and ^{39}Ar, again leading to marked improvements in accuracy and precision of measurement, as well as an extension of the ^{40}Ar/^{39}Ar technique at the younger end of the dating range. The striking advances that have been made in luminescence dating since the 1980s, most notably in the applications of IRSL to single grains of sediment, have produced a dating method that is beginning to rival radiocarbon over the last 50 000 years, with the added advantage of an age range that extends well beyond that of ^{14}C (Duller, 2004). Further analytical refinements in all of these areas might confidently be expected over the coming years.

Radiometric dates, however, are only as good as the materials that are being dated, and in recent years a concerted effort has been made to understand the physical, chemical and biological composition of sample material, and this too is likely to be a feature of dating programmes in the next few decades. Over the past ten years or so, much has been learned about post-mortem diagenesis in fossil assemblages, and the development of increasingly sophisticated models of uranium uptake, for example, are now enabling more reliable U-series and electron spin resonance (ESR) dates to be obtained from biochemically complex materials such as bones and teeth (Pike and Hedges, 2001; Pike and Pettitt, 2003). Similarly, careful screening of sample material and prior determination of key physical and/or chemical characteristics mean that it is now possible to obtain coherent U-series dates from, for instance, such hitherto problematical material as marine molluscs (Jedoui *et al.*, 2003). Improved methods of chemical extraction and a better knowledge of environmental biomolecular diagenesis are now enabling radiocarbon dates to be obtained on compound-specific molecules, such as lipids, and work is likely to continue in this area. Some success has been had in the breaking down of polymers,[1] such as insect chitin (Hodgins *et al.*, 2001), in order to generate ‘purer’ (and hence less contaminated) samples for radiocarbon dating. Similarly, considerable progress has been made in the extraction, purification and radiocarbon dating of residues from rock art sites (Hedges *et al.*, 1998b; Watchman and Jones, 2002), one of the greatest challenges for archaeology in the next decade. It is in these and other areas of sample chemistry that knowledge and capabilities are almost certain to increase in the near future (Hedges, 2001a).

Three radiometric techniques in particular perhaps merit further consideration: radiocarbon, electron spin resonance and cosmogenic nuclide dating. In **radiocarbon dating**, one of the disappointments of the past 20 years has been the difficulties that have been experienced in pushing the age range back beyond the 50 000-year ‘barrier’. There were great expectations of AMS, but a number of technical difficulties, including the necessity of keeping background contamination low enough, have meant that the upper age limit of the method is not a great deal different from what it was two decades ago. Refinements to the pretreatment process, such as the ABOX system, along with the stepped combustion approach to target preparation (Bird *et al.*, 2003), have succeeded in extending the dating range to 50 000–60 000 years, but that is still some way short of the 75 000-year range of the isotopic enrichment methods of the 1970s (section 2.3.3). Whether the next generation of AMS systems, many of which are likely to be on a much smaller scale (reflecting a combination of lower capital costs and reduced running/maintenance costs), will be capable of pressing the age range of radiocarbon back beyond, say, 80 000–90 000 years will only become apparent in due course.

One area where further developments are certain to occur is in radiocarbon calibration. The recently published calibration (CALIB 4.4) is the most comprehensive of the four calibrations now available, incorporating as it does data from dendrochronology, marine varve sequences, corals and speleothems. However, the most securely dated part of the curve, that based on dendrochronology, still only extends to ca. 12 000 cal. years BP, and while the combination of varve chronology and larger numbers of TIMS U-series coral and speleothem dates to calibrate older parts of the curve is a significant advance on previous calibrations, uncertainties in all of these areas mean that further refinements of the curve beyond the dendrochronology limit are bound to emerge over the course of the next decade or so. A key element here is the extension of the European dendrochronology record back into the Lateglacial (Friedrich *et al.*, 2001), and there are realistic expectations that this will be achieved within the next few years. Dendrochronological work in New Zealand (section 8.3) may provide additional important dendrochronological calibration beyond the range of European pine. A more securely based calibration curve should enable further applications of the 'wiggle-match' approach (section 2.6.5), although this will require much larger numbers of radiocarbon dates than have often been available from individual sites hitherto (Lowe *et al.*, 2001). The next significant development in radiocarbon calibration is likely to be the publication of INTCAL04, the successor to INTCAL98, but at the time of writing (June 2004), this program was still not available.

In some respects, the major advances that have occurred in luminescence dating (Chapter 4) have not been matched in **ESR dating**. A number of recent developments, however, might point to ESR becoming a more widely used technique over the coming decades. For example, problems of accurate determination of the equivalent dose rate, which had hitherto been a problem in the ESR dating of fossil coral, may soon be resolved (Schellmann and Radtke, 2001), while innovations in the use of ESR for the dating of sediments (Voinchet *et al.*, 2003), including the single-grain dating of quartz (Beerten *et al.*, 2003), could herald a new phase of applied ESR dating. Indeed, with a time range extending from 10 000 to more than 1 million years, this may prove to be a valuable ancillary, and in some cases an alternative to luminescence dating. The use of ESR, in combination with U-series, particularly for the dating of human tooth enamel is likely to become a routine method over the next few years (Grün, 2001).

The newest of the Quaternary radiometric dating family, **cosmogenic nuclide dating (CN)**, is certain to see major developments over the coming decades. Although important advances have been made during the past 15 years, further refinements are still needed in estimates of terrestrial production rates of different nuclides, and improved simulation models are required for CN production in shallow rocks (Gosse and Phillips, 2001). More empirical calibrations are also necessary at well-characterised sites in order to establish temporal variations in CN production, while better control is required on the parameters that affect nuclear interactions in rock. Developments in all of these aspects of CN dating might reasonably be anticipated in the next decade or so. Of particular significance in this respect is the recently established CRONUS (Cosmic-Ray Produced Nuclide Systematics) Project, an international research programme involving earth scientists, particle physicists and chemists from North America and Europe, which will investigate the sources of uncertainty in cosmogenic nuclide production in order to improve the accuracy and reliability of cosmogenic surface exposure applications. The aim is to reduce the current levels of uncertainty in CN measurements to around 5%, which would bring CN dating more into line with other radiometric techniques. In addition to these technical developments, the range of nuclides routinely used for dating is almost certain be increased to include, for example, ^{41}Ca and ^{14}C. The short half-lives of these two isotopes mean that they could potentially be used in the determinations of younger exposure ages. Successful measurements of *in situ* production of ^{14}C in such diverse media as desert sand, meteorites and glacier ice (Jull *et al.*, 2000; Lal and Jull, 2001) point to some of the potential future uses of this cosmogenic nuclide. CN dating is also likely to see an expansion beyond the present focus on geomorphology and landscape history into such diverse areas as environmental geology involving, for example, issues of tectonic stability and nuclear waste disposal (Gosse and Phillips, 2001), and archaeology, where initial attempts have already been made to date rock engravings (Phillips *et al.*, 1997) and stone artefacts (Ivy-Ochs *et al.*, 2001).

8.3 Annually Banded Records

Three areas of dating, involving the use of annually banded records, where there are likely to be significant advances over the coming years are dendrochronology, ice-core chronology and sclerochronology. In **dendrochronology**, one of the principal challenges is to extend the European pine record through the Younger Dryas Stadial to anchor the floating 1051-year chronology that has been developed for the Bølling-Allerød period (Friedrich *et al.*, 2001). This would enable a continuous dendrochronological series to be developed back into the early part of the Lateglacial and, as noted above, will provide a basis for dendrochronological calibration of the radiocarbon timescale back to 14 000 years. Beyond this date, the most important developments in dendrochronological radiocarbon calibration may well be in New Zealand, where the kauri (*Agathis australis*), a species that commonly lives up to 600 years and, in some cases, >1000 years (Ogden *et al.*, 1992; 1993), is likely to play a major role. This is because kauri wood, which is well preserved in peats, is frequently of considerable antiquity with ages in the range of ca. 20 000–40 000 ^{14}C years BP (Figure 8.1). The potential for developing a continuous dendrochronological series from these sub-fossil kauri is considerable and this, in turn, could form the basis for a new high-precision radiocarbon calibration. In addition to these long dendrochronological series, the coming years are likely to see the development of an increasing number of 'floating chronologies' associated, for example, with archaeological investigations, and both the anchoring of these tree-ring series in time and the integration of the records into regional master chronologies.

Although numerous drilling programmes have been mounted in both Greenland and Antarctica over the past 15 years, there are no signs of a slowing in the pace of **ice-core research**, despite the considerable costs that are involved in mounting and sustaining

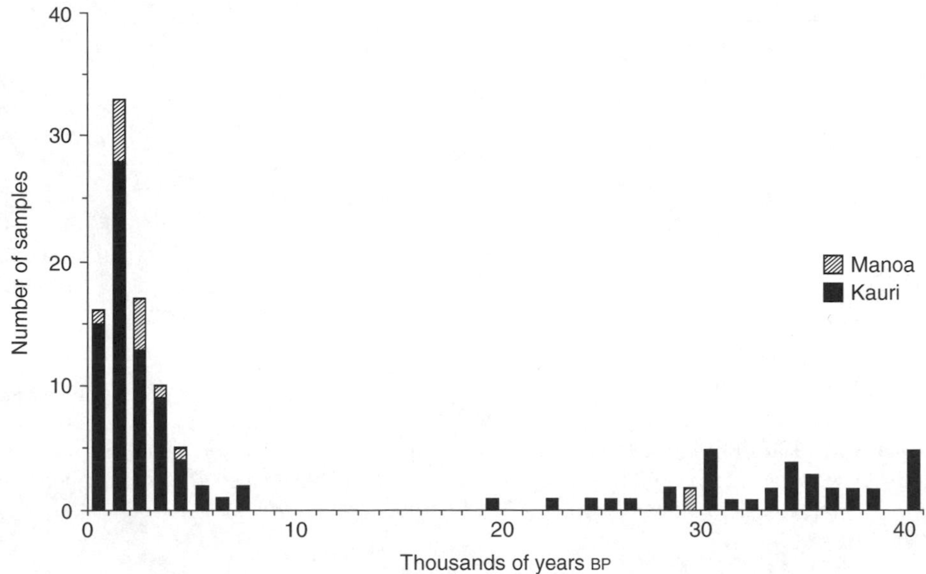

Figure 8.1 *Frequency distribution of radiocarbon dates on kauri* (A. australis). *Note the cluster of dates in the Mid- and Late Holocene, and also in the period from 20 000 to 40 000 ^{14}C years BP (after Ogden* et al., *1992). Reproduced by permission of Blackwell Publishing*

these programmes. The first detailed results of the new European ice-core project at North GRIP, Greenland, have recently appeared (North Greenland Ice Core Project Members, 2004) and further data from this deep ice core are likely to be published over the course of the next few years. Here, the development of multi-parameter dating approaches, including high-resolution visual stratigraphy derived from digital scanners (Svensson *et al.*, 2005), offers the potential for providing an ice-core timescale of remarkable accuracy and precision. In Antarctica, the EPICA coring programme is ongoing with an expectation of eventually reaching ice almost one million years old (EPICA Community Members, 2004). The EPICA core clearly offers a basis for the development of a long continuous chronology, although ice-flow modelling, which has hitherto been the principal method in the dating of deep Antarctic ice cores (Chapter 5), may not be capable of providing a timescale of sufficiently high resolution. An alternative could be to employ the technique of *orbital tuning*, in other words to adapt the approach that has been used for the dating of deep-ocean sediments where the oxygen isotope signal has been 'tuned' to the astronomical variables (section 7.2.2). Ruddiman and Raymo (2003) have shown how this method can be applied to the methane record from the Vostok ice core to generate a coherent timescale for the past 350 000 years (Figure 8.2), and it is possible that a similar approach might well be used for EPICA and for other cores that will be drilled over the course of the next two to three decades.

A third aspect of annually banded records where progress is likely to be made within the next few years is **sclerochronology**. In section 5.6.3 we saw how composite chronologies are being developed on the basis of certain long-lived molluscan species, and the prospect was raised of a 1000-year continuous or near-continuous chronology

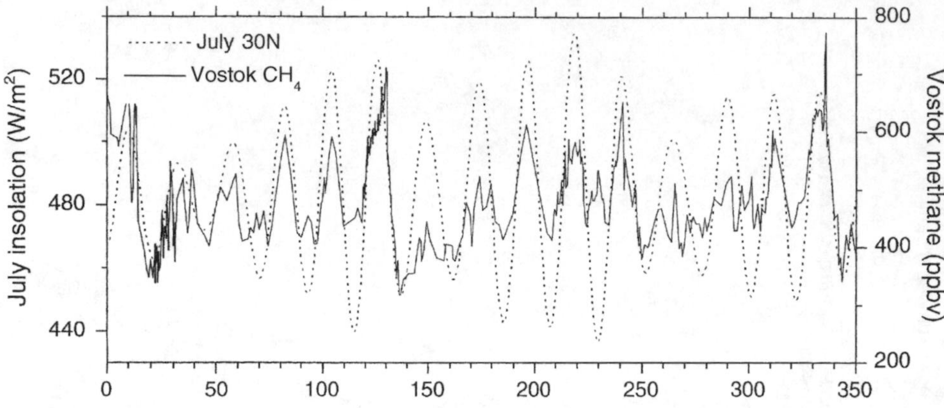

Figure 8.2 *The methane (CH_4) record from the Vostok ice core (solid line) for the past 350 000 years tuned to the July 30°C insolation curve (dotted line) derived from the astronomical variables (after Ruddiman and Raymo, 2003). Reproduced with permission of Elsevier*

being developed on the basis of *A. islandica* (Marchitto *et al.*, 2000). Recent work on specimens from the North Sea, involving the cross-matching of living and dead specimens of *Arctica*, underpinned by a programme of AMS radiocarbon dating, suggest that a 1000-year sclerochronology may indeed soon be attainable (Forsythe *et al.*, 2004). Indeed, the recovery of *Arctica* specimens from the North Sea dating to various phases of the Holocene and into the Lateglacial means that it may be possible, at some time in the future, to construct a continuous chronology of 10 000 years, or even more. This would have major implications for our understanding of spatial and temporal marine reservoir changes in coastal and shelf seas, as well as providing an absolute chronological template for geochemical proxies of marine environmental change. The prospects for the further development of chronologies based on coral records (section 5.6.2) are also good. Thus far, coral sclerochronologies have been confined largely to the past few hundred years, but recent work has shown that these records could well be extended back into the Late Quaternary through the sampling of dead (fossil) corals (Tudhope *et al.*, 2001). This offers the prospect for developing floating chronologies for raised interglacial reef sequences which can then be anchored by U-series dating. Such high-resolution chronologies could have far-reaching implications for studies of past sea-level change, sea-surface temperature fluctuations, and short-term climatic variability such as the El Niño Southern Oscillation.

8.4 Age Equivalence

In the field of age-equivalent dating, perhaps the greatest advances in recent years have been made in **tephrochronology**, most notably in the development of a tephrochronology for the eastern North Atlantic region during the Last Termination (section 7.3.4.2) and also during the Holocene (Hall and Pilcher, 2002; Van den Bogaard and Schmincke,

2002; Chambers *et al.*, 2004). Refinements in tephra extraction techniques, in geochemical fingerprinting and tephra provenancing, have led to the development of an outline tephrostratigraphic framework that is applicable at the continental scale. The coming years are likely to see an increasing fine-tuning of this scheme, as the ranges of individual tephras are more clearly defined, and new tephras are discovered and sourced (Davies *et al.*, 2003; 2004; Hunt, 2004). Progress towards a high-resolution tephrochronology for the North Atlantic prior to the Last Termination might also be anticipated, as recent work on North Atlantic Ash Zone II (ca. 53 000 years ago) has demonstrated (Austin *et al.*, 2004). There are certain to be further developments in tephra identification, perhaps through the use of minor, trace and rare element analysis (Pearce *et al.*, 2002), and through experimental work on the chemical stability of tephras in different depositional contexts (Pollard *et al.*, 2003). The ages of individual tephra horizons will almost certainly be more closely constrained by additional radiocarbon dates and also, perhaps, by the direct dating of tephras by $^{40}Ar/^{39}Ar$ (e.g. Van den Bogaard, 1995). In addition, the search for and detection of tephras in varved sediment sequences will provide a further chronological underpinning for this evolving tephrochronological framework (Litt *et al.*, 2003), as also will the identification of volcanic events in the Greenland ice-core records. Work currently in progress on the NorthGRIP, which is focussing not only on the detection of tephra shards in the ice but also on the proxy signals of volcanic events, notably sulphur and calcium peaks,[2] should, within the next few years, enable many of the major North Atlantic Tephra horizons to be dated with a very high degree of accuracy and precision.

Major advances in tephrochronology may also be anticipated in regions other than the North Atlantic. In New Zealand, for example, which was one of the first areas of the world to develop an integrated tephrochronology (Lowe, 1988), a very large number of Late Quaternary tephras have been identified and many have been dated (Froggatt and Lowe, 1990; Shane, 2000). Refinement of this tephrochronological framework might be anticipated in the coming years, notably in the extension of the range of key tephras through the identification of microtephras in both terrestrial and marine sediments (Newnham *et al.*, 2003). In South America, by contrast, despite the numerous active volcanoes in the mountains of the Andes, tephrochronology is in its infancy. However, recent work has shown the enormous potential for tephrochronology in this region of the world (Haberle and Lumley, 1998; Thouret *et al.*, 1999; 2002), and in the next decade or so it is likely that the outline tephrochronology for the past 15 000 years for southern Peru will eventually be extended in time back to ca. 50 000 years, and in space into adjacent areas of Bolivia and northern Chile.

8.5 Biomolecular Dating

Finally, one area that has not been covered in this book, but where there may be the potential for the development of a new field of dating, is biomolecular evidence. The last ten years have seen dramatic advances in this field, no more so than in the study of ancient DNA (Yang, 1997; Brown, 2001). The recent remarkable discoveries of plant and animal DNA preserved in the permafrost and in cave sediments, for example, have shown the potential for reconstructing palaeoenvironments from these genetic signals

(Willerslev *et al.*, 2003). Changing plant and animal distributions in response to the global climate changes of the Quaternary would also have had major genetic consequences, and the advent of DNA technology provides the basis for the development of suitable markers to examine these genetic responses (Hewitt, 2000). Most significant, perhaps, are the DNA data that are emerging from research into the evolution of anatomically modern humans. Molecular genetic and archaeological studies are now combining to elucidate the origin and spread of modern humans worldwide, with studies of genetic molecular diversity providing a basis for reconstructing human lineages, for example the relationship (or non-relationship) between Neanderthals and *Homo sapiens* (Krings *et al.*, 1997), and periods of population expansion (Jorde *et al.*, 1998).

Recent research has shown how the timing of genetic mutations can be determined by 'molecular clock analysis' (Chou *et al.*, 2002). Work linking gene mutation with anatomical changes in human lineage, specifically accelerated encephalisation (brain expansion) possibly associated with a reduction in size of jaw muscles, has enabled this key genetic change to be dated at 2.4 ± 0.3 million years (Stedman *et al.*, 2004). Of particular interest, however, are the small changes or mutations in mtDNA, the tiny pieces of DNA that are inherited only down the female line. The genetic clock based on changes in mtDNA suggests, for example, that modern humans migrated out of Africa into Asia 80 000 years ago, and then spread from South Asia through the Near East and into Europe between 54 000 and 50 000 years ago (Oppenheimer, 2003). Whether such an approach will eventually form the basis for a complete human evolutionary chronology remains to be seen, but as Hedges (2001b) among others has observed, it would be surprising if, in the very near future, much of the chronological evidence for human evolution and population dispersal is not based on genetic data.

Notes

1. **Polymers** are compounds with very large molecules made up of repeating molecular units. They may be natural substances, such as proteins, starch, cellulose and rubber, or synthetic materials such as nylon and plastic.
2. Large volumes of sulphur are often ejected into the atmosphere during the course of a volcanic eruption. This reacts with other atmospheric constituents to form sulphuric acid which is disseminated globally and, through precipitation, becomes incorporated into polar ice. Peaks in sulphur concentration in ice cores are therefore reflective of former episodes of volcanic activity. In addition, the interaction between the sulphur aerosol and calcium-rich dust circulating in the atmosphere over the polar ice sheets produces calcium sulphate (gypsum), which also finds its way into the ice via precipitation. Hence calcium peaks in polar ice cores may also provide a record of volcanic eruptions.

References

Adams, K.D., Locke, W.W. and Rossi, R., 1992, Obsidian-hydration dating of fluvially reworked sediments in the West Yellowstone region, Montana, *Quaternary Research*, **38**, 180–195.

Aguirre, E. and Carbonell, E., 2000, Early human expansions into Eurasia: the Atapuerca evidence, *Quaternary International*, **75**, 11–18.

Aitken, M.J., 1985, *Thermoluminescence Dating* (New York: Academic Press).

Aitken, M.J., 1990, *Science-Based Dating in Archaeology* (London: Longman).

Aitken, M.J., 1998, *An Introduction to Optical Dating* (Oxford: Oxford University Press).

Allen, J.R., 2004, Annual textural banding in Holocene estuarine silts, Severn Estuary Levels (SW Britain): patterns, cause and implications, *The Holocene*, **14**, 536–552.

Allen, J.R. and Rae, J., 1987, Late Flandrian shoreline oscillations in the Severn Estuary: a geomorphological and stratigraphical assessment, *Philosophical Transactions of the Royal Society, London*, **B315**, 185–230.

Allen, M.J., Godden, D., Matthews, C. and Powell, A.B., 2002, Mesolithic, Late Bronze Age and Medieval activity at Katherine's Farm, Avonmouth, 1998, *Archaeology in the Severn Estuary*, **13**, 89–105.

Alley, R.B., 2000a, *The Two-Mile Time Machine* (Princeton, New Jersey: Princeton University Press).

Alley, R.B., 2000b, The Younger Dryas cold interval as viewed from central Greenland, *Quaternary Science Reviews*, **19**, 213–226.

Alley, R.B., Meese, D.A., Shuman, C.A., Gow, A.J., Taylor, K.C., Grootes, P.M., Zielinski, G.A., Ram, M., Waddington, E.D., Mayewski, P.A. and Zielinski, G.A., 1993, Abrupt increase in snow accumulation at the end of the Younger Dryas event, *Nature*, **362**, 527–529.

Alley, R.B., Shuman, C.A., Meese, D.A., Gow, A.J., Taylor, K.C., Cuffey, K.M., Fitzpatrick, J.J., Grootes, P.M., Zielinski, G.A., Ram, M., Spinelli, G. and Elder, B., 1997a, Visual-stratigraphic dating of the GISP2 ice core: basis, reproducibility and application, *Journal of Geophysical Research*, **102**, 26 367–26 382.

Alley, R.B., Gow, A.J., Meese, D.A., Fitzpatrick, J.J., Waddington, E.D. and Bolzan, J.F., 1997b, Grain-scale processes, folding, and stratigraphic disturbances in the GISP2 ice core, *Journal of Geophysical Research*, **102**, 26 819–26 830.

Ambers, J. and Bowman, S., 2002, Letters to the Editor, *Radiocarbon*, **44**, 599.

Ambrose, W.R., 1994, Obsidian hydration dating of a Pleistocene age site from the Manus Islands, Papua New Guinea, *Quaternary Science Reviews*, **13**, 137–142.

Quaternary Dating Methods M. Walker

American Commission on Stratigraphic Nomenclature, 1961, Code of Stratigraphic Nomenclature, *American Association of Petroleum Geologists' Bulletin*, **45**, 645–665.

Andrén, T., Björck, J. and Johnsen, S., 1999, Correlation of Swedish glacial varves with the Greenland (GRIP) oxygen isotope record, *Journal of Quaternary Science*, **14**, 361–371.

Andrews, J.T., Austin, W.E.N., Bergsten, H. and Jennings, A.E. (eds), 1996, *Late Quaternary Palaeoceanography of the North Atlantic Margins*. Geological Society of London, Special Publication No. 111, London.

Anovitz, L.M., Elam, J.M., Riciputi, L.R. and Cole, D.R., 1999, The failure of obsidian hydration dating: sources, implications and new directions, *Journal of Archaeological Science*, **26**, 735–752.

Antevs, E., 1931, Late-glacial correlations and ice-recession in Manitoba, *Geological Society of Canada Memoir*, **168**, 1–76.

Appleby, P.G. and Oldfield, F., 1992, Applications of lead-210 to sedimentation studies, in M. Ivanovich, R.S. Harmon (eds), *Uranium-Series Disequilibrium: Applications to Earth, Marine and Environmental Sciences*, 731–779 (Oxford: Clarendon Press).

Appleby, P.G., Oldfield, F., Thompson, R., Huttunen, P. and Tolonen, K., 1979, ^{210}Pb dating of annually laminated lake sediments from Finland, *Nature*, **280**, 53–55.

Appleby, P.G., Sjotyk, W. and Fankhauser, A., 1997, Lead-210 age dating of three peat cores in the Jura mountains, Switzerland. *Water, Air and Soil Pollution*, **100**, 223–231.

Arribas, A. and Palmqvist, P., 1999, On the ecological connection between sabre-tooths and hominids: faunal dispersal events in the Lower Pleistocene and a review of the evidence for the first human arrival in Europe, *Journal of Archaeological Science*, **26**, 571–585.

Augustinus, P.C., Short, S.A. and Heijnis, H., 1997, Uranium/thorium dating of ferricretes from mid- to late-Pleistocene glacial sediments, western Tasmani, Australia, *Journal of Quaternary Science*, **12**, 295–308.

Austin, W.E.N., Bard, E., Hunt, J.B., Kroon, D. and Peacock, J., 1995, The ^{14}C age of the Icelandic Vedde Ash: implications for Younger Dryas marine reservoir corrections, *Radiocarbon*, **37**, 53–62.

Austin, W.E.N., Wilson, L.J. and Hunt, J.B., 2004, The age and chronostratigraphic significance of North Atlantic Ash Zone II, *Journal of Quaternary Science*, **19**, 137–146.

Bada, J.L., 1985, Aspartic acid racemisation ages of California Palaeoindian skeletons, *American Antiquity*, **50**, 645–647.

Bada, J.L., Schroeder, R.A. and Carter, G.F., 1974, New evidence for the antiquity of man in North America deduced from aspartic acid racemisation, *Science*, **184**, 791–793.

Baillie, M.G.L., 1982, *Tree-Ring Dating and Archaeology* (London: Croom Helm).

Baillie, M.G.L., 1995, *A Slice Through Time* (London: Batsford).

Baillie, M.G.L. and Brown, D.M., 1988, An overview of oak chronologies, *British Archaeological Reports (British Series)*, **196**, 543–548.

Baillie, M.G.L. and Munroe, M.A.R., 1988, Irish tree rings, Santorini and volcanic dust veils, *Nature*, **332**, 344–346.

Baker, A., Smart, P., Edwards, R.L. and Richards, D.A., 1993, Annual growth banding in cave stalagmite, *Nature*, **364**, 518–520.

Baker, A., Proctor, C.J. and Barnes, W.L., 1999, Variations in stalagmite luminescence laminae structure at Poole's Cavern, England, AD 1910–1996: calibration of a palaeoprecipitation proxy, *The Holocene*, **9**, 683–688.

Ballantyne, C.K. and Harris. C., 1994, *The Periglaciation of Great Britain* (Cambridge: Cambridge University Press).

Ballantyne, C.K., McCarroll, D., Nesje, A., Dahl, S. and Stone, J.O., 1998b, The last ice sheet in north-west Scotland: reconstruction and implications, *Quaternary Science Reviews*, **17**, 1149–1184.

Ballantyne, C.K., Stone, J.O. and Fifield, L.K., 1998a, Cosmogenic Cl-36 dating of postglacial landsliding at The Old Man of Storr, Isle of Skye, Scotland, *The Holocene*, **8**, 347–351.

Ballarini, M., Wallinga, J., Murray, A.S., van Heteren, S., Oost, A.P., Bos, A.J.J. and van Eijk, C.W.E., 2003, Optical dating of young coastal dunes on a decadal timescale, *Quaternary Science Reviews*, **22**, 1011–1018.

Balter, M., 2001, In search of the first Europcans, *Science*, **291**, 1722–1725.

Bannerjee, D., Murray, A.S. and Foster, I.D.L., 2001, Scilly Isles, UK: optical dating of a possible tsunami deposit from the 1755 Lisbon earthquake, *Quaternary Science Reviews*, **20**, 715–718.

Barber, K.E., Chambers, F.M., Maddy, D., Stoneman, R.E. and Brew, J.S., 1994, A sensitive high-resolution record of Late Holocene climatic change from a raised bog in northern England, *The Holocene*, **4**, 194–205.

Bard, E., 2001, Extending the calibrated radiocarbon record, *Science*, **292**, 244–245.

Bard, E., Arnold, M. and Duplessy, J.-C., 1991, Reconciling the sea level record of the last deglaciation with ^{18}O spectra from deep sea cores, in J.J. Lowe (ed.), *Radiocarbon Dating: Recent Applications and Future Potential. Quaternary Proceedings* **1**, 67–73 (Cambridge: Quaternary Research Association).

Bard, E., Arnold, M., Mangerud, J., Paterne, M., Labeyrie, L., Duprat, J., Mélières, M.-A., Sønstergaard, E., and Duplessy, J.-C., 1994, The North Atlantic atmospheric-sea surface ^{14}C gradient during the Younger Dryas climatic event, *Earth and Planetary Science Letters*, **126**, 275–287.

Barrow, J., 1988, *The World within the World* (Oxford: Clarendon).

Barrows, T.T., Stone, J.O. and Fifield, L.K., 2004, Exposure ages for Pleistocene periglacial deposits in Australia, *Quaternary Science Reviews*, **23**, 697–708.

Bartsiokas, A. and Middleton, A.P., 1992, Characterisation and dating of recent and fossil bone by X-ray diffraction, *Journal of Archaeological Science*, **19**, 63–72.

Bassinot, F.C., Labeyrie, L., Vincent, E., Quidelleur, X., Shackleton, N.J. and Lancelot, Y., 1994, The astronomical theory of climate and the age of the Brunhes-Matuyama magnetic reversal, *Earth and Planetary Science Letters*, **126**, 91–108.

Bateman, M.D., Frederick, C.D., Jaiswal, M.K. and Singhvi, A.K., 2003, Investigations into the potential effects of pedoturbation on luminescence dating, *Quaternary Science Reviews*, **22**, 1169–1176.

Bates, M.R., 1993, Quaternary aminostratigraphy in northwestern France, *Quaternary Science Reviews*, **12**, 793–809.

Bates, M.R., Keen, D.H. and Lautridou, J.P., 2003, Pleistocene marine and periglacial deposits of the English Channel, *Journal of Quaternary Science*, **18**, 319–337.

Batt, C.M., 1997, The British archaeomagnetic calibration curve: an objective treatment, *Archaeometry*, **39**, 153–168.

Beck, C. and Jones, G.T., 1994, Dating surface assemblages using obsidian hydration, in C. Beck (ed.), *Dating in Exposed and Surface Contexts*, 47–75 (Albuquerque: University of New Mexico Press).

Beck, J.W., Edwards, R.L., Ito, E., Taylor, F.W., Recy, J., Rougerie, F., Joannot, P. and Hennin, C., 1992, Sea-surface temperature from coral skeletal Sr/Ca ratios, *Science*, **257**, 644–647.

Beck, J.W., Donahue, D.J., Jull, A.J.T., Burr, G., Broecker, W.S., Bonani, G., Hajdas, I. and Malotki, E., 1998, Ambiguities in direct dating of rock surfaces using radiocarbon measurements, *Science*, **280**, 2132–2135.

Beck, J.W., Richards, D.A., Edwards, R.L., Silverman, B.W., Smart, P.L., Donahue, D.J., Hererra-Osterheld, S., Burr, G.S., Calsoyas, L., Jull, A.J.T. and Biddulph, D., 2001, Extremely large variations of atmospheric ^{14}C concentration during the last glacial period, *Science*, **292**, 2453–2458.

Beerten, K., Pierreux, D. and Stesmans, A., 2003, Towards single grain ESR dating of sedimentary quartz: first results, *Quaternary Science Reviews*, **22**, 1329–1334.

Behre, K.-E. and van der Plicht, J., 1992, Towards an absolute chronology for the last glacial period in Europe: radiocarbon dates from Oerel, northern Germany, *Vegetation History and Archaeobotany*, **1**, 111–117.

Bell, M. and Walker, M.J.C., 2005, *Late Quaternary Environmental Change: Physical and Human Perspectives*. 2nd edition (London: Pearson International).

Bell, M., Caseldine A.C. and Neumann, H. (eds), 2000, *Prehistoric Intertidal Archaeology in the Welsh Severn Estuary*. Council for British Archaeology, Research Report **120**, York.

Benedict, J.B., 1990, Experiments on lichen growth. 1. Seasonal patterns and environmental controls, *Arctic and Alpine Research*, **22**, 244–253.

Benson, L., Liddicoat, J., Smoot, J., Sarna-Wojcicki, A., Negrini, R. and Lund, S., 2003, Age of the Mono Lake excursion and associated tephra, *Quaternary Research*, **22**, 135–140.

Benson, L., Madole, R., Phillips, W., Landis, G., Thomas, T. and Kubik, P., 2004, The probable importance of snow and sediment shielding on cosmogenic ages of north-central Colorado Pinedale and pre-Pinedale moraines, *Quaternary Science Reviews*, **23**, 193–206.

Berger, G.W., 1988, Dating Quaternary events by luminescence, in D.J. Easterbrook (ed.), *Dating Quaternary Sediments*. Geological Society of America, Special Paper **227**, 13–50.

Berger, G.W., 1995, Progress in luminescence dating methods for Quaternary sediments, in N.W. Rutter, N.R. Catto (eds), *Dating Methods for Quaternary Deposits*. Volume 2. Geological Association of Canada, GEOtext, 81–104.

Bergman, J., Wastegård, S., Hammarlund, D., Wohlfarth, B. and Roberts, S.J., 2004, Holocene Tephra horizons at Klocka Bog, west-central Sweden: aspects of reproducibility in subarctic peat deposits, *Journal of Quaternary Science*, **19**, 241–249.

Berry, M., 1994, Soil-geomorphic analysis of Late Pleistocene glacial sequences in the McGee, Pine and Bishop Creek drainages, east-central Sierra Nevada, California, *Quaternary Research*, **41**, 160–175.

Beschel, E.E., 1973, Lichens as a measure of the age of recent moraines, *Arctic and Alpine Research*, **5**, 303–309 (translated from 'Flechten als Altersmasstab rezenter Moränen', *Zeitschrift für Gletscherkunde und Glazialgeologie*, **1**, 152–161).

Betts, M.W. and Latta, M.A., 2000, Rock surface hardness as an indication of exposure age: an archaeological application of the Schmidt hammer, *Archaeometry*, **42**, 209–223.

Bickerton, R.W. and Matthews, J.A., 1992, On the accuracy of lichenometric dates: an assessment based on the 'Little Ice Age' moraine sequence of Nigardsbreen, southern Norway, *The Holocene*, **2**, 227–237.

Bickerton, R.W. and Matthews, J.A., 1993, 'Little Ice Age' variations of outlet glaciers from the Jostedalsbreen ice-cap, southern Norway: a regional lichenometric-dating study of ice-marginal moraine sequences and their climatic significance, *Journal of Quaternary Science*, **8**, 45–66.

Bierman, P.R., 1994, Using *in situ* cosmogenic isotopes to estimate rates of landscape evolution: a review from the geomorphic perspective, *Journal of Geophysical Research*, **99**, 13 885–13 896.

Bierman, P.R. and Gillespie, R.A., 1994, Evidence suggesting that methods of rock-varnish cation-ratio dating are neither comparable nor consistently reliable, *Quaternary Research*, **41**, 82–90.

Bird, M.I., Fifield, L.K., Santos, G.M., Beaumont, P.B., Zhou, Y., di Tada, M.L. and Hausladen, P.A., 2003, Radiocarbon dating from 40 to 60 ka BP at Border Cave, South Africa, *Quaternary Science Reviews*, **22**, 943–947.

Birkeland, P.W., 1999, *Soils and Geomorphology* (Oxford: Oxford University Press).

Birkeland, P.W., Burke, R.M. and Benedict, J.B., 1989, Pedogenic gradients for iron and aluminium accumulation and phosphorus depletion in arctic and alpine soils as a function of time and climate, *Quaternary Research*, **32**, 193–204.

Birks, H.H., Gulliksen, S., Haflidason, H., Mangerud, J. and Possnert, G., 1996, New radiocarbon dates for the Vedde Ash from western Norway, *Quaternary Research*, **45**, 119–127.

Björck, J., 1999, Event stratigraphy for the Last Glacial-Holocene transition in eastern middle Sweden, *Quaternaria, Stockholm University A*, **6**, 48 pp.

Björck, S., Koç, N. and Skog, G., 2003, Consistently large marine reservoir ages in the Norwegian Sea during the last deglaciation, *Quaternary Science Reviews*, **22**, 429–435.

Björck, S., Walker, M.J.C., Cwynar, L.C., Johnsen, S., Knudsen, K.-L., Lowe, J.J., Wohlfarth, B. and INTIMATE members, 1998, An event stratigraphy for the Last Termination in the North Atlantic region based on the Greenland ice-core record: a proposal by the INTIMATE group, *Journal of Quaternary Science*, **13**, 283–292.

Blaauw, M., Heuvenlink, G.B.M., Mauquoy, D., van der Plicht, J. and van Geel, B., 2003, A numerical approach to ^{14}C wiggle-match dating of organic deposits: best fits and confidence intervals, *Quaternary Science Reviews*, **22**, 1485–1500.

Blaauw, M., van Geel, B., Mauquoy, D. and van der Plicht, J., 2004, Carbon-14 wiggle-match dating of peat deposits: advantages and limitations, *Journal of Quaternary Science*, **19**, 177–182.

Blackwell, B.A.B., 2001a, Electron spin resonance (ESR) dating in lacustrine environments, in W.M. Last, J.P. Smol (eds), *Tracking Environmental Change using Lake Sediments: Volume 1. Basin Analysis, Coring and Chronological Techniques*, 283–369 (Dordrecht: Kluwer).

Blackwell, B.A.B., 2001b, Amino acid racemisation (AAR) dating and analysis in lacustrine environments, in W.M. Last, J.P. Smol (eds), *Tracking Environmental Change using Lake Sediments: Volume 1. Basin Analysis, Coring and Chronological Techniques*, 391–450 (Dordrecht: Kluwer).

Blockley, S.P.E., Lowe, J.J., Walker, M.J.C., Asioli, A., Trincardi, F., Coope, G.R., Donahue, R.E. and Pollard, A.M., 2004, Bayesian analysis of radiocarbon chronologies: examples from the European Late-glacial, *Journal of Quaternary Science*, **19**, 159–175.

Bluth, G.J.S., Schnetzler, C.C., Krueger, D.A.J. and Walter, L.S., 1993, The contribution of explosive volcanism to global sulfur dioxide concentrations, *Nature*, **366**, 327–330.

Boardman, J. (ed), 1985, *Soils and Quaternary Landscape Evolution* (Chichester and New York: John Wiley).

Boaretto, E., Bryant, C., Carmi, I., Cook, G., Gulliksen, S., Harkness, D, Heinemeier, J., McClure, J., McGee, E., Naysmith, P., Possnert, G., Scott, M., van der Plicht, H. and van Strydonck, M., 2002, Summary findings of the fourth international radiocarbon intercomparison (FIRI) (1998–2001), *Journal of Quaternary Science*, **17**, 633–637.

Bond, G., Showers, W., Cheseby, M., Lotti, R., Alamisi, P. de Menocal, P., Priore, P., Cullen, H., Hajdas, I. and Bonani, G., 1997, A pervasive millennial-scale cycle in North Atlantic Holocene and glacial climates, *Science*, **278**, 1257–1266.

Bond, G.C., Mandeville, C. and Hoffman, S., 2001, Were rhyolitic glasses in the Vedde Ash and in the North Atlantic's Ash Zone 1 produced by the same volcanic eruption? *Quaternary Science Reviews*, **20**, 1189–1199.

Bonnichsen, R., Hodges, L., Ream, W., Field, K.G., Kirner, D.L., Selsor, K. and Taylor, R.E., 2001, Methods for the study of ancient hair: radiocarbon dates and gene sequences from individual hairs, *Journal of Archaeological Science*, **28**, 775–785.

Bourdon, B., Henderson, G.M., Lundstrom, C.C. and Turner, S. (eds), 2003a, *Uranium Series Geochemistry, Reviews in Mineralogy and Geochemistry*, **52**, 1–656.

Bourdon, B., Turner, S., Henderson, G.M. and Lundstrom, C.C., 2003b, Introduction to U-series geochemistry, in B. Bourdon, G.M. Henderson, C.C. Lundstrom, S. Turner (eds), *Uranium Series Geochemistry, Reviews in Mineralogy and Geochemistry*, **52**, 1–22.

Bowen, D.Q., 1991, Time and space in the glacial sediment systems of the British Isles, in J. Ehlers, P.L. Gibbard, J. Rose (eds.), *Glacial Deposits in Great Britain and Ireland*, 3–11 (Rotterdam: Balkema).

Bowen, D.Q., 1999, On the correlation and classification of Quaternary deposits and land–sea correlations, in D.Q. Bowen (ed.), *A Revised Correlation of Quaternary Deposits in the British Isles*. Geological Society Special Report No. 23, London, 1–9.

Bowen, D.Q., Hughes, S., Sykes, G.A. and Miller, G.H., 1989, Land–sea correlations in the Pleistocene based on isoleucine epimerisation in non-marine molluscs, *Nature*, **340**, 49–51.

Bowen, D.Q., Pillans, B., Sykes, G.A., Beu, A.G., Edwards, A.R., Kamp, P.J.J. and Hull, A.G., 1998, Amino acid geochronology of Pleistocene marine sediments in the Wanganui Basin:

a New Zealand framework for correlation and dating, *Journal of the Geological Society, London*, **155**, 439–446.

Bowman, S. 1990, *Radiocarbon Dating* (London: British Museum).

Brauer, A., Endres, C., Günter, C., Litt, T., Stebich, M. and Negendank, J.F.W., 1999, High resolution sediment and vegetation responses to Younger Dryas climate change in varved lake sediments from Meerfelder Maar, Germany, *Quaternary Science Reviews*, **18**, 321–329.

Bridgland, D., Maddy, D. and Bates, M., 2004, River terrace sequences: templates for Quaternary geochronology and marine–terrestrial correlation, *Journal of Quaternary Science*, **19**, 203–218.

Briffa, K.R., 2000, Annual climatic variability in the Holocene: interpreting the message of ancient trees, *Quaternary Science Reviews*, **19**, 87–105.

Briffa, K.R. and Matthews, J.A. (eds), 2002, Analysis of dendrochronological variability and associated natural climates in Eurasia (ADVANCE-10k), *The Holocene*, **12**, 639–794.

Briffa, K.R., Jones, P.D., Schweingruber, F.H. and Osborn, T.J., 1998, Influence of volcanic eruptions on northern hemisphere summer temperatures over the past 600 years, *Nature*, **393**, 450–455.

Briner, J.P., Swanson, T.W. and Caffee, M., 2001, Late Pleistocene cosmogenic ^{36}Cl glacial chronology of the southwestern Ahklun Mountains, Alaska, *Quaternary Research*, **56**, 148–154.

Broadbent, N.D. and Bergqvist, K.I., 1986, Lichenometric chronology and archaeological features on raised beaches: preliminary results from the Swedish north Bothnian coastal region, *Arctic and Alpine Research*, **18**, 297–306.

Broecker, W.S., 1984, Terminations, in A. Berger, J. Imbrie, J. Hays, G. Kukla, B. Saltzman (eds), *Milanovitch and Climate*, 687–698 (Dordrecht: Reidel).

Bronk Ramsay, C., 1995, Radiocarbon calibration and analysis of stratigraphy: the OxCal program, *Radiocarbon*, **37**, 425–430.

Bronk Ramsay, C., 1998, Probability and dating, *Radiocarbon*, **40**, 461–474.

Bronk Ramsay, C., van der Plicht, J. and Weninger, B., 2001, 'Wiggle matching' radiocarbon dates, *Radiocarbon*, **43**, 381–389.

Brooks, A.S., Hare, P.E., Kokis, J.E., Miller, G.H., Ernst, R.D. and Wendorf, F., 1990, Dating Pleistocene archaeological sites by protein diagenesis in ostrich eggshell, *Science*, **248**, 60–64.

Brown, T.A., 2001, Ancient DNA, in D.R. Brothwell, A.M. Pollard (eds), *Handbook of Archaeological Sciences*, 301–311 (Chichester and New York: John Wiley).

Brown, D.M. and Baillie, M.G.L., 1992, Construction and dating of a 5000-year English bog oak tree-ring chronology, *Lundqua Report*, **34**, 72–75.

Brunelle, A. and Anderson, R.S., 2003, Sedimentary charcoal as an indicator of late-Holocene drought in the Sierra Nevada, California, and its relevance to the future, *The Holocene*, **13**, 21–28.

Brunnberg, L., 1995, Clay-varve chronology and deglaciation during the Younger Dryas and Pre-boreal in the easternmost part of the Middle Swedish Ice Marginal Zone, *Quaternaria, Stockholm University A*, **2**, 94 pp.

Brunning, R., 2000, Wood as an archaeological resource, *Archaeology in the Severn Estuary*, **11**, 175–185.

Bryson, B., 2003, *A Short History of Nearly Everything* (London: Doubleday).

Buck, C.E., 2001, Applications of the Bayesian statistical paradigm, in D.R. Brothwell, A.M. Pollard (eds), *Handbook of Archaeological Sciences*, 695–702 (Chichester and New York: John Wiley).

Buck, C.E., Cavanagh, W.G. and Litton, C.D., 1996, *The Bayesian Approach to Interpreting Archaeological Data* (Chichester and New York: John Wiley).

Buck, C.E., Higham, T.F.G. and Lowe, D.J., 2003, Bayesian tools for tephrochronology, *The Holocene*, **13**, 639–647.

Bursik, M.I. and Gillespie, A.R., 1993, Late Pleistocene glaciation of Mono Basin, California, *Quaternary Research*, **39**, 24–35.

Cande, S.C. and Kent, D.V., 1995, Revised calibration of the geomagnetic polarity timescale for the Late Cretaceous and Cenozoic, *Journal of Geophysical Research*, **100**, 6093–6095.

Carew, J.L. and Mylroie, J.E., 1995, Quaternary tectonic stability of the Bahamian Archipelago: evidence from fossil coral reefs and flank margin caves, *Quaternary Science Reviews*, **14**, 145–153.

Cato, I., 1985, The definitive connection of the Swedish geochronological time scale with the present, and the new date for the zero year in Döviken, northern Sweden, *Boreas*, **14**, 117–122.

Catt, J.A., 1986, *Soils and Quaternary Geology* (Oxford: Clarendon Press).

Catt, J.A., 1988, Soils of the Plio-Pleistocene: do they distinguish types of interglacial? *Philosophical Transactions of the Royal Society, London*, **B318**, 539–557.

Catt, J.A., 1990, Palaeopedology Manual, *Quaternary International*, **6**, 1–95.

Catto, N., 1987, Lacustrine sedimentation in a proglacial environment, Caribou River Valley, Yukon, Canada, *Boreas*, **16**, 197–206.

Cerling, T.E. and Craig, H., 1994, Geomorphology and *in situ* cosmogenic isotopes, *Annual Review of Earth and Planetary Science*, **22**, 273–317.

Chalmers, A.F., 1999, *What is This Thing Called Science?* 3rd edition (Buckingham: Open University Press).

Chambers, F.M., Barber, K.E., Maddy, D. and Brew, J.S., 1997, A 5500-year proxy-climate and vegetational record from a blanket mire at Talla Moss, Borders, Scotland, *The Holocene*, **7**, 391–399.

Chambers, F.M., Daniell, J.R.G., Hunt, J.B., Molloy, K. and O'Connell, M., 2004, Tephrostratigraphy of An Loch Mór, Inis Oírr, western Ireland: implications for Holocene tephrochronology in the northeastern Atlantic region, *The Holocene*, **14**, 703–720.

Charles, C.D., Hunter, D.E. and Fairbanks, R.G., 1997, Interaction between the ENSO and the Asian Monsoon in a coral record of tropical climate, *Science*, **277**, 925–928.

Charman, D., 2003, *Peatlands and Environmental Change* (Chichester and New York: John Wiley).

Charman, D.J., Hendon, D. and Woodland, W.A., 2000, *The Identification of Testate Amoebae (Protozoa: Rhizopoda) in Peats*. Technical Guide **9** (London: Quaternary Research Association).

Chesner, C.A., Rose, W.L., Deino, A., Drake, R. and Westgate, J.A., 1991, Eruptive history of Earth's largest Quaternary caldera (Toba, Indonesia) clarified, *Geology*, **10**, 200–203.

Chlachula, J., 2003, The Siberian loess record and its significance for reconstruction of Pleistocene climatic change in north-central Asia, *Quaternary Science Reviews*, **22**, 1879–1906.

Chou, H.-H., Hayakawa, T., Diaz, S., Krings, M., Indriati, E., Leakey, M., Paabo, S., Satta, Y., Takahata, N. and Varki, A., 2002, Inactivation of CMP-*N*-acetylneuraminic acid hydrozylase occurred prior to brain expansion during human evolution, *Proceedings of the National Academy of Sciences*, **99**, 11736–11741.

Christen, J.A. and Litton, C.D., 1995, A Bayesian approach to wiggle-matching, *Journal of Archaeological Science*, **22**, 719–725.

Christen, J.A., Clymo, R.S. and Litton, C.D., 1995, A Bayesian approach to the use of ^{14}C dates in the estimation of the age of peat, *Radiocarbon*, **37**, 431–442.

Christiansen, H.H., Bennike, O., Böcher, J., Elberling, B., Humlum, O. and Jakobsen, B.H., 2002, Holocene environmental reconstruction from deltaic deposits in northeast Greenland, *Journal of Quaternary Science*, **17**, 145–160.

Clarke, R.J., 1998, First ever discovery of a well-preserved skull and associated skeleton of *Australopithecus*, *South African Journal of Science*, **94**, 460–463.

Clark, A.J., Tarling, D.H. and Nöel, M., 1988, Developments in archaeomagnetic dating in Britain, *Journal of Archaeological Science*, **15**, 645–667.

Clark, D.H., Clark, M.M. and Gillespie, A.R., 1994, Debris-covered glaciers in the Sierra Nevada, California, and their implications for snowline reconstruction, *Quaternary Research*, **41**, 39–163.

Clark, J.D., Beyen, Y., Woldegabriel, G., Hart, W.K., Renne, P.R., Gilbert, H., Defleur, A., Suwa, G., Katoh, S., Ludwig, K.R., Bolserrie, J.-R., Asfaw, B. and White, T.D., 2003, Stratigraphic, chronological and behavioural contexts of Pleistocene *Homo sapiens* from Middle Awash, Ethiopia, *Nature*, **423**, 747–752.

Clausen, H.B., 1973, Dating of polar ice by ^{32}Si, *Journal of Glaciology*, **66**, 411–416.

Close, F., Marten, M. and Sutton, C., 1987, *The Particle Explosion* (Oxford: Oxford Unversity Press).

Cockburn, H.A.P. and Summerfield, M.A., 2004, Geomorphological applications of cosmogenic isotope analysis, *Progress in Physical Geography*, **28**, 1–42.

Colgan, P.M., 1999, Early Middle Pleistocene glacial sediments (780 000–620 000 BP) near Kansas City, northeastern Kansas and northwestern Missouri, USA, *Boreas*, **28**, 477–489.

Collins, M.J., Waite, E.R. and van Duin, A.C.T., 1999, Predicting protein decomposition: the case of aspartic acid racemization kinetics, *Philosphical Transactions of the Royal Society, London*, **B354**, 51–64.

Colman, S.M. and Pierce, K.L., 1986, Glacial sequences near McCall, Idaho: weathering rinds, soil development, morphology and other relative age criteria, *Quaternary Research*, **25**, 25–42.

Cook, A.C., Wadsworth, J. and Southon, J.R., 2001, AMS radiocarbon dating of ancient iron artifacts: a new carbon extraction method in use at LLNL, *Radiocarbon*, **43**, 221–227.

Cook, A.C., Wadsworth, J., Southon, J.R. and van der Merwe, N.J., 2003, AMS radiocarbon dating of rusty iron, *Journal of Archaeological Science*, **30**, 95–101.

Cook, E.R. and Kariukistis, L.A. (eds), 1990, *Methods of Dendrochronology* (Dordrecht: Kluwer).

Cowan, E.A., Cai, J., Powell, R.D., Clark, J.D. and Pitcher, J.N., 1997, Temperate glacimarine varves: an example from Disenchantment Bay, southern Alaska, *Journal of Sedimentary Research*, **67**, 536–549.

Cresswell, R.G., 1991, The radiocarbon dating of iron artifacts using accelerator mass spectrometry, *Historical Metallurgy*, **25**, 76–85.

Cresswell, R.G., 1992, Radiocarbon dating of iron artifacts, *Radiocarbon*, **34**, 898–905.

Crook, R., 1986, Relative dating of Quaternary deposits based on P-wave velocities in weathered granitic clasts, *Quaternary Research*, **25**, 281–292.

Dahms, D.E., 1994, Mid-Holocene erosion of soil catenas on moraines near the type Pinedale Till, Wind River Range, Wyoming, *Quaternary Research*, **42**, 41–48.

Dalsgaard, K. and Odgaard, B.V., 2001, Dating sequences of buried horizons of podzols developed in wind-blown sand at Ulfborg, western Jutland, *Quaternary International*, **78**, 53–60.

Damon, P.E., Donahue, D.J., Gore, B.H., Hatheway, A.L., Jull, A.J.T., Linick, T.W., Sercel, P.J., Toolin, L.J., Bronk, C.R., Hall, E.T., Hedges, R.E.M., Housley, R., Law, I.A., Perry, C., Bonani, G., Trumbore, S., Wölfli, W., Ambers, J.C., Bowman, S.G.E., Leese, M.N. and Tite, M.S., 1989, Radiocarbon dating of the Shroud of Turin, *Nature*, **337**, 611–615.

Davidson, D.A., Grieve, I.C., Tyler, A.N., Barclay, G.J. and Maxwell, G.S., 1998, Archaeological sites: assessment of erosion risk, *Journal of Archaeological Science*, **25**, 857–860.

Davidson, D.A. and Simpson, I.A., 2001, Archaeology and soil micromorphology, in D.E. Brothwell, A.M. Pollard (eds), *Handbook of Archaeological Sciences*, 167–177 (Chichester and New York: John Wiley).

Davies, G.L., 1969, *The Earth in Decay* (London: Macdonald).

Davies, S.M., Turney, C.S.M. and Lowe, J.J., 2001, Identification and significance of a visible basalt-rich Vedde Ash layer in a Late-glacial sequence on the Isle of Skye, Inner Hebrides, Scotland, *Journal of Quaternary Science*, **16**, 99–104.

Davies, S.M., Branch, N.P., Lowe, J.J. and Turney, C.S.M., 2002, Towards a European tephrochronological framework for Termination 1 and the Early Holocene, *Philosophical Transactions of the Royal Society, London*, **A360**, 767–802.

Davies, S.M., Wastegård, S. and Wohlfarth, B., 2003, Extending the limits of the Borrobol Tephra to Scandinavia and detection of new Early Holocene tephras, *Quaternary Research*, **59**, 345–352.

Davies, S.M., Wohlfarth, B., Wastegård, S., Blockley, S. and Possnert, G., 2004, Were there two Borrobol Tephras during the early Lateglacial period: implications for tephrochronology?, *Quaternary Science Reviews*, **23**, 581–590.

Davis, R. and Schaeffer, O.A., 1955, Chlorine-36 in nature, *Annals of the New York Academy of Sciences*, **62**, 105–122.

Deacon, H.J. and Deacon, J., 1999, *Human Beginnings in South Africa* (Cape Town and Johannesburg: David Phillip).

Dearing, J., 1999, Magnetic susceptibility, in J. Walden, F. Oldfield, J. Smith (eds), *Environmental Magnetism: A Practical Guide*. Technical Guide **6**, Quaternary Research Association, London, 35–62.

De Geer, G., 1912, A geochronology of the last 12,000 years, *XIth International Geological Congress, Stockholm*, **1**, 241–253.

De Vries, H., 1958, Variations in concentration of radiocarbon with time and location on earth, *Koninkijk Nederlandse Akademie von Wetenschappen, Amsterdam, Proceedings*, **B61**, 94–102.

Demetsopoulos, J.C., Burleigh, R. and Oakley, K.P., 1983, Relative and absolute dating of the human skeleton from Galley Hill, Kent, *Journal of Archaeological Science*, **10**, 129–134.

Derbyshire, E. (ed.), 2001, Loess and palaeosols: characteristics, stratigraphy, chronology and climate. A contribution to IGCP 413, *Quaternary International*, **76/77**, 1–260.

Derbyshire, E., 2003, Loess, and the dust indicators and records of terrestrial and marine palaeoenvironments (DIRTMAP) database, *Quaternary Science Reviews*, **22**, 1813–2052.

Dillehay, T.D., 1997, *Monte Verde, a Late Pleistocene settlement in Chile. Volume 2. The Archaeological Context and Interpretation* (Washington, DC: Smithsonian Institute Press).

Dillehay, T.D., 2003, Tracking the first Americans, *Science*, **425**, 24–25.

Ding, Z., Rutter, N. and Liu, T.S., 1993, Pedostratigraphy of Chinese loess deposits and climatic cycles in the last 2.5 Ma, *Catena*, **20**, 73–91.

Ding, Z., Yu, Z., Rutter, N. and Liu, T.S., 1994, Towards an orbital time scale for Chinese loess deposits, *Quaternary Science Reviews*, **13**, 39–70.

Donahue, D.J., Olin, J.S. and Harbottle, G., 2002, Determination of the radiocarbon age of parchment of the Vinland Map, *Radiocarbon*, **44**, 45–52.

Dorighel, O., Poupeau, G., Bellot-Gurlet L. and Labrin, E., 1998, Fission track dating and provenience of archaeological obsidian artefacts in Colombia and Ecuador, in P. van den Haute, F. de Corte (eds), *Advances in Fission-Track Geochronology*, 313–324 (Dordrecht: Kluwer).

Dorn, R.I., 1994, Surface exposure dating with rock varnish, in C. Beck (ed.), *Dating in Exposed and Surface Contexts*, 77–113 (Albuquerque: University of New Mexico Press).

Dorn, R.I., 1998, Ambiguities in direct dating of rock surfaces using radiocarbon measurements: Response, *Science*, **280**, 2135–2139.

Dorn, R.I., Clarkson, P.B., Nobbs, M.F., Loendorf, L.L. and Whitley, D.S., 1992, Radiocarbon dating inclusions of organic matter in rock varnish, with examples from drylands, *Annals of the Association of American Geographers*, **82**, 136–151.

Douglass, A.E., 1919, *Climatic Cycles and Growth 1*, (Washington, DC: Carnegie Institute).

Dragovich, D., 2000, Rock engraving chronologies and accelerator mass spectrometry radiocarbon ages of desert varnish, *Journal of Archaeological Science*, **27**, 871–876.

Dugmore, A.J. and Newton, A.J., 1992, Thin tephra layers in peat revealed by X-radiography, *Journal of Archaeological Science*, **19**, 163–170.

Duller, G.A.T., 2004, Luminescence dating of Quaternary sediments: recent advances, *Journal of Quaternary Science*, **19**, 183–192.

Dunai, T.J., 2000, Scaling factors for production rates of *in situ* produced cosmogenic nuclides: a critical reevaluation, *Earth and Planetary Science Letters*, **176**, 157–169.

Dutta, K., Bushan, R. and Somayajulu, B.L.K., 2001, ΔR correction values for the northern Indian Ocean, *Radiocarbon*, **43**, 483–488.

Edwards, R.L., Gallup, C.D. and Cheng, H., 2003, Uranium-series dating of marine and lacustrine carbonates, in B. Bourdon, G.M. Henderson, C.C. Lundstrom, S. Turner (eds), *Uranium Series Geochemistry, Reviews in Mineralogy and Geochemistry*, **52**, 363–405.

Eggins, S., Grün, R., Pike, A.W.L., Sheley, M., and Taylor, L., 2003, ^{238}U, ^{232}Th profiling and U-series isotope analysis of fossil teeth by laser ablation-ICPMS, *Quaternary Science Reviews*, **22**, 1373–1382.

Eighmy, J.L. and Sternberg, R.S., 1990, *Archaeomagnetic Dating* (Tucson: University of Arizona Press).

Eikenberg, J., Vezzu, G., Zumsteg, I., Bajo, S., Ruethi, M. and Wyssling, G., 2001, Precise two chronometer dating of Pleistocene travertine: the $^{230}Th/^{234}U$ and $^{226}Ra_{ex}/^{226}Ra(0)$ approach, *Quaternary Science Reviews*, **20**, 1935–1953.

Einarsson, T., 1986, Tephrochronology, in B.E. Berglund (ed.), *Handbook of Holocene Palaeoecology and Palaeohydrology*, 329–342 (Chichester and New York: John Wiley).

England, J., 1999, Coalescent Greenland and Innuitian ice during the last glacial maximum: revising the Quaternary of the Canadian High Arctic, *Quaternary Science Reviews*, **18**, 421–456.

EPICA Community Members (2004), Eight glacial cycles from an Antarctic ice core, *Nature*, **429**, 823–828.

Eronen, M., Zetterberg, P., Briffa, K.R., Lindholm, M., Mariläinen, J. and Timonen, M., 2002, The supra-long Scots pine tree-ring record for Finnish Lapland: Part 1, Chronology construction and initial inferences, *The Holocene*, **12**, 673–680.

Espizua, L.E. and Bigazzi, G., 1998, Fission-track dating of the Punta de Vacas glaciation in the Río Mendoza Valley, Argentina, *Quaternary Science Reviews*, **17**, 755–760.

Espizua, L.E., Bigazzi, G., Junes, P.J., Hadler, J.C. and Osorio, A.M., 2002, Fission-track dating of a tephra layer related to Poti-Malal and Seguro drifts in the Río Grande basin, Mendoza, Argentina, *Journal of Quaternary Science*, **17**, 781–788.

Esposito, M., Reyss, J.-L., Chaimanee, Y. and Jaeger, J.J., 2002, U-series dating of fossil teeth and carbonates from Snake Cave, Thailand, *Journal of Archaeological Science*, **29**, 341–350.

Evans, D.J.E., Archer, S. and Wilson, D.J.H., 1999, A comparison of the lichenometric and Schmidt hammer dating techniques based on data from the proglacial areas of some Icelandic glaciers, *Quaternary Science Reviews*, **18**, 13–41.

Evans, M.E. and Heller, F., 2003, *Environmental Magnetism: Principles and Applications of Enviromagnetics* (New York: Academic Press).

Ferguson, C.W. and Graybill, D.A., 1983, Dendrochronology of bristlecone pine: a progress report, *Radiocarbon*, **25**, 287–288.

Fiedel, S.J., 2002, Initial human colonisation of the Americas: an overview of the issues and evidence, *Radiocarbon*, **44**, 407–436.

Fjeldskaar, W., Lindholm, C., Dehls, J.F. and Fjeldskaar, I., 2000, Postglacial uplift, neotectonics and seismicity in Fennoscandia, *Quaternary Science Reviews*, **19**, 1413–1422.

Folk, R.L. and Valastro, S. Jr, 1976, Successful technique for dating of lime mortar by carbon-14, *Journal of Field Archaeology*, **3**, 203–208.

Follmer, L., 1983, Sangamon and Wisconsinan Pedogenesis in the Midwestern United Staes, in S.C. Porter (ed.), *Late Quaternary Environments of the United States. Volume 1: The Late Pleistocene*, 138–144 (London: Longman).

Forsythe, G.T.W., Scourse, J., Harris, I., Richardson, C.A., Jones, P., Briffa, K. and Heinemeier, J., 2004, Towards an absolute chronology for the marine environment: the development of a 1000 year record from *Arctica islandica*, in J.D. Scourse, L. Clarke, C. Richardson, F. Marret (eds), *Annually-Banded Records in the Quaternary*, 27 (London: Quaternary Research Association).

Forbes, J.D., 1853, Norway and its Glaciers Visited in 1851 (London: Adam and Charles Black).

Friedman, I. and Obradovich, J., 1981, Obsidian hydration dating of volcanic events, *Quaternary Research*, **16**, 37–47.

Friedrich, M., Kromer, B., Spurk, M., Hoffmann, J. and Kaiser, K.F., 1999, Palaeo-environment and radiocarbon calibration as derived from Lateglacial/Early Holocene tree-ring chronologies, *Quaternary International*, **61**, 27–29.

Friedrich, M., Kromer, B., Kaiser, K.F., Spurk, M., Hughen, K.A. and Johnsen, S., 2001, High-resolution climate signals in the Bølling-Allerød Interstadial (Greenland Interstadial 1) as reflected in European tree-ring chronologies compared to marine varves and ice-core records, *Quaternary Science Reviews*, **20**, 1223–1232.

Fritts, H.C., 1976, *Tree-Rings and Climate* (London: Academic Press).

Froggatt, P.C. and Lowe, D.J., 1990, A review of late Quaternary silicic and some other tephra formations from New Zealand: their stratigraphy, nomenclature, distribution, volume and age, *New Zealand Journal of Geology and Geophysics*, **33**, 89–109.

Frumkin, A., Ford, D.C. and Schwarcz, H.P., 1999, Continental oxygen isotopic record of the last 170,000 years in Jerusalem, *Quaternary Research*, **51**, 317–327.

Fuchs, M. and Lang, A., 2001, OSL dating of coarse-grain fluvial quartz using single-aliquot protocols on sediments from NE Peleponnese, Greece, *Quaternary Science Reviews*, **20**, 783–787.

Fuchs, M., Lang. A. and Wagner, G.A., 2004, The history of Holocene soil erosion in the Phlious Basin, NE Peleponnese, Greece, based on optical dating, *The Holocene*, **14**, 334–345.

Funnell, B.M., 1995, Global sea level and the (pen)insularity of late Cenozoic Britain, in R.C. Preece (ed.), *Island Britain: A Quaternary Perspective*. Geological Society of London, Special Publication 96, 3–14.

Gabunia, L., Vekua, A., Lordkipanidze, D., Swisher III, C.C., Ferring, R., Justus, A., Nioradze, M., Tvalchrelidze, M., Anton, S.C., Bosinski, G., Jöris, O., Lumley, M.-A.-de, Majsuradze, G. and Mouskhelishvili, A., 2000, Earliest Pleistocene hominid cranial remains from Dmanisi, Republic of Georgia: taxonomy, geological setting and age, *Science*, **288**, 1019–1025.

Garnett, E.R., Gilmour, M.A., Rowe, P.J., Andrews, J.E. and Preece, R.C., 2004, ^{230}Th/^{234}U dating of Holocene tufas: possibilities and problems, *Quaternary Science Reviews*, **23**, 947–958.

Gernaey, A.M., Waite, E.R., Collins, M.J., Craig, O.E. and Sokol, R.J., 2001, Survival and interpretation of archaeological proteins, in D.R. Brothwell, A.M. Pollard (eds), *Handbook of Archaeological Sciences*, 323–329 (Chichester and New York: John Wiley).

Geyh, M., Schotterer, U. and Grosjean, M., 1998, Temporal changes of the ^{14}C reservoir effect in lakes, *Radiocarbon*, **40**, 921–931.

Gibbard, P.L., Boreham, S., Cohen, K.M. and Moscariello, A., 2004, *Global Chronostratigraphical Correlation Table for the Last 2.7 Million Years*. Subcommission on Quaternary Stratigraphy, Cambridge.

Godfrey-Smith, D.I. and Casey, J.L., 2003, Direct thermoluminescence chronology for Early Iron Age smelting technology on the Gambaga Escarpment, northern Ghana, *Journal of Archaeological Science*, **30**, 1037–1050.

Godwin, H., 1962, Half-life of radiocarbon, *Nature*, **195**, 944.

Goh, K.M. and Molloy, B.J.P., 1979, Contaminants in charcoals used for radiocarbon dating, *New Zealand Journal of Soil Science*, **27**, 89–100.

Goldstein, S.J. and Stirling, C.H., 2003, Techniques for measuring uranium-series nuclides: 1992–2002, in B. Bourdon, G.M. Henderson, C.C. Lundstrom and S. Turner (eds), *Uranium Series Geochemistry, Reviews in Mineralogy and Geochemistry*, **52**, 23–57.

Gonzalez, S., Concepción J.-L., Hedges, R., Huddart, D., Ohman, J.C., Turner, A. and Pompa y Padilla, J.A., 2003, Earliest humans in the Americas: new evidence from Mexico, *Journal of Human Evolution*, **44**, 379–387.

Goodfriend, G.A., 1987, Radiocarbon age anomalies in shell carbonate of land snails from semi-arid areas, *Radiocarbon*, **29**, 159–167.

Goodfriend, G.A., 1992, The use of land snails in palaeoenvironmental reconstruction, *Quaternary Science Reviews*, **11**, 665–685.

Goodfriend, G., Collins, M.J., Fogel, M.L., Macko, S.A. and Wehmiller, J.F., 2000, *Perspectives in Amino Acid and Protein Geochemistry* (Oxford: Oxford University Press).

Goodman, A.Y., Rodbell, D.T., Seltzer, G.O. and Mark, B.G., 2001, Subdivision of glacial deposits in southeastern Peru based on pedogenic development and radiometric ages, *Quaternary Research*, **56**, 31–50.

Goodwin, D.H., Flessa, K.W., Schöne, B. and Lettman, D.L., 2001, Cross-calibration of daily growth increments, stable isotope variation, and temperature in the Gulf of California bivalve mollusk *Chione coretzi*: implications for paleoenvironmental analysis, *Palaios*, **16**, 387–398.

Gorter, S.C., 1936, Paramagnetic relaxation in a transversal magnetic field, *Physica*, **3**, 1006–1008.

Gorter, S.C. and Kronig, R. de L., 1936, On the theory of absorption and dispersion in paramagnetic and dielectric media, *Physica*, **3**, 1009–1020.

Goslar, T., Arnold, M., Bard, E., Kuc, T., Pazdur, M.F., Ralska-Jasiewiczowa, M., Rozanski, K., Tisneret, N., Walanus, A., Wicik, B., and Więckowski, K., 1995, High concentration of atmospheric ^{14}C during the Younger Dryas cold episode, *Nature*, **377**, 414–417.

Goslar, T., Arnold, M., Tisneret-Laborde, N., Czernik, J. and Więckowski, K., 2000, Variations of Younger Dryas atmospheric radiocarbon explicable without ocean circulation changes, *Nature*, **403**, 877–880.

Gosse, J.C. and Phillips, F.M., 2001, Terrestrial *in situ* cosmogenic nuclides: theory and applications, *Quaternary Science Reviews*, **20**, 1475–1560.

Gove, H.E., Mattingly, S.J., David, A.R. and Garza-Valdes, L.A., 1997, A problematic source of organic contamination of linen, *Nuclear Instruments and Methods in Physics Research*, B, **123**, 504–507.

Gribbin, J., 1984, *In Search of Shrödinger's Cat* (New York: Bantam, and London: Black Swan).

Gribbin, J., 1995, *Schrödinger's Kittens and the Search for Reality* (London: Weidenfeld & Nicholson).

Grimley, D.A., Follmer, L., Hughes, R.E. and Solheid, P.A., 2003, Modern, Sangamon and Yarmouth soil development in loess of unglaciated southwestern Illinois, *Quaternary Science Reviews*, **22**, 225–244.

Grissino-Mayer, H.D. and Fritts, H.C., 1997, The International Tree-Ring Data Bank: an enhanced global database serving the global scientific community, *The Holocene*, **7**, 235–239.

Grönvold, K., Oskarsson, N., Johnsen, S.J., Clausen, H.B., Hammer, C.U., Bond, G. and Bard, E., 1995, Ash layers from Iceland in the GRIP ice core correlated with oceanic and land sediments, *Earth and Planetary Science Letters*, **135**, 149–155.

Grudd, H., Briffa, K.R., Karlén, W., Bartholin, T.S., Jones, P.D. and Kromer, B., 2002, A 7400-year tree-ring chronology in northern Swedish Lapland: natural climatic variability expressed on annual to millennial timescales, *The Holocene*, **12**, 657–665.

Grün, R., 1997, Electron spin resonance dating, in R.E. Taylor, M.J. Aitken (eds), *Chronometric and Allied Dating in Archaeology*, 217–261 (New York: Plenum).

Grün, R., 2001, Trapped charge dating (ESR, TL, OSL), in D.E. Brothwell, A.M. Pollard (eds), *Handbook of Archaeological Sciences*, 47–62 (Chichester and New York: John Wiley).

Grün, R. and Schwarcz, H.P., 2000, Revised open-system U-series/ESR age calculations for teeth from Stratum C at the Hoxnian Interglacial type locality, England, *Quaternary Science Reviews*, **19**, 1151–1154.

Grün, R. and Wintle, A.G. (eds), 2001, Proceedings of the Ninth International Conference on Luminescence and Electron Spin Resonance Dating LED99, *Quaternary Science Reviews*, **20**, 683–1061.

Grün, R. and Wintle, A.G. (eds), 2003, Proceedings of the Tenth International Conference on Luminescence and Electron Spin Resonance Dating – LED 02, *Quaternary Science Reviews*, **22**, 951–1382.

Grün, R., Beaumont, P.B., Tobias, P.V. and Eggins, S., 2003, On the age of Border Cave 5 human mandible, *Journal of Human Evolution*, **45**, 155–167.

Haberle, S.G. and Lumley, S.H., 1998, Age and origin of the tephras recorded in postglacial lake sediments to the west of the southern Andes, *Journal of Volcanology and Geothermal Research*, **84**, 239–256.

Haflidason, H., Eiriksson, J. and van Kreveld, S., 2000, The tephrochronology of Iceland and the North Atlantic region during the Middle and Late Quaternary: a review, *Journal of Quaternary Science*, **15**, 3–22.

Hall, R.D. and Anderson, A.K., 2000, Comparative soil development of Quaternary paleosols of the central United States, *Palaeogeography, Palaeoclimatology, Palaeoecology*, **158**, 109–145.

Hall, V.A. and Pilcher, J.R., 2002, Late-Quaternary Icelandic tephras in Ireland and Great Britain: detection, characterisation and usefulness, *The Holocene*, **12**, 223–230.

Hammer, C.U., Mayewski, P.A., Peel, D. and Stuiver, M. (eds), 1997, Greenland Summit Ice Cores. Greenland Ice Sheet Project 2/Greenland Ice Core Project, *Journal of Geophysical Research*, **102**, 26 315–26 886.

Harden, G.W. and Taylor, E.M., 1983, A quantitative comparison of soil development in four climatic regimes, *Quaternary Research*, **28**, 342–359.

Harkness, D.D., 1979, Radiocarbon dates from Antarctica, *British Antarctic Survey Bulletin*, **47**, 43–59.

Harle, K.J., Heijnis, H., Chisari, R., Kershaw, A.P., Zoppi, U. and Jacobsen, G., 2002, A chronology for the long pollen record from Lake Wangoom, western Victoria (Australia) as derived from uranium/thorium disequilibrium dating, *Journal of Quaternary Science*, **17**, 707–720.

Hearty, P.J., 2003, Stratigraphy and timing of aeolianite deposition on Rottnest Island, western Australia, *Quaternary Research*, **60**, 211–222.

Hearty, P.J. and Kaufman, D.S., 2000, Whole-rock aminstratigraphy and Quaternary sea-level history of the Bahamas, *Quaternary Research*, **54**, 163–173.

Hearty, P.J., Vacher, H.L. and Mitterer, R.M., 1992, Aminostratigraphy and ages of Pleistocene limestones of Bermuda, *Geological Society of America Bulletin*, **104**, 471–480.

Hedges, R.E.M., 2001a, The future of the past, *Radiocarbon*, **43**, 141–148.

Hedges, R.E.M., 2001b, Dating in archaeology: past, present and future, in D.E. Brothwell, A.M. Pollard (eds), *Handbook of Archaeological Sciences*, 3–8 (Chichester and New York: John Wiley).

Hedges, R.E.M. and Law, I.A., 1989, The radiocarbon dating of bone, *Applied Geochemistry*, **4**, 249–253.

Hedges, R.E.M., Tiemei, C. and Housley, R.A., 1992, Results and methods in the radiocarbon dating of pottery, *Radiocarbon*, **34**, 906–915.

Hedges, R.E.M., Bronk Ramsay, C. and van Klinken, G.J., 1998a, An experiment to refute the likelihood of cellulose carboxylation, *Radiocarbon*, **40**, 59–60.

Hedges, R.E.M., Bronk Ramsay, C., van Klinken, G.J., Pettit, P.B., Nielsen-Marsh, C., Etchegoyen, A., Fernandez Niello, J.O., Boschin, M.T. and Llamazares, A.M., 1998b, Methodological issues in the ^{14}C dating of rock paintings, *Radiocarbon*, **40**, 35–44.

Hedges, R.E.M., Millard, A.R. and Pike, A.W.G., 1995, Measurements and relationships of diagenetic alteration of bone from three archaeological sites, *Journal of Archaeological Science*, **22**, 201–209.

Heijnis, H., 1995, *Uranium/Thorium Dating of Late Pleistocene Peat Deposits in N.W. Europe* (The Netherlands: Rijksuniversiteit Groningen).

Heijnis, H. and van der Plicht, J., 1992, Uranium/thorium dating of Late Pleistocene peat deposits in N.W. Europe, uranium/thorium isotope systematics and open-system behaviour of peat layers, *Chemical Geology*, **94**, 161–171.

Heinemeier, J., Jungner, H., Lindroos, A., Ringbom, Å., von Konow, T. and Rud, N., 1997, AMS ^{14}C dating of lime mortar, *Nuclear Instruments and Methods in Physics Research*, B, **123**, 487–495.

Henderson, G.M. and Slowey, N.C., 2000, Evidence from U-Th dating against northern Hemisphere forcing of the penultimate deglaciation, *Nature*, **404**, 61–66.

Hendon, D. and Charman, D., 2004, High-resolution peatland water-table changes for the past 200 years: the influence of climate and implications for management, *The Holocene*, **14**, 125–134.

Hendy, E.J., Gagan, M.K., Alibert, C.A., McCullough, M.T., Lough, J.M. and Isdale, P.J., 2002, Abrupt decrease in tropical Pacific sea surface salinity at the end of the Little Ice Age, *Science*, **295**, 1511–1514.

Hendy, E.J., Gagan, M.K. and Lough, J.M., 2003, Chronological control of coral records using luminescent lines and evidence for non-stationary ENSO teleconnections in northeastern Australia, *The Holocene*, **13**, 187–199.

Henshilwood, C.S., d'Errico, F., Marean, C.W., Milo, R.G. and Yates, R., 2001, An early bone tool industry from the Middle Stone Age at Blombos Cave, South Africa: implications for the origins of modern human behaviour, symbolism and language, *Journal of Human Evolution*, **41**, 631–678.

Henshilwood, C., d'Errico, F., Vanhaeren, M., van Niekerk, K. and Jacobs, Z., 2004, Middle Stone Age shell beads from South Africa, *Science*, **304**, 404.

Hewitt, G., 2000, The genetic legacy of the Quaternary ice ages, *Nature*, **405**, 907–913.

Higgitt, D.L., Walling, D.E. and Haigh, M.J., 1994, Estimating rates of ground retreat on mining spoils using caesium-137, *Applied Geography*, **14**, 294–307.

Hillam, J., 1992, *Dendrochronology in England: The Dating of a Wooden Causeway from Lincolnshire and a Logboat from Humberside*. Proceedings of the 13th Colloquium AFEAF, Guerat 1989, 137–141.

Hillam, J., 2000, Dendrochronological dating, in M. Bell, A.E. Caseldine, Neumann, H. (eds), *Prehistoric Intertidal Archaeology in the Welsh Severn Estuary*. Council for British Archaeology, Research Report **120**, York, 159–168.

Hillam, J., Groves, C.M., Brown, D.M., Baillie, M.G.L., Coles, J.M. and Coles, B.J., 1990, Dendrochronology of the British Neolithic, *Antiquity*, **64**, 210–220.

Hillman, G., Hedges, R., Moore, A., Colledge, S. and Pettitt, P., 2001, New evidence of Lateglacial cereal cultivation at Abu Hureyra on the Euphrates, *The Holocene*, **11**, 383–393.

Hodgins, G.W.L., Thorpe, J.L., Coope, G.R. and Hedges, R.E.M., 2001, Protocol development for purification and characterisation of sub-fossil insect chitin for stable isotopic analysis and radiocarbon dating, *Radiocarbon*, **43**, 199–208.

Hogg, A.G., Higham, T.F.G., Lowe, D.J., Palmer, J.G., Reimer, P.J. and Newnham, R.M., 2003, A wiggle-match date for Polynesian settlement of New Zealand, *Antiquity*, **77**, 116–125.

Holmes, A., 1915, Radioactivity and the measurement of geological time, *Proceedings of the Geologists' Association*, **26**, 289–309.

Horrocks, M., Deng, Y., Ogden, J., Alloway, B.V., Nichol, S.L. and Sutton, D.G., 2001, High spatial resolution of pollen and charcoal in relation to the c. 600 year BP Kaharoa Tephra: implications for Polynesian settlement of Gt. Barrier Island, northern New Zealand, *Journal of Archaeological Science*, **28**, 153–168.

Houghton, J.T., Ding, Y., Griggs, D.J., Noguer, M., van der Linden, P.J. and Xiaosu, D. (eds), 2001, *Climate Change 2001: The Scientific Basis* (Cambridge: Cambridge University Press).

Huber, U.M. and Markgraf, V., 2003, European impact on fire regimes and vegetation dynamics at the steppe-forest ecotone of southern Patagonia, *The Holocene*, **13**, 567–579.

Huffman, O.F., 2001, Geologic context and age of the Perning/Mojokerto *Homo erectus*, East Java, *Journal of Human Evolution*, **40**, 353–362.

Huggett, R.J., 1998, Soil chronosequences, soil development, and soil evolution: a critical review, *Catena*, **32**, 155–172.

Hughen, K.A., Overpeck, J.T., Peterson, L.C. and Anderson, R.F., 1996, The nature of varved sedimentation in the Cariaco Basin, Venezuela, and its palaeoclimatic significance, in A.E.S. Kemp (ed.), *Palaeoclimatology and Palaeoceanography from Laminated Sediments*, Geological Society Special Publication No. 116, 171–183.

Hughen, K.A., Overpeck, J.T., Lehman, S.J., Kashgarian, M., Southon, J. and Peterson, L.C., 1998, A new ^{14}C calibration dataset for the Last Deglaciation based on marine varves, *Radiocarbon*, **40**, 483–494.

Hughen, K.A., Overpeck, J.T. and Anderson, R.F., 2000, Recent warming in a 500-year palaeotemperature record from varved sediments, Upper Soper Lake, Baffin Island, Canada, *The Holocene*, **10**, 9–19.

Hughen, K., Lehman, S., Southon, J., Overpeck, J., Marchal, O., Herring, C. and Turnbull, J., 2004, ^{14}C activity and global carbon cycle changes over the past 50 000 years, *Science*, **303**, 202–207.

Hughes, P.D.M., Mauquoy, D., Barber, K.E. and Langdon, P.G., 2000, Mire-development pathways and palaeoclimatic records from a full Holocene peat archive at Walton Moss, Cumbria, England, *The Holocene*, **10**, 465–479.

Hull, K.L., 2001, Reasserting the utility of obsidian hydration dating: a temperature-dependent empirical approach to practical temporal resolution with archaeological obsidians, *Journal of Archaeological Science*, **28**, 1025–1040.

Hunt, J.B., 2004, Tephrostratigraphical evidence for the timing of Pleistocene explosive volcanism at Jan Mayen, *Journal of Quaternary Science*, **19**, 121–136.

Hunt, J.B. and Hill, P.G., 1993, Tephra geochemistry: a discussion of some persistent analytical problems, *The Holocene*, **3**, 271–278.

Hunt, J.B. and Hill, P.G., 1996, An inter-laboratory comparison of the electron probe microanalysis of glass geochemistry, *Quaternary International*, **34–36**, 229–241.

Hunt, J.B., Clift, P.D., Lacasse, C., Vallier, T.L. and Werner, R., 1998, Standardisation of electron probe microanalysis of glass geochemistry, in A.D. Saunders, H.C. Larsen, S.W. Wise Jr (eds), *Proceedings of the Ocean Drilling Program, Scientific Results*, **152**, 85–91.

Huntley D.J. and Lamothe, M., 2001, Ubiquity of anomalous fading in K-feldspars, and the measurement and correction for it in optical dating, *Canadian Journal of Earth Sciences*, **38**, 1093–1106.

Huntley, D.J., Godfrey-Smith, D.I. and Thewalt, M.L.W., 1985, Optical dating of sediments, *Nature*, **313**, 105–107.

Huntley, D.J., Hutton, J.T. and Prescott, J.R., 1993a, The stranded beach-dune sequence of south-east South Australia: a test of thermoluminescence dating, 0–800 ka, *Quaternary Science Reviews*, **12**, 1–20.

Huntley, D.J., Hutton, J.T. and Prescott, J.R., 1993b, Optical dating using inclusions within quartz grains, *Geology*, **21**, 1087–1910.

Hurford, A.J., 1991, Fission track dating, in P.L. Smart and P.D. Frances (eds), *Quaternary Dating Methods – A User's Guide*. Technical Guide **4**, Quaternary Research Association, London, 84–107.

Hütt, G., Jack, I. and Tchonka, J., 1988, Optical dating: K-feldspars optical response stimulation spectra, *Quaternary Science Reviews*, **7**, 381–385.

Ikeya, M., 1975, Dating a stalactite by electron paramagnetic resonance, *Nature*, **255**, 48–50.

Imbrie, J., Hays, J.D., Martinson, D.G., McIntyre, A., Mix, A.C., Morley J.J., Pisias, N.J., Prell, W.L. and Shackleton, N.J., 1984, The orbital theory of Pleistocene climate: support from a revised chronology of the marine $\delta^{18}O$ record, in A. Berger, J. Imbrie, J. Hays, G. Kukla, B. Saltzman (eds), *Milanovitch and Climate*, 269–306 (Dordrecht: Reidel).

Ingólfsson, O., Norddahl, H. and Haflidason, H., 1995, Rapid isostatic rebound in southwestern Iceland at the end of the last glaciation, *Boreas*, **24**, 245–259.

Innes, J.L., 1985, An examination of some factors affecting the largest lichens on a substrate, *Arctic and Alpine Research*, **17**, 99–106.

Ivanovich, M. and Harmon, R.S. (eds), 1995, *Uranium-Series Disequilibrium: Applications to Earth, Marine and Environmental Sciences*. 2nd edition (Oxford: Clarendon Press).

Ivy-Ochs, S., Wüst, R., Kubik, P.W., Müller-Beck, H. and Schlüchter, C., 2001, Can we use cosmogenic isotopes to date stone artifacts? *Radiocarbon*, **43**, 759–764.

Jacobs, Z., Duller, G.A.T. and Wintle, A.G., 2003a, Optical dating of dune sands from Blombos Cave, South Africa: II – single grain data, *Journal of Human Evolution*, **44**, 613–625.

Jacobs, Z., Wintle, A.G. and Duller, G.A.T., 2003b, Optical dating of dune sands from Blombos Cave, South Africa: I – multiple grain data, *Journal of Human Evolution*, **44**, 599–612.

Jacobsson, S., 1979, Outline of the petrology of Iceland, *Jökull*, **29**, 57–73.

Jansen, E., 1989, The use of stable oxygen and carbon isotope stratigraphy as a dating tool, *Quaternary International*, **1**, 151–166.

Jensen, K.G., Kuipers, A., Koç, N. and Heinemeier, J., 2004, Diatom evidence of hydrographic changes and ice conditions in Igaliku Fjord, South Greenland, during the past 1500 years, *The Holocene*, **14**, 152–164.

Jedoui, Y., Reyss, J.-L., Kallel, N., Montacer, M., Ismail, H.B., Davaud, E., 2003, U-series evidence for two high Late Interglacial sea levels in southeastern Tunisia, *Quaternary Science Reviews*, **22**, 343–351.

Johnsen, S., Dahl-Jensen, D., Gundestrup, N., Steffensen, J.P., Clausen, H.B., Miller, H., Masson-Delmotte, V., Sveinbjörnsdottir, A.E. and White, J., 2001, Oxygen isotope and palaeotemperature records from six Greenland ice-core stations: Camp Century, Dye-3, GRIP, GISP2, Renland and northGRIP, *Journal of Quaternary Science*, **16**, 299–308.

Johnson, A., 1999, *The Ancient Bristlecone Pine Forest* (Bishop, California: Community Printing and Publishing).

Johnsson, K., 1997, Chemical dating of bones based on diagenetic changes in bone apatite, *Journal of Archaeological Science*, **24**, 431–437.

Jones, M.K. and Colledge, S., 2001, Archaeobotany and the transition to agriculture, in D.R. Brothwell, A.M. Pollard (eds), *Handbook of Archaeological Sciences*, 393–401 (Chichester and New York: John Wiley).

Jones, R.T., Marshall, J.D., Crowley, S.F., Bedford, A., Richardson, N., Bloemendal, J. and Oldfield, F., 2002, A high resolution, multiproxy late-glacial record of climate change and intrasystem responses in northwest England, *Journal of Quaternary Science*, **17**, 329–340.

Jorde, L.B., Barnshad, M., and Rogers, A.R., 1998, Using mitochondrial and nuclear DNA markers to reconstruct human evolution, *BioEssays*, **20**, 126–136.

Jull, A.J.T., Donahue, D.J., Broshi, M. and Tov, E., 1995, Radiocarbon dating of scrolls and linen fragments from the Judean Desert, *Radiocarbon*, **37**, 11–19.

Jull, A.J.T., Donahue, D.J. and Damon, P.E., 1996, Factors affecting the apparent radiocarbon age of textiles: a comment on 'Effects of fires and biofractionation of carbon isotopes on results of radiocarbon dating of old textiles: The Shroud of Turin', by D.A. Kouznetzov *et al.*, *Journal of Archaeological Science*, **23**, 157–160.

Jull, A.J.T., Lal, D., Burr, G.S., Bland, P.A., Bevan, A.W.R. and Beck, J.W., 2000, Radiocarbon beyond this world, *Radiocarbon*, **42**, 151–172.

Keen, D.H., 1995, Raised beaches and sea-levels in the English Channel in the Middle and Late Pleistocene: problems of interpretation and implications for the isolation of the British Isles, in R.C. Preece (ed.), *Island Britain: A Quaternary Perspective*. Geological Society Special Publication No. **96**, London, 63–74.

Kelley, S., Williams-Thorpe, O. and Thorpe, R.S., 1994, Laser argon dating and geological provenancing of a stone axe from the Stonehenge environs, *Archaeometry*, **36**, 209–216.

Kelly, M., Kubik, P.W., von Blanckenburg, F. and Schlüchter, C., 2004, Surface exposure dating of the Great Aletsch Egesen moraine system, western Swiss Alps, using the cosmogenic nuclide ^{10}Be, *Journal of Quaternary Science*, **19**, 431–442.

Kemp, A.E.S. (ed.), 1996, *Palaeoclimatology and Palaeoceanography From Laminated Sediments*. Geological Society Special Publication No. 116, London.

Kemp, R.A., 1998, Role of micromorphology in palaeopedological research, *Quaternary International*, **51/52**, 133–141.

Kemp, R.A. and Faulkner, J.S., 1998, Short-range variations in the micromorphology of a palaeosol from Wivenhoe in southeast England, *Journal of Quaternary Science*, **13**, 233–243.

Kemp, R.A., Whiteman, C.A. and Rose, J., 1993, Palaeoenvironmental and stratigraphic significance of the Valley Farm and Barham Soils in eastern England, *Quaternary Science Reviews*, **12**, 833–848.

King, J. and Peck, J., 2001, Use of palaeomagnetism in studies of lake sediments, in W.M. Last, J.P. Smol (eds), *Tracking Environmental Change using Lake Sediments: Volume 1. Basin Analysis, Coring and Chronological Techniques*, 371–389 (Dordrecht: Kluwer).

Kitagawa, H. and van der Plicht, J., 1998, A 40 000-year varve chronology from Lake Suigetsu, Japan: extension of the ^{14}C calibration curve, *Radiocarbon*, **40**, 505–515.

Kitagawa, H. and van der Plicht, J., 2000, Atmospheric radiocarbon calibration beyond 11 900 cal BP from Lake Suigetsu laminated sediments, *Radiocarbon*, **42**, 369–380.

Knuepfer, P.L.K., 1994, Use of rock weathering rinds in dating geomorphic surfaces, in C. Beck (ed.), *Dating in Exposed and Surface Contexts*, 15–28 (Albuquerque: University of New Mexico Press).

Konrad, S.K. and Clark, D.H., 1998, Evidence for an early Neoglacial glacier advance from rock glaciers and lake sediments in the Sierra Nevada, California USA, *Arctic and Alpine Research*, **30**, 272–284.

Kouznetzov, D.A., Ivanov, A.A. and Veletsky, P.R., 1996, Effects of fires and biofractionation of carbon isotopes on results of radiocarbon dating of old textiles: the Shroud of Turin, *Journal of Archaeological Science*, **23**, 109–121.

Krings, M., Stone, A., Schmitz, R.W. and Krainitzki, H., Stoneking, M. and Pääbo, S., 1997, Neanderthal DNA sequences and the origin of modern humans, *Cell*, **90**, 19–30.

Kromer, B. and Spurk, M., 1998, Revision and tentative extension of the tree-ring based ^{14}C calibration, 9200–11 855 cal BP, *Radiocarbon*, **40**, 1117–1125.

Kukla, G. and An, Z.S., 1989, Loess stratigraphy in central China, *Palaeogeography, Palaeoclimatology, Palaeoecology*, **72**, 203–255.

Kuman, K. and Clark, R.J., 2000, Stratigraphy, artefact industries and hominid associations for Sterkfontein Member 5, *Journal of Human Evolution*, **38**, 827–847.

Kurz, M.D. and Brook, E.J., 1994, Surface exposure dating with cosmogenic nuclides, in C. Beck (ed.), *Dating in Exposed and Surface Contexts*, 139–159 (Albuquerque: University of New Mexico Press).

Kuzmin, Y.V., Hall, S., Tite, M.S., Bailey, R., O'Malley, J.M. and Medvedev, V.E., 2001, Radiocarbon and thermoluminescence dating of the pottery from the early Neolithic site of Gasya (Russian Far East): initial results, *Quaternary Science Reviews*, **20**, 945–948.

Lai, Z.-P., Stokes, S., Bailey, R., Fattahi, M. and Arnold, L., 2003, Infrared stimulated red luminescence from Chinese loess: basic observations, *Quaternary Science Reviews*, **22**, 961–966.

Lal, D. and Jull, A.J.T., 2001, *In-situ* cosmogenic ^{14}C: production and examples of its unique applications in studies of terrestrial and extraterrestrial processes, *Radiocarbon*, **43**, 731–742.

LaMarche, V.C. Jr, 1974, Palaeoclimatic inferences from long tree-ring records, *Science*, **183**, 1043–1048.

Lamb, H.H., 1995, *Climate, History and the Modern World*. 2nd edition (London: Routledge).

Lamoreux, S., 2001, Varve chronology, in W.M. Last, J.P. Smol (eds), *Tracking Environmental Change Using Lake Sediments: Volume 1. Basin Analysis, Coring and Chronological Techniques*, 247–260 (Dordrecht: Kluwer).

Lamoureux, S.F., England, J.H., Sharp. M.J. and Bush, A.B.G., 2001, A varve record of increased 'Little Ice Age' rainfall associated with volcanic activity, Arctic Archipelago, Canada, *The Holocene*, **11**, 243–249.

Langbroek, M. and Roebroeks, W., 2000, Extraterrestrial evidence on the age of the hominids from Java, *Journal of Human Evolution*, **38**, 595–600.

Langdon, P.E., Barber, K.E. and Hughes, P.D.M., 2003, A 7500-year peat-based palaeoclimatic reconstruction and evidence for an 1100-year cyclicity in bog surface wetness from Temple Hill Moss, Pentland Hills, southeast Scotland, *Quaternary Science Reviews*, **22**, 259–274.

Lantin, J., Aerts-Bijma, A.E. and van der Plicht, J., 2001, Dating of cremated bones, *Radiocarbon*, **43**, 249–254.

Latham, A.G., 2001, Uranium-series dating, in D.E. Brothwell, A.M. Pollard (eds), *Handbook of Archaeological Sciences*, 63–72 (Chichester and New York: John Wiley).

Lauritzen, S.-E., Haugen, J.E., Løvlie, R. and Gilje-Nielsen, H., 1994, Geochronological potential of isoleucine epimerization in calcite speleothems, *Quaternary Research*, **41**, 52–58.

Lauritzen, S.-E. and Lundberg, J., 1999, Calibration of the speleothem delta function: an absolute temperature record for the Holocene in northern Norway, *The Holocene*, **9**, 659–669.

Lengyel, S.N. and Eighmy, J.L., 2002, A revision to the U.S. southwest archaeomagnetic master curve, *Journal of Archaeological Science*, **29**, 1423–1433.

Leonard, E., 1986, Varve studies at Hector Lake, Alberta, Canada, and the relationship between glacier activity and sedimentation, *Quaternary Research*, **25**, 199–214.

Lian, O.B. and Huntley, D.J., 2001, Luminescence dating, in W. Last, J. Smol (eds), *Tracking Environmental Change using Lake Sediments. Volume 1. Basin Analysis, Coring and Chronological Techniques*, 261–282 (Dordrecht: Kluwer).

Lian, O. and Brooks, G.R., 2004, Optical dating studies of mud-dominated alluvium and buried hearth-like features from Red River Valley, southern Manitoba, Canada, *The Holocene*, **14**, 570–578.

Lian, O.B., Hu, J., Huntley, D.J. and Hicock, S.R., 1995, Optical dating studies of Quaternary organic-rich sediments from southwestern British Columbia and northwestern Washington State, *Canadian Journal of Earth Sciences*, **32**, 1194–1207.

Libby, W.F., 1952 (1955), *Radiocarbon Dating*. 2nd edition (Chicago: Chicago University Press).

Lillieskóld, M. and Sundqvist, B., 1994, Weathering of surface clasts as an indicator of the relative age of glacial deposits on Jameson Land, East Greenland, *Boreas*, **23**, 473–478.

Lindeberg, G. and Ringberg, B., 1999, Image analysis of rhythmites in proximal varves in Blekinge, southeastern Sweden, *Geologiska Föreningens i Stockholm Förhandlingar*, **121**, 182–186.

Litt, T., Schmincke, H.-U. and Kromer, B., 2003, Environmental response to climatic and volcanic events in central Europe during the Weichselian Lateglacial, *Quaternary Science Reviews*, **22**, 7–32.

Liu, B., Phillips, F.M., Pohl, M.M. and Sharma, P., 1996, An alluvial surface chronology based on cosmogenic ^{36}Cl dating, Ajo Mountains (Organ Pipe Cactus National Monument), southern Arizona, *Quaternary Research*, **45**, 30–37.

Liu, T., Ding, Z. and Rutter, N., 1999, Comparison of Milankovitch periods between continental loess and deep sea records over the last 2.5 Ma, *Quaternary Science Reviews*, **18**, 1205–1212.

Locock, M., 1999, Buried soils of the Wentlooge Formation, *Archaeology in the Severn Estuary*, **10**, 1–10.

Locock, M., Robinson, S. and Yates, A., 1998, Late Bronze Age sites at Cabot Park, Avonmouth, *Archaeology in the Severn Estuary*, **9**, 31–36.

Long, A., 1998, Attempt to affect the apparent ^{14}C age of cotton by scorching in a CO_2 environment, *Radiocarbon*, **40**, 57–58.

Lotter, A.F. and Lemcke, G., 1999, Methods for preparing and counting biochemical varves, *Boreas*, **28**, 243–252.

Lowe, D.J., 1988, Late Quaternary volcanism in New Zealand: towards an integrated record using distal airfall tephras in lakes and bogs, *Journal of Quaternary Science*, **3**, 111–120.

Lowe, D.J., McFadgen, B.G., Higham, T.F.G., Hogg, A.G., Froggatt, P.C. and Nairn, I.A., 1998, Radiocarbon age of the Kaharoa Tephra, a key marker for Late-Holocene stratigraphy and archaeology in New Zealand, *The Holocene*, **8**, 487–495.

Lowe, J.J., 2001, Abrupt climatic changes in Europe during the last glacial–interglacial transition; the potential for testing hypotheses on the synchroneity of climatic events using tephrochronology, *Global and Planetary Change*, **30**, 73–84.

Lowe, J.J. and Walker, M.J.C., 1997, *Reconstructing Quaternary Environments*. 2nd edition (London: Pearson International).

Lowe, J.J., Lowe, S., Fowler, A.J., Hedges, R.E.M. and Austin T.J.F., 1988, Comparison of accelerator and radiometric age measurements obtained from Late Devensian Lateglacial lake deposits from Llyn Gwernan, North Wales, *Boreas*, **17**, 355–369.

Lowe, J.J., Hoek, W. and INTIMATE group, 2001, Inter-regional correlation of palaeoenvironmental changes during the last glacial–interglacial transition: a protocol for improved precision recommended by the INTIMATE project group, *Quaternary Science Reviews*, **20**, 1175–1187.

Lücke, A., Schleser, G.H., Zolitschka, B. and Negendank, J.F.W., 2003, A Lateglacial and Holocene organic carbon isotope record of lacustrine palaeoproductivity and climate change derived from varved lake sediments of Lake Holzmaar, Germany, *Quaternary Science Reviews*, **22**, 569–580.

Ludwig, K.R. and Paces, J.B., 2002, Uranium-series dating of pedogenic-silica and carbonate, Crater Flat, Nevada, *Geochimica et Cosmochimica Acta*, **66**, 487–506.

Lund, S.P., 1996, A comparison of palaeomagnetic secular variation records from North America, *Journal of Geophysical Research*, **101B**, 8008–8024.

Lundqvist, J. and Wohlfarth, B., 2001, Timing and east–west correlation of south Swedish ice marginal lines during the Late Weichselian, *Quaternary Science Reviews*, **20**, 1127–1148.

Machida, H., 1981, Tephrochronology and Quaternary studies in Japan, in S. Self, R.S.J. Sparks (eds), *Tephra Studies*, 161–191 (Dordrecht: Reidel).

Mackie, E.A.V., Davies, S.M., Turney, C.S.M., Dobby, K., Lowe, J.J. and Hill, P.G.L., 2002, The use of magnetic separation techniques to detect basaltic microtephra in last glacial–interglacial transition (LGIT; 15–10 ka cal. BP) sediment sequences in Scotland, *Scottish Journal of Geology*, **38**, 21–30.

Mahaney, W.C., Halvorsen, D.L., Piegat, J. and Sanmugadas, K., 1984, Evaluation of dating methods used to assign ages in the Wind River and Teton Ranges, western Wyoming, in W.C. Mahaney (ed.), *Quaternary Dating Methods*, 355–374 (Amsterdam: Elsevier).

Mangerud, J., Løvlie, R., Gulliksen, S., Hufthammer, A.-K., Larsen, E. and Valen, V., 2003, Palaeomagnetic correlations between Scandinavian ice-sheet fluctuations and Greenland Dansgaard-Oeschger events, 45,000–25,000 yr BP, *Quaternary Research*, **59**, 213–222.

Mankinen, E.A. and Dalrymple, G.B., 1979, Revised geomagnetic polarity time scale for the interval 0–5 M.y BP, *Journal of Geophysical Research*, **84**, 615–626.

Marchitto, T.A., Jones, G.A., Goodfriend, G.A. and Weidman, C.R., 2000, Precise temporal correlation of Holocene mollusk shells using sclerochronology, *Quaternary Research*, **53**, 236–246.

Marcolini, F., Bigazzi, G., Bonadonna, F.P., Centamore, E., Ciomi, R. and Zanchetta, G., 2003, Tephrochronology and tephrostratigraphy of two Pleistocene continental fossiliferous successions from central Italy, *Journal of Quaternary Science*, **18**, 545–556.

Markewich, H.W., Wysocki, D.A., Pavich, M.J., Rutledge, E.M., Millard, H.T., Rich, F.J., Maat, P.B., Rubin, M. and McGeehin, J.P., 1998, Paleopedology plus TL, ^{10}Be and ^{14}C dating as tools in stratigraphic and paleoclimatic investigations, Mississippi River valley, USA, *Quaternary International*, **51/52**, 143–167.

Marshall, E., 1990, Racemisation dating: great expectations, *Science*, **247**, 799.

Marshall, E., 2001, Pre-Clovis sites fight for acceptance, *Science*, **291**, 1730–1732.

Martinson, D.G., Pisias, N.G., Hayes, J.D., Imbrie, J., Moore, T.C. and Shackleton, N.J., 1987, Age dating and the orbital theory of ice ages: development of a high-resolution 0–300 000 year chronostratigraphy, *Quaternary Research*, **27**, 1–29.

Matthews, J.A., 1985, Radiocarbon dating of surface and buried soils: principles, problems and prospects, in K.S. Richards, R.R. Arnett, S. Ellis (eds), *Geomorphology and Soils*, 271–288 (London: Allen & Unwin).

Matthews, J.A., 1991, The late Neoglacial ('Little Ice Age') glacier maximum in southern Norway: new ^{14}C-dating evidence and climatic implications, *The Holocene*, **1**, 219–233.

Matthews, J.A., 1992, *The Ecology of Recently Deglaciated Terrain* (Cambridge: Cambridge University Press).

Matthews, J.A., 1994, Lichenometric dating: a review with particular reference to 'Little Ice Age' moraines in southern Norway, in C. Beck (ed.), *Dating in Exposed and Surface Contexts*, 185–212 (Albuquerque: University of New Mexico Press).

Mayewski, P.A. and White, F., 2002, *The Ice Chronicles* (Hanover and London: University Press of New England).

Mazeaud, A., Laj, C., Bard, E., and Tric, E., 1992, A geomagnetic calibration of the radiocarbon timescale, in E. Bard and W. Broecker (eds), *The Last Deglaciation: Absolute and Relative Chronologies, NATO ASI Series 1,2*, 163–169 (Berlin: Springer-Verlag).

McBrearty, S. and Brooks, A.S., 2000, The revolution that wasn't; a new interpretation of modern human behaviour, *Journal of Human Evolution*, **39**, 453–563.

McCarroll, D., 1991, Relative-age dating of inorganic deposits: the need for a more critical approach, *The Holocene*, **1**, 174–180.

McCarroll, D., 1994, The Schmidt hammer as a measure of degree of rock surface weathering and terrain age, in C. Beck (ed.), *Dating in Exposed and Surface Contexts*, 29–45 (Albuquerque: University of New Mexico Press).

McCarroll, D., 2002, Amino-acid geochronology and the British Pleistocene: secure stratigraphical framework or a case of circular reasoning, *Journal of Quaternary Science*, **17**, 647–651.

McCarroll, D., Shakesby, R.A. and Matthews, J.A., 1998, Spatial and temporal patterns of Late Holocene rockfall activity on a Norwegian talus slope a lichenometric and simulation-modelling approach, *Arctic and Alpine Research*, **30**, 51–60.

McCarroll, D., Jalkanen, R., Hicks, S., Tuovinen, M., Gagen, M., Pawellek, F., Eckstein, D., Schmitt, U., Autio, J. and Heikkinen, O., 2003, Multiproxy dendroclimatology: a pilot study in northern Finland, *The Holocene*, **13**, 829–838.

McDougall, I. and Harrison, T.M., 1999, *Geochronology and Thermochronology by the* $^{40}Ar/^{39}Ar$ *Method*. 2nd edition (Oxford: Oxford University Press).

McGarry, S. and Baker, A., 2000, Organic acid fluorescence: applications to speleothem palaeoenvironmental reconstruction, *Quaternary Science Reviews*, **19**, 1087–1101.

McLaren, S.J. and Rowe, P.J., 1996, The reliability of uranium-series mollusc dates from the western Mediterranean basin, *Quaternary Science Reviews*, **15**, 709–717.

McLaren, S. and Gardner, R., 2000, New radiocarbon dates from a Holocene aeolianite, Isla Cancun, Quintana Roo, Mexico, *The Holocene*, **10**, 757–761.

Meese, D.A., Gow, A.J., Alley, R.B., Zielinski, G.A., Grootes, P.M., Ram, M., Taylor, K.C., Mayewski, P.A. and Bolzan, J.F., 1997, The Greenland Ice Sheet Project 2 depth-age scale: methods and results, *Journal of Geophysical Research*, **102**, 26 411–26 423.

Mellars, P. and Dark, P., 1998, *Star Carr in Context* (Cambridge: McDonald Institute Monograph).

Mensing, S.A. and Southon, J.R., 1999, A simple method to separate pollen for AMS radiocarbon dating and its application to lacustrine and marine sediments, *Radiocarbon*, **41**, 1–8.

Mercier, N., Valladas, H. and Valladas, G., 1995, Flint thermoluminescence dates from the CFR Laboratory at Gif: contributions to the study of the chronology of the Middle Pleistocene, *Quaternary Science Reviews*, **14**, 351–364.

Meyer, E., Sarna-Wojcicki, A.M., Hillhouse, J.W., Woodward, M.J., Slate, J.L. and Sorg, D.H., 1991, Fission-track age (400,000 yr) of the Rockland Tephra, based on inclusions of zircon grains lacking fossil fission tracks, *Quaternary Research*, **35**, 367–382.

Meyrick, R., 2003, Holocene molluscan faunal history and environmental change at Kloster Mühle, Rheinland-Pfalz, western Germany, *Journal of Quaternary Science*, **18**, 121–132.

Millard, A., 2001, The deterioration of bone, in D.E. Brothwell, A.M. Pollard (eds), *Handbook of Archaeoological Science*, 637–647 (Chichester and New York: John Wiley).

Millard, A.R. and Hedges, R.E.M., 1995, The role of the environment in the uranium uptake by buried bone, *Journal of Archaeological Science*, **22**, 239–250.

Millard, A.R. and Hedges, R.E.M., 1996, A diffusion-absorption model of uranium uptake by archaeological bone, *Geochimica et Cosmochimica Acta*, **60**, 2139–2152.

Miller, G.H. and Mangerud, J., 1985, Aminostraigraphy of European marine interglacial deposits, *Quaternary Science Reviews*, **4**, 279–318.

Miller, G.H., Beaumont, P.B., Deacon, H.J., Brooks, A.S., Hare, P.E. and Jull, A.J.T., 1999, Earliest modern humans in southern Africa dated by isoleucine epimerization in ostrich eggshell, *Quaternary Science Reviews*, **18**, 1537–1548.

Miracle, P., 2002, Mesolithic middens, in P. Miracle, N. Milner (eds), *Consuming Passions and Patterns of Consumption*, 65–88 (Cambridge: McDonald Institute Monograph).

Mithen, S. and Reed, M., 2002, Stepping out: a computer simulation of hominid dispersal from Africa, *Journal of Human Evolution*, **43**, 433–462.

Molodkov, A., 2001, ESR dating evidence for early man at a Lower Palaeolithic cave-site in the northern Caucasus as derived from terrestrial mollusc shells, *Quaternary Science Reviews*, **20**, 1051–1055.

Molloy, K. and O'Connell, M., 1993, Early land use and vegetation history at Derryinver Hill, Renvyle Peninsula, Co. Galway, Ireland, in F.M. Chambers (ed.), *Climatic Change and Human Impact on the Landscape*, 185–199 (London: Chapman & Hall).

Mook, W.G., 1986, Recommendations/resolutions adopted by the Twelfth International Radiocarbon Conference, *Radiocarbon*, **28(2A)**, 799.

Morgenstern, U., Taylor, C.B., Parrat, Y., Gäggeler, H.W. and Eicher, B., 1996, ^{32}Si in precipitation: evaluation of temporal and spatial variation and as a dating tool for glacier ice, *Earth and Planetary Science Letters*, **144**, 289–296.

Morgenstern, U., Fifield, L.K. and Zondervan, A., 2000, New frontiers in glacial ice dating: measurement of natural ^{32}Si by AMS, *Nuclear Instruments and Methods*, **B172**, 605–609.

Morgenstern, U., Geyh, M.A., Kudrass, H.R., Ditchburn, R.G. and Graham, I.J., 2001, ^{32}Si dating of marine sediments from Bangladesh, *Radiocarbon*, **43**, 909–916.

Morrison, R.B., 1978, Quaternary soil stratigraphy – concepts, methods and problems, in W.C. Mahaney (ed.), *Quaternary Soils*, 77–108 (Norwich: GeoAbstracts).

Muhs, D.R., Simmons, K.R. and Steinke, B., 2002, Timing and warmth of the Last Interglacial period: new U-series evidence from Hawaii and Bermuda and a new fossil compilation for North America, *Quaternary Science Reviews*, **21**, 1355–1383.

Muhs, D.R., Ager, T.A., Bettis III, E.A., McGeehin, J., Been, J.M., Begét, J.E., Pavich, M.J., Stafford, T.W. Jr, and Stevens, D.A.S.P., 2003, Stratigraphy and palaeoclimatic significance of Late Quaternary loess-palaeosol sequences of the Last Interglacial–Glacial cycle in central Alaska, *Quaternary Science Reviews*, **22**, 1947–1986.

Murray, A.S. and Funder, S., 2003, Optically stimulated luminescence dating of a Danish Eemian coastal marine deposit: a test of accuracy, *Quaternary Science Reviews*, **22**, 1177–1183.

Murray-Wallace, C.V., Belperio, A.P., Picker, K. and Kimber, R.W.L., 1991, Coastal aminostratigraphy of the Last Interglaciation in southern Australia, *Quaternary Research*, **35**, 63–71.

Nakamura, T., Taniguchi, Y., Tsuji, S. and Oda, H., 2001, Radiocarbon dating of charred residues on the earliest pottery in Japan, *Radiocarbon*, **43**, 1129–1138.

Nayling, N., 1995, The excavation, recovery and provisional analysis of a medieval wreck from Magor Pil, South Wales, *Archaeology in the Severn Estuary*, **6**, 85–96.

Newnham, R.M., Eden, D.N., Lowe, D.J. and Hendy, C.H., 2003, Rerewhakaaitu Tephra a land-sea marker for the Last Termination in New Zealand, with implications for global climate change, *Quaternary Science Reviews*, **22**, 289–308.

Niedermann, S., 2002, Cosmic-ray-produced noble gases in terrestrial rocks: dating tools for surface processes, *Reviews in Mineralogy and Geochemistry*, **47**, 731–784.

Nijampurkar, V.N. and Rao, D.K., 1992, Accumulation and flow rates of ice on Chota Shigri glacier, central Himalaya, using radioactive and stable isotopes, *Journal of Glaciology*, **38**, 43–50.

Nijampurkar, V.N., Rao, D.K., Oldfield., F. and Renberg, I., 1998, The half-life of ^{32}Si: a new estimate based on varved lake sediments, *Earth and Planetary Science Letters*, **163**, 191–196.

Nöel, M. and Batt, C.M., 1990, A method for correcting geographically separated remanence directions for the purpose of archaeomagnetic dating, *Geophysical Journal International*, **102**, 753–756.

North American Commission on Stratigraphic Nomenclature, 1983, North American Stratigraphic Code, *American Association of Petroleum Geologists' Bulletin*, **67**, 841–875.

North Greenland Ice Core Project Members, 2004, High-resolution record of northern hemisphere climate extending into the last interglacial period, *Nature*, **431**, 147–151.

Oakley, K.P., 1969, Analytical methods of dating bones, in D.E. Brothwell, E. Higgs (eds), *Science in Archaeology*, 35–45 (London: Thames & Hudson).

Ó Cofaigh, C. and Dowdeswell, J.A., 2001, Laminated sediments in glacimarine environments: diagnostic criteria for their interpretation, *Quaternary Science Reviews*, **20**, 1411–1436.

Oda, H. and Nakamura, T., 1998, ^{14}C dating of ancient Japanese documents, *Radiocarbon*, **40**, 701–705.

Oeschger, H. and Langway, C.C. Jr (eds), 1989, *The Environmental Record in Ice Sheets and Glaciers* (Chichester and New York: John Wiley).

Ogden, J., Wilson, A., Hendy, C. and Newnham, R.M., 1992, The late Quaternary history of kauri (*Agathis australis*) in New Zealand and its climatic significance, *Journal of Biogeography*, **19**, 611–622.

Ogden, J., Newnham, R.M., Palmer, J.G., Serra, R.G. and Mitchell, N., 1993, Climatic implications of macro- and microfossil assemblages from Late Pleistocene deposits in northern New Zealand, *Quaternary Research*, **39**, 107–119.

Ojala, A.E.K. and Francus, P., 2002, X-ray densitrometry vs BSE-image analysis of thin sections; a comparative study of varved sediments of Lake Nautajärvi, Finland, *Boreas*, **31**, 57–64.

Ojala, A.E.K. and Saarinen, T., 2002, Palaeosecular variation of the earth's magnetic field during the last 10000 years based on the annually laminated sediment of Lake Nautajärvi, central Finland, *The Holocene*, **12**, 391–400.

Ojala, A.E.K. and Tiljander, M., 2003, Testing the fidelity of sediment chronology: comparison of varve and palaeomagnetic results from Holocene lake sediments from central Finland, *Quaternary Science Reviews*, **22**, 1787–1803.

Oldfield, F., Appleby, P.G. and Battarbee, R.W., 1978, Alternative ^{210}Pb dating: results from New Guinea Highlands and Lough Erne, *Nature*, **271**, 339–342.

Oldfield, F., Richardson, N. and Appleby, P.G., 1995, Radiometric dating (^{210}Pb, ^{137}Cs, ^{241}Am) of recent ombrotrophic peat accumulation and evidence for changes in mass balance, *The Holocene*, **5**, 141–148.

Oldfield, F., Wake, R., Boyle, J., Jones, R., Nolan, S., Gibbs, Z., Appleby, P., Fisher, E. and Wolff, G., 2003, The Late-Holocene history of Gormire Lake (NE England) and its catchment: a multiproxy reconstruction of past human impact, *The Holocene*, **13**, 677–690.

Olsson, I.U., 1986, Radiometric dating, in B.E. Berglund (ed.), *Handbook of Holocene Palaeoecology and Palaeohydrology*, 273–312 (Chichester and New York: John Wiley).

O'Neal, M.A. and Schoenenberger, K.R., 2003, A *Rhizocarpon geographicum* growth curve for the Cascade Range of Washington and northern Oregon, USA, *Quaternary Research*, **60**, 233–241.

Oppenheimer, C., 2002, Limited global change due to the largest known Quaternary eruption: Toba ~74 kyr BP? *Quaternary Science Reviews*, **21**, 1593–1609.

Oppenheimer, S., 2003, *Out of Eden* (London: Constable and Robinson).

O'Sullivan, P.E., 1983, Annually-laminated lake sediments and the study of Quaternary environmental changes – a review, *Quaternary Science Reviews*, **1**, 245–313.

Owen, L.A., Spencer, J.Q., Haizhou, M., Barnard, P.L., Derbyshire, E., Finkel, R.C., Caffee, M.W. and Nian, Y., 2003, Timing of Late Quaternary glaciation along the southwestern slopes of the Qilian Shan, Tibet, *Boreas*, **32**, 281–291.

Parés, J.M. and Pérez-González, A., 1999, Magnetochronology and stratigraphy at Gran Dolina section, Atapuerca (Burgos, Spain), *Journal of Human Evolution*, **37**, 325–342.

Partridge, T.C., Shaw, J., Heslop, D. and Clarke, R.J., 1999, The new hominid skeleton from Sterkfontein, South Africa: age and preliminary assessment, *Journal of Quaternary Science*, **14**, 293–298.

Patience, A.J. and Kroon, D., 1991, Oxygen isotope chronostratigraphy, in P.L. Smart, P.D. Frances (eds), *Quaternary Dating Methods: A User's Guide*. Technical Guide **4**, Quaternary Research Association, Cambridge, 199–228.

Paulsen, D.E., Li, H.-C. and Ku, T.-L., 2003, Climate variability in central China over the last 1270 years revealed by high-resolution stalagmite records, *Quaternary Science Reviews*, **22**, 691–701.

Peacock, J.D., 1996, Marine molluscan proxy data applied to Scottish late glacial and Flandrian sites: strengths and limitations, in J.T. Andrews, W.E.N. Austin, H. Bergsten, A.E. Jennings (eds), *Late Quaternary Palaeoceanography of the North Atlantic Margins*. Geological Society of London Special Publication, **111**, 215–228.

Peacock, J.D., 1999, The pre-Windermere Interstadial (Late Devensian) raised marine strata of eastern Scotland and their macrofauna: a review, *Quaternary Science Reviews*, **18**, 1655–1680.

Peacock, J.D. and Harkness, D.D., 1990, Radiocarbon ages and the full-glacial to Holocene transition in seas adjacent to Scotland and southern Scandinavia: a review, *Transactions of the Royal Society of Edinburgh: Earth Science*, **81**, 385–396.

Pearce, N.J.G., Eastwood, A.J., Westgate, J.A. and Perkins, W.T., 2002, Trace-element composition of single glass shards in distal Minoan Tephra from SW Turkey, *Journal of the Geological Society, London*, **159**, 545–556.

Peglar, S.M., Fritz, S.C. and Birks, H.J.B., 1989, Vegetation and land-use history at Diss, Norfolk, *Journal of Ecology*, **77**, 203–222.

Penck, A. and Bruckner, E., 1909, *Die Alpen im Eiszeitalter* (Leipzig: Tachnitz).

Penrose, R., 1999, *The Emperor's New Mind* (Oxford: Oxford University Press).

Peteet, D.M., Vogel, J.S., Nelson, D.E., Southon, J.R., Nickmann, R.J. and Heusser, L.E., 1990, Younger Dryas climatic reversal in northeastern USA? AMS ages for an old problem, *Quaternary Research*, **33**, 219–230.

Peterson, L.C., Huang, G.H., Hughen, K.A. and Röhl, U., 2001, Rapid changes in the hydrological cycle of the tropical Atlantic during the last glacial, *Science*, **290**, 1947–1951.

Petit, J.R., Jouzel, J., Raynaud, D., Barkov, N.I., Basile, I., Bender, M., Chapellaz, J., Davis, J., Delaygue, G., Delmotte, M., Kotlyakov, V.M., Legrand, M., Lipkenov, V., Lorius, C., Pépin, L., Ritz, C., Saltzman, E. and Stievenard, M., 1999, 420,000 years of climate and atmospheric history revealed by the Vostok deep Antarctic ice core, *Nature*, **399**, 429–436.

Phillips, F.M., Zreda, M.G., Smith, S.S., Elmore, D., Kubik, P.W. and Sharma, P., 1990, Cosmogenic chlorine-36 chronology for glacial deposits at Bloody Canyon, eastern Sierra Nevada, *Science*, **248**, 1529–1532.

Phillips, F.M., Flinsch, M., Elmore, D. and Sharma, P., 1997, Maximum ages of the Côa valley (Portugal) engravings measured with chlorine-36, *Antiquity*, **71**, 100–104.

Phillips, W.M., Sloan, V.F., Shroder, J.F., Sharma, P., Clarke, M.L. and Rendell, H.M., 2000, Asynchronous glaciation at Nanga Parbat, northwestern Himalaya Mountains, Pakistan, *Geology*, **28**, 431–434.

Pierce, K.L., Obradovich, J.D. and Friedman, I., 1976, Obsidian hydration correlation and dating of Bull Lake and Pinedale Glaciations near West Yellowstone, Montana, *Geological Society of America Bulletin*, **87**, 703–710.

Pike, A.W.G. and Hedges, R.E.M., 2001, Sample geometry and U uptake in archaeological teeth: implications for U-series and ESR dating, *Quaternary Science Reviews*, **20**, 1021–1025.

Pike, A.W.G. and Pettitt, P.B., 2003, U-series dating and human evolution, in B. Bourdon, G.M. Henderson, C.C. Lundstrom and S. Turner (eds), *Uranium Series Geochemistry, Reviews in Mineralogy and Geochemistry*, **52**, 607–630.

Pike, A.W.G., Hedges, R.E.M. and van Calsteren, P., 2002, U-series dating of bone using the diffusion-absorption model, *Geochimica et Cosmochimica Acta*, **66**, 4273–4286.

Pilcher, J.R. and Hall, V.A., 1992, Towards a Holocene tephrochronology for the north of Ireland, *The Holocene*, **2**, 255–259.

Pirazzoli, P.A., Radtke, U., Hantoro, W.S., Jouannic, C., Hoang, C.T., Causse, C. and Borel-Best, M., 1991, Quaternary raised coral reef terraces on Sumba Island, Indonesia, *Science*, **252**, 1834–1836.

Pollard, A.M., Blockley, S.P.E. and Ward, K.R., 2003, Chemical alteration of tephra in the depositional environment: theoretical stability modelling, *Journal of Quaternary Science*, **18**, 385–394.

Preece, R.C., Kemp, R.A. and Hutchinson, J.A., 1995, A Late-glacial colluvial sequence at Watcombe Bottom, Ventnor, Isle of Wight, England, *Journal of Quaternary Science*, **10**, 107–121.

Preece, S.J., Westgate, J.A., Alloway, B.V. and Milner. M.W., 2000, Characterisation, identity, distribution, and source of late Cenozoic Tephra beds in the Klondike district of the Yukon, Canada, *Canadian Journal of Earth Sciences*, **37**, 983–996.

Prescott, J.R. and Hutton, J.T., 1994, Cosmic ray contributions to dose rates for luminescence and ESR dating: large depths and long-time variations, *Radiation Measurements*, **23**, 497–500.

Prescott, J.R., Huntley, D.J. and Hutton, J.T., 1993, Estimation of equivalent dose in thermoluminescence dating – *the Australian slide method, Ancient TL*, **11**, 1–5.

Putkonen, J. and Swanson, T., 2003, Accuracy of cosmogenic ages for moraines, *Quaternary Research*, **59**, 255–261.

Qin, X., Tan, M., Liu, T., Wang, X., Li, T. and Lu, J., 1999, Spectral analysis of a 1000-year stalagmite lamina thickness record from Shihua Cavern, Beijing, China, and its climatic significance, *The Holocene*, **9**, 689–694.

Quinn, T.M., Taylor, F.W. and Crowley, T.J., 1993, A 173 year stable isotope record from a tropical south Pacific coral, *Quaternary Science Reviews*, **12**, 407–418.

Radtke, U., Schellman, G., Scheffers, A., Kelletat, D., Kromer, B. and Kasper, H.U., 2003, Electron spin resonance and radiocarbon dating of coral deposited by Holocene tsunami events on Curaçao, Bonaire and Aruba (Netherlands Antilles), *Quaternary Science Reviews*, **22**, 1309–1315.

Raynaud, D., Barnola, J.-M., Chapellaz, J., Blunier, T., Indermühle, A. and Stauffer, B., 2000, The ice core record of greenhouse gases: a view in the context of future changes, *Quaternary Science Reviews*, **19**, 9–17.

Rees, M., 2000, *Just Six Numbers* (London: Phoenix).

Reider, R.G., 1983, A soil catena in the Medicine Bow Mountains, Wyoming, USA, with reference to palaeoenvironmental influences, *Arctic and Alpine Research*, **15**, 181–192.

Renberg, I., 1981, Formation, structure and visual appearance of iron-rich, varved lake sediments, *Verhandlungen Internationalen Vereinigung für Limnologie*, **21**, 94–101.

Reneau, S.L. and Raymond, R.J., 1991, Cation-ratio dating of rock varnish: why does it work? *Geology*, **19**, 937–940.

Renfrew, C., 1973, *Before Civilisation: The Radiocarbon Revolution and Prehistoric Europe*, (London: Jonathan Cape).

Renne, P.R., Sharp, W.D., Deino, A.L., Orsi, G. and Civetta, L., 1997, $^{40}Ar/^{39}Ar$ dating into the Historical Realm: calibration against Pliny the Younger, *Science*, **277**, 1279–1280.

Richards, D.A. and Smart, P.L., 1991, Potassium–argon and argon–argon dating, in P.L. Smart, P.F.D. Frances (eds), *Quaternary Dating Methods – A User's Guide*. Technical Guide **4**, Quaternary Research Association, Cambridge, 37–44.

Richards, D.A. and Beck, J.W., 2001, Dramatic shifts in atmospheric radiocarbon during the Last Glacial period, *Antiquity*, **75**, 482–485.

Richards, D.A. and Dorale, J.A., 2003, U-series chronology and environmental applications of speleothems, in B. Bourdon, G.M. Henderson, C.C. Lundstrom, S. Turner (eds), *Uranium Series Geochemistry, Reviews in Mineralogy and Geochemistry*, **52**, 407–460.

Richmond, G.M. and Fullerton, D.S., 1986, Summation of Quaternary glaciations in the United States of America, *Quaternary Science Reviews*, **5**, 183–196.

Riciputi, L.R., Elam, J.M., Anovitz, L.M. and Cole, D.R., 2002, Obsidian diffusion dating by secondary ion mass spectrometry: a test using results from Mound 65, Chalco, Mexico, *Journal of Archaeological Science*, **19**, 1055–1075.

Ridings, R., 1996, Where in the world does obsidian hydration work? *Antiquity*, **61**, 136–148.

Ringberg, B., 1984, Cyclic lamination in proximal varves reflecting the length of summer during the Late Weichsel in southernmost Sweden, in N.A. Mörner, W. Karlén (eds), *Climatic Changes on a Yearly to Millennial Basis*, 57–62 (Dordrecht: Reidel).

Rink, W.J., 1997, Electron spin resonance (ESR) dating and ESR applications in Quaternary science and archaeometry, *Radiation Measurements*, **27**, 975–1025.

Rippon, S., 1996, *The Gwent Levels: The Evolution of a Wetland Landscape.* Council for British Archaeology, Research Report 1105, York.

Rittenour, T.M., Goble, R.J. and Blum, M.D., 2003, An optical age chronology of Late Pleistocene fluvial deposits in the northern lower Mississippi Valley, *Quaternary Science Reviews*, **22**, 1105–1110.

Roberts, N., 1998, *The Holocene*. 2nd edition (Oxford: Blackwell).

Roberts, R.G., 1997, Luminescence dating in archaeology: from origins to optical, *Radiation Measurements*, **27**, 819–892.

Roberts, R.G., Bird, M., Olley, J., Galbraith, R., Lawson, E., Laslett, G., Yoshida, H., Jones, R., Fullager, R., Jacobsen, G. and Hau, Q., 1998, Optical and radiocarbon dating at Jinmium rock shelter in northern Australia, *Nature*, **393**, 358–362.

Roberts, R.G., Galbraith, R., Olley, J.M., Yoshida, H. and Laslett, G.M., 1999, Optical dating of single and multiple grains of quartz from Jinmium rock shelter, northern Australia. Part II. Results and implications, *Archaeometry*, **41**, 365–395.

Robinson, D.A. and Williams, R.B.G. (eds), 1994, *Rock Weathering and Landform Evolution* (Chichester and New York: John Wiley).

Rodbell, D.T., 1990, Soil–age relationships on Late Quaternary moraines, Arrowsmith Range, Southern Alps, New Zealand, *Arctic and Alpine Research*, **22**, 355–365.

Rodbell, D.T., 1993, Subdivision of Late Pleistocene moraines in the Cordillera Blanca, Peru, based on rock-weathering features, soils and radiocarbon dates, *Quaternary Research*, **39**, 133–143.

Roe, F., 1990, Comments on non-local stone, in J. Richards (ed.), *The Stonehenge Environs Project*. Report No. 16, HMSO, London, 229–231.

Roebroecks, W. and von Kolfschoten, T., 1994, The earliest occupation of Europe: a short chronology, *Antiquity*, **68**, 489–503.

Roos-Barraclough, F., van der Knaap, W.O., van Leeuwen, J.F.N. and Shotyk, W., 2004, A Late-glacial and Holocene record of climate change from a Swiss peat humification profile, *The Holocene*, **14**, 7–20.

Rose, J., Boardman, J., Kemp, R.A. and Whiteman, C.A., 1985, Palaeosols and the interpretation of the British Quaternary stratigraphy, in K.S. Richards, R.R. Arnett, S. Ellis (eds), *Geomorphology and Soils*, 348–375 (London: George Allen & Unwin).

Rose, N.L., Golding, P.N.E. and Battarbee, R.W., 1996, Selective concentration and enumeration of tephra shards from lake sediment cores, *The Holocene*, **6**, 243–246.

Roucoux, K.H., Shackleton, N.J., Abreu, L., Schönield, J. and Tzedakis, P.C., 2001, Combined marine proxy and pollen analyses reveal rapid Iberian vegetation response to North Atlantic, *Quaternary Research*, **56**, 128–132.

Rowe, P.J., Atkinson, T.C. and Turner, C., 1999, U-series dating of Hoxnian interglacial deposits at Marks Tey, Essex, England, *Journal of Quaternary Science*, **14**, 693–702.

Ruddiman, W.F. and Raymo, M.E., 2003, A methane-based time scale for Vostok ice, *Quaternary Science Reviews*, **22**, 141–155.

Ruddiman, W.F. and Thompson, J.S., 2001, The case for human causes of increased atmospheric CH_4 over the last 5000 years, *Quaternary Science Reviews*, **20**, 1769–1777.

Saarinen, T., 1998, High-resolution palaeosecular variation in northern Europe during the last 3200 years, *Physics of the Earth and Planetary Interiors*, **106**, 301–311.

Saarinen, T., 1999, Palaeomagnetic dating of Late Holocene sediments in Fennoscandia, *Quaternary Science Reviews*, **18**, 889–897.

Saarnisto, M., 1986, Annually laminated lake sediments, in B.E. Berglund (ed.), *Handbook of Holocene Palaeoecology and Palaeohydrology*, 343–370 (Chichester and New York: John Wiley).

Saarnisto, M. and Kahra, A. (eds), 1992, *Laminated Sediments*. Geological Society of Finland, Special Paper 14, Helsinki.

Scharpenseel, H.W. and Becker-Heidemann, P., 1992, Twenty-five years of radiocarbon dating soils: paradigm erring and learning, *Radiocarbon*, **34**, 541–549.

Schellmann, G. and Radtke, U., 2001, Progress in ESR dating of Pleistocene corals – a new approach, *Quaternary Science Reviews*, **20**, 1015–1020.

Schöne, B.R., 2003, A 'clam-ring' master-chronology constructed from a short-lived bivalve mollusc from the northern Gulf of California, USA, *The Holocene*, **13**, 39–50.

Schöne, B.R., Lega, J., Flessa, K.W., Goodwin, D.H. and Dettman, D.L., 2002, Reconstructing daily temperatures from growth rates of the intertidal bivalve mollusk *Chione cortezi* (northern Gulf of California, Mexico), *Palaeogeography, Palaeoclimataology, Palaeoecology*, **184**, 131–146.

Schöne, B.R., Dunca, E., Mutvei, H. and Norlund, U., 2004, A 217-year record of summer air temperature reconstructed from freshwater pearl mussels (*M. margarifitera*, Sweden), *Quaternary Science Reviews*, **23**, 1803–1816.

Schreve, D.C., 2001, Differentiation of the British late Middle Pleistocene interglacials: the evidence from mammalian biostratigraphy, *Quaternary Science Reviews*, **20**, 1693–1716.

Schulman, E., 1956, *Dendroclimatic Changes in Semiarid America* (Arizona: University of Tucson).

Schwander, J., Jouzel, J., Hammer, C.U., Petit, J.-R., Udisti, R. and Wolff, E., 2001, A tentative chronology for the EPICA Dome Concordia, *Geophysical Research Letters*, **28**, 4243–4246.

Schweingruber, F.H., 1988, *Tree Rings. Basics and Applications of Dendrochronology* (Dordrecht: Reidel).

Schweingruber, F.H., 1996, *Tree Rings and Environment: Dendroecology*. (Bern: Paul Haupt).

Schwarcz, H.P., 1989, Uranium series dating of Quaternary deposits, *Quaternary International*, **1**, 7–17.

Schwarcz, H.P. and Latham, A.G., 1989, Dirty calcites 1. Uranium series dating of contaminated calcite using leachates alone, *Geochemical Geology (Isotope Geoscience)*, **80**, 35–43.

Scott, E.M. (ed.), 2003, The Fourth International Radiocarbon Intercomparison (FIRI), *Radiocarbon*, **45**, 135–291.

Scourse, J.D., 1997, Transport of the Stonehenge Bluestones: testing the glacial hypothesis, *Proceedings of the British Academy*, **92**, 271–314.

Scuderi, L.A., 1987, Glacier variations in the Sierra Nevada, California, as related to a 1200-year tree-ring chronology, *Quaternary Research*, **27**, 220–231.

Scuderi, L.A., 1990, Tree-ring evidence for climatically-effective volcanic eruptions, *Quaternary Research*, **34**, 67–85.

Sejrup, H.-P. and Haugen, J.-E., 1994, Amino-acid diagenesis in the marine bivalve *Arctica islandica* Linné from northwest European sites: only time and temperature? *Journal of Quaternary Science*, **9**, 301–309.

Sémah, F., Salekei, H. and Falguères, C., 2000, Did early man reach Java during the Late Pliocene? *Journal of Archaeological Science*, **27**, 763–769.

Severinghaus, J.P., Sowers, T., Brook, E.J., Alley, R.B. and Bender, M.L., 1998, Timing of abrupt climate change at the end of the Younger Dryas interval from thermally fractionated gases in polar ice, *Nature*, **391**, 141–146.

Shackleton, N.J. and Opdyke, N.D., 1973, Oxygen isotope and palaeomagnetic stratigraphy of Pacific core V28-238: oxygen isotope temperatures and ice volumes on a 10^5 and 10^6 year scale, *Quaternary Research*, **3**, 39–55.

Shackleton, N.J., Duplessy, J.-C., Arnold, M., Maurice, P., Hall, M.A. and Cartlidge, J., 1988, Radiocarbon ages of last glacial Pacific deep water, *Nature*, **335**, 708–711.

Shackleton, N.J., Berger, A. and Peltier, W.R., 1990, An alternative astronomical calibration of the lower Pleistocene timescale based on ODP Site 677, *Transactions of the Royal Society of Edinburgh: Earth Sciences*, **81**, 251–261.

Shackleton, N.J., Fairbanks, R.G., Chiu, T.-C. and Parrenin, F., 2004, Absolute calibration of the Greenland time scale: implications for Antarctic time scales and for $\Delta^{14}C$, *Quaternary Science Reviews*, **23**, 1513–1522.

Shane, P., 2000, Tephrochronology: a New Zealand case study, *Earth Science Reviews*, **49**, 223–259.

Sharp, W.D., Ludwig, K.R., Chadwick, O.A., Amundson, R. and Glaser, L.I., 2003, Dating fluvial terraces by $^{230}Th/U$ on pedogenic carbonate, Wind River Basin, Wyoming, *Quaternary Research*, **59**, 139–150.

Shen, G., Ku, T.-L., Cheng, H., Edwards, R.L., Yuan, Z. and Wang, Q., 2001, High-precision U-series dating of Locality 1 at Zhoukoudian, China, *Journal of Human Evolution*, **41**, 679–688.

Shiraiwa, T. and Watanabe, T., 1991, Late Quaternary glacial fluctuations in the Langtang Valley, Nepal Himalaya, reconstructed by relative dating methods, *Arctic and Alpine Research*, **23**, 404–416.

Siegert, M.J., 2001, *Ice Sheets and Late Quaternary Environmental Change* (Chichester and New York: John Wiley).

Sillen, A. and Parkington, J., 1996, Diagenesis of bones fron Eland's Bay Cave, *Journal of Archaeological Science*, **23**, 535–542.

Simola, H., 1992, Structural elements in varved lake sediments, in M. Saarnisto, A. Kahra (eds), *Laminated Sediments*. Geological Society of Finland, Special Paper 14, Helsinki, 5–10.

Singer, M.J., Fine, P., Verosub, K.L. and Chadwick, O.A., 1992, Time dependence of magnetic susceptibility of soil chronosequences on the Californian coast, *Quaternary Research*, **37**, 323–332.

Singer, R., Gladfelder, B.G. and Wymer, J.J., 1993, *The Lower Palaeolithic Site at Hoxne, England* Chicago: University of Chicago Press.

Singer, B.S., Ackert, R.P. and Hervé Guillou, 2004, $^{40}Ar/^{39}Ar$ and K-Ar chronology of Pleistocene glaciations in Patagonia, *Geological Society of America Bulletin*, **116**, 434–450.

Singhvi, A.K., Sharma, Y.P., Agrawal, D.P. and Dhir, R.P., 1982, Thermoluminescence dating of dune sands in Rajasthan, India, *Nature*, **295**, 313–315.

Smart, P.L., 1991a, Uranium series dating, in P.L. Smart, P.D. Frances (eds), *Quaternary Dating Methods – A User's Guide*. Technical Guide **4**, Quaternary Research Association, Cambridge, 45–83.

Smart, P.L., 1991b, Electron spin resonance (ESR) dating, in P.L. Smart, P.D. Frances (eds), *Quaternary Dating Methods – A User's Guide*. Technical Guide **4**, Quaternary Research Association, London, 128–160.

Smith, J., 1999, An introduction to the magnetic properties of natural minerals, in J. Walden, F. Oldfield, J. Smith (eds), *Environmental Magnetism: A Practical Guide*. Technical Guide No. **6**, Quaternary Research Association, London, 5–25.

Snowball, T. and Thompson, R., 1992, A mineral magnetic study of Holocene sediment yields and deposition patterns in the Llyn Geirionydd catchment, North Wales, *The Holocene*, **2**, 238–248.

Snowball, I. and Sandgren, P., 2002, Geomagnetic field variations in northern Sweden during the Holocene quantified from varved lake sediments and their implications for cosmogenic nuclide production rates, *The Holocene*, **12**, 517–530.

Snowball, I., Sandgren, P. and Petterson, G., 1999, The mineral magnetic properties of an annually laminated Holocene lake sediment sequence in northern Sweden, *The Holocene*, **9**, 353–362.

Sonnett, C.P. and Finney, S.A., 1990, The spectrum of radiocarbon, *Philosophical Transactions of the Royal Society*, **A330**, 413–426.

Sonninen, E. and Jungner, H., 2001, An improvement in preparation of mortar for radiocarbon dating, *Radiocarbon*, **43**, 271–273.

Spencer, J.Q. and Owen, L.A., 2004, Optically stimulated luminescence dating of Late Quaternary glaciogenic sediments in the upper Hunza Valley: validating the timing of glaciation and assessing dating methods, *Quaternary Science Reviews*, **23**, 175–191.

Spurk, M., Friedrich, M., Hofmann, J., Remmele, S., Frenzel, B., Leuschner, H.H. and Kromer, B., 1998, Revision and extensions of the Hohenheim Oak and Pine Chronologies – new evidence about the timing of the Younger Dryas/Preboreal-Transition, *Radiocarbon*, **40**, 1107–1116.

Stallings, W.S., 1937, Some early papers on tree-rings, *Tree-Ring Bulletin*, **3**, 27–28.

Stauffer, B., 1989, Dating of ice by radioactive isotopes, in H. Oeschger, C.C. Langway, Jr (eds), *The Environmental Record in Glaciers and Ice Sheets*, 123–139 (Chichester and New York: John Wiley).

Stedman, H.H., Kozyak, B.W., Nelson, A., Thesier, D.M., Su, L.T., Low, D.W., Bridges, C.R., Shrager, J.B., Minugh-Purvis, N. and Mitchell, M.A., 2004, Myosin gene mutation correlates with anatomical changes in the human lineage, *Nature*, **428**, 415–418.

Sternberg, R.S., 1997, Archaeomagnetic dating, in R.E. Taylor, M.J. Aitken (eds), *Chronometric Dating in Archaeology*, 323–356 (New York: Plenum Press).

Sternberg, R.S., 2001, Magnetic properties and archaeomagnetism, in D.R. Brothwell, A.M. Pollard (eds), *Handbook of Archaeological Sciences*, 73–79 (Chichester and New York: John Wiley).

Stevenson, C.M., Sheppard, P.J. and Sutton, D.G., 1996, Advances in the hydration dating of New Zealand obsidian, *Journal of Archaeological Science*, **23**, 233–242.

Stone, J.O., 2000, Air pressure and cosmogenic isotope production, *Journal of Geophysical Research*, **105**, 23, 723–753, 759.

Stott, A.W., Berstan, R., Evershed, P., Hedges, R.E.M., Bronk Ramsay, C. and Humm, M.J., 2001, Radiocarbon dating of single compounds isolated from pottery cooking vessel residues, *Radiocarbon*, **43**, 191–197.

Strauss, L.G. and Bar-Yosef, O. (eds), 2001, Out of Africa in the Pleistocene, *Quaternary International*, **75**, 1–130.

Street-Perrott, F.A., Huang, Y., Perrott, R.A., Eglinton, N.G., Barker, P., Khelifa, L.B., Harkness, D.D. and Olago, D.O., 1997, Impact of lower atmosphere carbon dioxide on tropical mountain ecosystems, *Science*, **278**, 1422–1426.

Stringer, C., 2003, Out of Ethiopia, *Nature*, **423**, 692–694.

Strömberg, B., 1994, Younger Dryas deglaciation at Mt Billingen, and clay varve dating of the Younger Dryas/Preboreal transition, *Boreas*, **23**, 177–193.

Stuart, F.M., 2001, *In situ* cosmogenic isotopes: principles and potential for archaeology, in D.E. Brothwell, A.M. Pollard (eds), *Handbook of Archaeological Sciences*, 93–100 (Chichester and New York: John Wiley).

Stuiver, M. and van der Plicht, J. (eds), 1998, INTCAL98 Calibration issue, *Radiocarbon*, **40**, 1041–1164.

Stuiver, M. and Grootes, P.M., 2000, GISP2 oxygen isotope ratios, *Quaternary Research*, **53**, 266–284.

Stuiver, M., Heusser, C.J. and Yang, C., 1978, North American glacial history extended to 75,000 years ago, *Science*, **200**, 16–21.

Stuiver, M., Brazunias, T.F., Becker, B. and Kromer, B., 1991, Climatic solar, oceanic and geomagnetic influences on Late-Glacial and Holocene atmospheric $^{14}C/^{12}C$ change, *Quaternary Research*, **35**, 1–24.

Stuiver, M., Reimer, P.J., Bard, E., Beck, J.W., Burr, G.S., Hughen, K.A., Kromer, B., McCormac, G., van der Plicht, M. and Spurk, M., 1998, INTCAL98 radiocarbon age calibration, 24,000-0 cal BP, *Radiocarbon*, **40**, 1041–1084.

Suckow, A., Morgenstern, U. and Kudrass, H.-R., 2001, Absolute dating of recent sediments in the cyclone-influenced shelf area off Bangladesh: comparison of gamma spectrometric (^{137}Cs, ^{210}Pb, ^{228}Ra), radiocarbon and ^{32}Si ages, *Radiocarbon*, **43**, 917–927.

Suess, H.E., 1970, Bristlecone pine calibration of the radiocarbon time-scale 5000 BC to the present, in I.U. Olsson (ed.), *Radiocarbon Variations and Absolute Chronology*, 303–311 (Chichester and New York: John Wiley).

Svensson, A., Nielsen, S.W., Kipfstuhl, S., Johnsen, S.J., Steffensen, J.P., Bigler, R., Ruth, R. and Röthlisberger, R., 2005, Visual stratigraphy of the NorthGRIP ice core during the last glacial period, *Journal of Geophysical Research*, **110**, D02108, 10.1029/2004JD005134.

Swart, P.K., Dodge, R.E. and Hudson, H.J., 1996, A 240-year stable oxygen and carbon isotope record in a coral from south Florida: implications for the prediction of precipitation in southern Florida, *Palaios*, **11**, 362–375.

Swisher, C.C. III, Curtis, G.H., Jacob, T., Getty, A.G., Suprijo, A. and Widiasmoro, 1994, Age of earliest known hominids in Java, Indonesia, *Science*, **263**, 1118–1121.

Sykes, G., 1991, Amino-acid dating, in P.L. Smart, P.D. Francis (eds), *Quaternary Dating Methods: A User's Guide*. Technical Guide **4**, Quaternary Research Association, London, 161–176.

Szabo, B.J., Ludwig, K.R., Muhs, D.R. and Simmons, K.R., 1994, Thorium-230 ages of corals and duration of the last interglacial sea-level high stand on Oahu, Hawaii, *Science*, **266**, 93–96.

Szabo, B.J., Bush, C.A. and Benson, L.V., 1996, Uranium-series dating of carbonate (tufa) deposits associated with Quaternary fluctuations of Pyramid Lake, Nevada, *Quaternary Research*, **45**, 271–281.

Tankersley, K.B., Schlecht, K.D. and Laub, R.S., 1998, Fluoride dating of mastodon bone from an early palaeoindian spring site, *Journal of Archaeological Science*, **25**, 805–811.

Tarling, D.H., 1983, *Palaeomagnetism: Principles and Applications in Geology, Geophysics and Archaeology* (London: Chapman & Hall).

Taylor, K.C., Lamorey, G.W., Doyle, G.A., Alley, R.B., Grootes, P.M., Mayewski, P.A., White, J.W.C. and Barlow, L.K., 1993, The 'flickering switch' of late Pleistocene climate change, *Nature*, **361**, 432–436.

Taylor, R.E., 1987, *Radiocarbon Dating: An Archaeological Perspective* (Orlando: Academic Press).

Taylor, R.E., 1992, Radiocarbon dating of bone: to collagen and beyond, in R.E. Taylor, R. Kra, A. Long (eds), *Radiocarbon After Four Decades: An Interdisciplinary Perspective*, 375–402 (New York: Springer-Verlag).

Taylor, R.E., 1997, Radiocarbon dating, in R.E. Taylor, M.J. Aitken (eds), *Chronometric Dating in Archaeology*, 65–96 (New York: Plenum Press).

Taylor, R.E., 2000, The contribution of radiocarbon dating to New World archaeology, *Radiocarbon*, **42**, 1–21.

Taylor, R.E., 2001, Radiocarbon Dating, in D.R. Brothwell, A.M. Pollard (eds), *Handbook of Archaeological Sciences*, 23–34 (Chichester and New York: John Wiley).

Taylor, R.E., Long, A. and Kra, R.S. (eds), 1992, *Radiocarbon Dating After Four Decades: An Interdisciplinary Perspective* (New York: Springer-Verlag).

Taylor, R.E., Hare, P.E., Prior, C.A., Kirner, D.L., Wan, L. and Burky, R.R., 1995, Radiocarbon dating of biochemically characterised hair, *Radiocarbon*, **37**, 319–330.

Thompson, L.G., 2000, Ice core evidence for climate change in the Tropics: implications for our future, *Quaternary Science Reviews*, **19**, 19–36.

Thompson, R. and Oldfield, F., 1986, *Environmental Magnetism* (London: Allen & Unwin).

Thompson, L.G., Yao, T., Davis, M.E., Henderson, K.A., Mosley-Thompson, E., Lin P.-L., Beer, J., Sybal, H-Al., Dai-Cole, J. and Bolzan, J.F., 1997, Tropical climate instability: the last glacial cycle from a Qinghai-Tibetan ice core, *Science*, **276**, 1821–1825.

Thorarinsson, S., 1981, The application of tephrochronology in Iceland, in S. Self, R.S.J. Sparks (eds), *Tephra Studies*, 109–134 (Dordrecht: Reidel).

Thouret, J.-C., Davila, J. and Eissen J-Ph., 1999, Largest historic explosive eruption in the Andes at Huaynaputina volcano, south Peru, *Geology*, **27**, 435–438.

Thouret, J.-C., Juvigné, E., Marino, J., Moscol, M., Loutch, I., Davila, J., Legeley-Padovani, A., Lamadon, S. and Rivera, M., 2002, Late Pleistocene and Holocene tephro-stratigraphy and chronology in southern Peru, *Boletin Sociedad Geologica del Peru*, **93**, 45–61.

Tinkler, K.J., 1985, *A Short History of Geomophology* (London: Croom Helm).

Tryon, C.A. and McBrearty, S., 2002, Tephrostratigraphy and the Acheulian to Middle Stone Age transition in the Kapthurin Formation, Kenya, *Journal of Human Evolution*, **42**, 211–235.

Tschudi, S., Ivy-Ochs, S., Schlüchter, C., Kubik, P. and Rainio, H., 2000, ^{10}Be dating of Younger Dryas Salpausselkä I formation in Finland, *Boreas*, **29**, 287–293.

Tudhope, A.W., Chilcott, C.P., McCulloch, M.T., Cook, E.R., Chappell, J., Ellam, R.M., Lea, D.W., Lough, J.M. and Shimmield, G.B., 2001, Variability in the El Niño-Southern Oscillation through a glacial–interglacial cycle, *Science*, **291**, 1511–1517.

Turner, G.M. and Thompson, R., 1981, Lake sediment record of the geomagnetic secular variation record in Britain during Holocene times, *Geophysical Journal of the Royal Astronomical Society*, **65**, 703–725.

Turney, C.S.M., 1998, Extraction of rhyolitic ash from minerogenic lake sediments, *Journal of Palaeolimnology*, **19**, 199–206.

Turney, C.S.M. and Lowe, J.J., 2001, Tephrochronology, in W.M. Last, J.P. Smol (eds), *Tracking Environmental Change using Lake Sediments: Volume 1. Basin Analysis, Coring and Chronological Techniques*, 451–471 (Dordrecht: Kluwer).

Turney, C.S.M., Lowe, J.J., Davies, S.M., Hal, V., Lowe, D.J., Wastegård, S., Hoek, W.Z., Alloway, B. and SCOTAV and INTIMATE members, 2004, Tephrochronology of Last Termination sequences in Europe: a protocol for improved analytical precision and robust correlation procedures (a joint SCOTAV-INTIMATE proposal), *Journal of Quaternary Science*, **19**, 111–120.

Valladas, H., Tisnérat-Laborde, N., Cachier, C., Arnodl, M., Bernaldo de Quirós, F., Cabrera-Valdés, V., Clottes, J., Courtin, J., Fortes-Pérez, J.J., Gonzáles-Sainz, C. and Moure-Romanillo, A., 2001, Radiocarbon AMS dates for Palaeolithic cave paintings, *Radiocarbon*, **43**, 977–986.

Valladas, H., Mercier, N., Joron, J.L., McPherron, S.P., Dibble, H.L. and Lenoir, M., 2003a, TL dates for the Middle Palaeolithic site of Combe-Capelle Bas, France, *Journal of Archaeological Science*, **30**, 1443–1450.

Valladas, H., Mercier, N., Michab, M., Joron, J.L., Reyss, J.L. and Guidon, N., 2003b, TL age-estimates of burnt quartz pebbles from the Toca do Boqueirão da Pedra Furada (Piaui, Northeastern Brazil), *Quaternary Science Reviews*, **22**, 1257–1263.

Van den Bogaard, C. and Schmincke, H.-U., 2002, Linking the North Atlantic to central Europe: a high-resolution Holocene tephrochronological record from northern Germany, *Journal of Quaternary Science*, **17**, 3–20.

Van den Bogaard, P., 1995, ^{40}Ar/^{39}Ar ages of sanidine phenocrysts from Laacher See Tephra (12,9000 yrs BP): chronostratigraphic and petrologic significance, *Earth and Planetary Science Letters*, **133**, 163–174.

Van den Haute, P. and Corte, F., 1998, *Advances in Fission Track Geochronology* (Dordrecht: Kluwer).

Van der Kaars, S. and van den Bergh, G.D., 2004, Anthropogenic changes in the landscape of West Java (Indonesia) during historic times, inferred from a sediment and pollen record from Teluk Basin, *Journal of Quaternary Science*, **19**, 229–239.

van der Merwe, N., 1969, *The Carbon-14 Dating of Iron* (Chicago: University of Chicago Press).

van der Plicht, J., 1993, The Groningen Radiocarbon Calibration Program, *Radiocarbon*, **35**, 231–237.

van der Plicht, J., 2002, Calibration of the ^{14}C timescale: towards the complete dating range, *Netherlands Journal of Geosciences*, **81**, 85–96.

van der Plicht, J., van der Sanden, W.A.B., Aerts, A.T. and Streurman, H.J., 2004, Dating bog bodies by means of ^{14}C-AMS, *Journal of Archaeological Science*, **31**, 471–491.

Van Geel, B., van der Plicht, J. and Renssen, H., 2003, Major Δ^{14}C excursions during the late glacial and early Holocene: changes in ocean ventilation or solar forcing of climate change, *Quaternary International*, **105**, 71–76.

Van Strydonck, M., Dupas, M. and Dauchot-Dehon, M., 1983, Radiocarbon dating of old mortars, *PACT Journal*, **8**, 337–343.

Vandergroes, M.J. and Prior, C.A., 2003, The AMS dating of pollen concentrates: a methodological study of Late Quaternary sediments from South Westland, New Zealand, *Radiocarbon*, **45**, 479–491.

Varvas, M. and Punning, J.-M., 1993, Use of the ^{210}Pb method in studies of the development and human-impact history of some Estonian lakes, *The Holocene*, **3**, 34–44.

Voinchet, P., Falguères, C., Laurent, M., Toyoda, S., Bahain, J.J. and Dolo, J.M., 2003, Artificial optical bleaching of the aluminium center in quartz: implications to ESR dating of sediments, *Quaternary Science Reviews*, **22**, 1335–1338.

Von Grafenstein, U., Erlenkauser, H., Brauer, A., Jouzel, J. and Johnsen, S.J., 1999, A mid-European decadal isotope-climate record from 15 500 to 5000 years BP, *Science*, **284**, 654–657.

Waelbroeck, C., Duplessy, J.-C., Michel, E., Labeyrie, L., Paillard, D. and Duprat, J., 2001, The timing of the last deglaciation in North Atlantic climate records, *Nature*, **412**, 724–727.

Wagner, G.A., 1998, *Age Determination of Young Rocks and Artefacts* (Heidelberg, Berlin and New York: Springer-Verlag).

Wagner, G.A. and van den Haute, P., 1992, *Fission Track Dating* (Dordrecht: Kluwer).

Walden, J., 1999, Remanence measurements, in J. Walden, F. Oldfield, J. Smith (eds), *Environmental Magnetism: A Practical Guide*. Technical Guide **6**, Quaternary Research Association, London, 63–88.

Walden, J., Oldfield, F. and Smith, J. (eds), 1999, *Environmental Magnetism: A Practical Guide*. Technical Guide **6**, Quaternary Research Association, London.

Walker, M.J.C., Björck, S., Lowe, J.J., Cwynar, L.C., Johnsen, S., Knudsen, K.-L., Wohlfarth, B. and INTIMATE group, 1999, Isotopic 'events' in the GRIP ice core: a stratotype for the Late Pleistocene, *Quaternary Science Reviews*, **18**, 1143–1150.

Walker, M.J.C., Bryant, C., Coope, G.R., Harkness, D.D., Lowe, J.J. and Scott, E.M., 2001, Towards a radiocarbon chronology for the Lateglacial: sample selection strategies, *Radiocarbon*, **43**, 1007–1020.

Walker, M.J.C., Coope, G.R., Sheldrick, C., Turney, C.S.M., Lowe, J.J., Blockley, S.P.E. and Harkness, D.D., 2003, Devensian Lateglacial environmental changes in Britain: a multi-proxy environmental record from Llanilid, South Wales, UK, *Quaternary Science Reviews*, **22**, 475–520.

Wallinga, J., Murray, A. and Duller, G., 2000, Underestimation of equivalent dose in single-aliquot optical dating of feldspars caused by preheating, *Radiation Measurements*, **32**, 691–695.

Wang, Y., Amundson, R. and Trumbore, S., 1996, Radiocarbon dating of soil organic matter, *Quaternary Research*, **45**, 282–288.

Wastegård, S., Björck, S., Possnert, G. and Wohlfarth, B., 1998, Evidence for the occurrence of Vedde Ash in Sweden: radiocarbon and calendar ages, *Journal of Quaternary Science*, **13**, 271–274.

Wastegård, S. and Rasmussen, T., 2001, New tephra horizons from Oxygen Isotope Stage 5 in the North Atlantic: correlation potential for terrestrial, marine and ice-core archives, *Quaternary Science Reviews*, **20**, 1587–1593.

Watanabe, S., Ayta, W.E.F., Hamaguci, H., Guidon, N., La Savia, E.S., Maranca, S. and Filho, O.B., 2003, Some evidence of a date of first humans to arrive in Brazil, *Journal of Archaeological Science*, **30**, 351–354.

Watchman, A. and Jones, R., 2002, An independent confirmation of the 4 ka antiquity of a beeswax figure in western Arnhem Land, Northern Territory, *Archaeometry*, **44**, 145–153.

Watchman, A., Ward, I., Jones, R. and O'Connor, S., 2001, Spatial and compositional variations within finely laminated mineral crusts at Carpenter's Gap, an archaeological site in tropical Australia, *Geoarchaeology*, **16**, 803–824.

Watson, E. and Luckman, B.H., 2001, Dendroclimatic reconstruction of precipitation for sites in the southern Canadian Rockies, *The Holocene*, **11**, 203–213.

Wehmiller, J.F. and Miller, G.H., 2000, Aminostratigraphic dating methods in Quaternary geology, in J.S. Noller, J.M. Sowers, W.R. Letts (eds), *Quaternary Geochronology, Methods and Applications*. American Geophysical Union, Washington, DC, Reference Shelf 4, 187–222.

Weidman, C.R., Jones, G.A. and Lohmann, K.C., 1994, The long-lived mollusc *Arctica islandica*: a new paleoceanographic tool for the reconstruction of bottom temperatures for the continental shelves of the northern North Atlantic Ocean, *Journal of Geophysical Research*, **99**, 18 305–18 314.

West, R.G., 1977, *Pleistocene Geology and Biology*. 2nd edition (London: Longman).

Westgate, J., 1989, Isothermal plateau fission track ages of hydrated glass shards from Silicic Tephra beds, *Earth and Planetary Science Letters*, **95**, 226–234.

Westgate, J.A., Preece, S.J., Froese, D.G., Walter, R.C., Sandhu, A.S. and Schweger, C.E., 2001, Dating Early and Middle (Reid) Pleistocene glaciations in central Yukon by tephrochronology, *Quaternary Research*, **56**, 335–348.

White, T.D., Asfaw, B., DeGusta, D., Gilbert, H., Richards, G.D., Suwa, G. and Howell, C., 2003, Pleistocene *Homo sapiens* from Middle Awash, Ethiopia, *Nature*, **423**, 742–747.

Wild, M.T., Tabner, B.J. and Macdonald, R., 1999, ESR dating of quartz phenocrysts in some rhyolitic extrusive rocks using Al and Ti impurity centres, *Quaternary Science Reviews*, **18**, 1507–1514.

Willerslev, E., Hansen, A.J., Binladen, J., Brand, T.B., Gilbert, M.T.B., Shapiro, B., Bunce, M., Wluf, C., Gilichinsky, D.A. and Cooper, A., 2003, Diverse plant and animal genetic records from Holocene and Pleistocene sediments, *Science*, **300**, 791–795.

Williams, M., Dunkerley, D., De Deckker, P., Kershaw, P. and Chappell, J., 1998, *Quaternary Environments*. 2nd edition (London: Arnold).

Williams, P.W., King, D.N.T., Zhao, J.-X. and Collerson, K.D., 2004, Speleothem master chronologies: combined Holocene ^{18}O and ^{13}C records from the North Island of New Zealand and their palaeoenvironmental implications, *The Holocene*, **14**, 194–208.

Wilson, R.C.L., Drury, S.A. and Chapman, J.L., 2000, *The Great Ice Age* (London: Routledge).

Winchester, S., 2002, *The Map that Changed the World* (London: Penguin Books).

Winkler, S., Matthews, J.A., Shakesby, R.A. and Dresser, P.Q., 2003, Glacier variations in Breheimen, southern Norway: dating Little Ice Age moraine sequences at seven low-altitude glaciers, *Journal of Quaternary Science*, **18**, 395–413.

Wintle, A.G., 1991, Luminescence dating, in P.L. Smart, P.D. Frances (eds), *Quaternary Dating Methods – A User's Guide*. Technical Guide **4**, Quaternary Research Association, London, 108–127.

Wintle, A.G., 1996, Archaeologically-relevant dating techniques for the next century, *Journal of Archaeological Science*, **23**, 123–138.

Wintle, A.G. and Huntley, D.J., 1980, Thermoluminescence dating of ocean sediments, *Canadian Journal of Earth Sciences*, **17**, 348–360.

Wintle, A.G. and Murray, A.S., 2000, Quartz OSL: effects of thermal treatment and their relevance to laboratory dating procedures, *Radiation Measurements*, **32**, 387–400.

Witbaard, R., Duineveld, G.C.A. and De Wilde, P.A.W.J., 1997, A long-term growth record derived from *Arctica islandica* (mollusc, bivalvia) from the Fladen Ground (northern North Sea), *Journal of the Marine Biological Association of the United Kingdom*, **77**, 801–816.

Wohlfarth, B., Björck, S., Cato, N. and Possnert, G., 1997, A new middle Holocene varve diagram from the river Ångersmanälven, northern Sweden: indications for a possible error in the Holocene varve chronology, *Boreas*, **26**, 347–353.

Wohlfarth, B., Skog, G., Possnert, G. and Holmquist, B., 1998a, Pitfalls in the AMS radiocarbon-dating of terrestrial macrofossils, *Journal of Quaternary Science*, **13**, 137–145.

Wohlfarth, B., Björck, S. and Possnert, G., 1998b, A 80-year long, radiocarbon-dated varve chronology from south-eastern Sweden, *Boreas*, **27**, 243–257.

Wolfman, D., 1990, Mesoamerican chronology and archaeomagnetic dating, in J.L. Eighmy, R.S. Sternberg (eds), *Archaeomagnetic Dating* (Tucson: University of Arizona Press).

Wooller, M.J., Swain, D.L., Ficken, K.J., Agnew, A.D.Q., Street-Perrott, F.A. and Eglinton, G., 2003, Late Quaternary vegetation changes around Lake Rutundu, Mount Kenya, East Africa: evidence from grass cuticles, pollen and stable carbon isotopes, *Journal of Quaternary Science*, **18**, 3–15.

Wykoff, R.W.G., 1980, Collagen in fossil bones, in P.E. Hare, T.C. Hoering, K. King Jr (eds), *Biogeochemistry of Amino Acids*, 17–22 (Chichester and New York: John Wiley).

Wymer, J.J., 1982, *The Palaeolithic Age* (London: Croom Helm).

Yang, H., 1997, Ancient DNA from Pleistocene fossils: preservation, recovery, and utility of ancient genetic information for Quaternary research, *Quaternary Science Reviews*, **16**, 1145–1161.

Yiou, F., Raisbeck, G.M., Baumgartner, S., Beer, J., Hammer, C., Johnsen, S., Jouzel, J., Kubik, P.W., Lestringuez, J., Stiévenard, M., Suter, M. and Yiou, P., 1997, Beryllium 10 in the Greenland Ice Core Project ice core at Summit, Greenland, *Journal of Geophysical Research*, **102**, 26 783–26 794.

Zagwijn, W., 1996, The Cromerian complex stage of the Netherlands and correlation with other areas in Europe, in C. Turner (ed.), *The Early Middle Pleistocene in Europe*, 145–172 (Rotterdam: Balkema).

Zeuner, F.E., 1959, *The Pleistocene Period* (London: Hutchinson).

Zhou, S., Li, J. and Zhang, S., 2002, Quaternary glaciation of the Bailang River Valley, Qilian Shan, *Quaternary International*, **97–98**, 103–110.

Zielinski, G.A., 2000, Use of paleo-records in determining variability within the volcanism-climate system, *Quaternary Science Reviews*, **19**, 417–438.

Zillen, L.M., Wastegård, S. and Snowball, I.F., 2002, Calendar ages of three mid-Holocene Tephra layers identified in varved lake sediments in west central Sweden, *Quaternary Science Reviews*, **21**, 1583–1591.

Zolitschka, B., Behre, K.-E. and Schneider, J., 2003, Human and climatic impact on the environment as derived from colluvial, fluvial and lacustrine archives – examples from the Bronze Age to the Migration period, Germany, *Quaternary Science Reviews*, **22**, 81–100.

Zreda, M.G. and Phillips, F.M., 1994, Surface exposure dating by cosmogenic chlorine-36 accumulation, in C. Beck (ed.), *Dating in Exposed and Surface Contexts*, 161–183 (Albuquerque: University of New Mexico Press).

Index

Quaternary Dating Methods M. Walker